Reliability Based Analysis and Design for Civil Engineers

Reliability Based Analysis and Design for Civil Engineers

Devaraj V
Professor of Civil Engineering, Former Chairman,
UVCE, Bangalore University, Bengaluru

Ravindra R
Associate Professor
R.V. College of Engineering, Bengaluru

TECHSAR™
NEW DELHI

Published and Printed by

TechSar Pvt. Ltd.
4435-36/7, Ansari Road, Daryaganj
New Delhi–110 002 (India)
E-mail: tech_info@techsarworld.com
Website: www.techsarworld.com

ISBN: 978-93-85909-80-1

Reprint 2025

Dedicated

to

Goddess Bhuvaneshwari

and

Our Parents

Foreword

The concern of the mankind for safety of its living environment is as old as the birth of civilization. Civil engineering practices have constantly evolved from rudimentary methods to the modern methods always keeping safety as the primary underlying criteria. The contemporary design practices employ partial safety factors to account for uncertainties in load and resistance parameters to realise safe designs for both concrete and steel structures. The design philosophies are gradually progressing towards more rational methods based on reliability concepts in which the goal is to realise the structures with desired level of reliability. Although the concept of reliability is not new in many fields of engineering such as mechanical, electrical, electronic, etc., its realisation in the field of civil engineering is gaining acceptance recently. This calls for adequate training of the next generation of civil engineers in this field. This requires sufficient numbers of textbooks and other teaching aids. In this context, this book *Reliability Based Analysis and Design for Civil Engineers* written by Prof. Devaraj V. is an excellent effort. He has covered all the basics required for understanding the subject adequately without compromising on the rigour. The best part of the book is the innumerable number of examples provided for explaining the concepts. Notwithstanding this, the book also provides enough number of problems for exercising the concepts learnt. I wish the book will go a long way in helping the students and professional engineers to learn and adopt the subject in civil engineering practice.

Dr M. Manjuprasad
Chief Scientist and Professor (AcSIR)
Structural Technologies Division
CSIR-National Aerospace Laboratories
Bangalore – 560017

Preface

Civil Engineering systems are designed for loads which act on them. The environmental actions such as earthquake, wind snow, temperature loads; also the materials used for construction display scatter in mechanical properties cannot be treated as deterministic. However they are random and uncertain.

Problems commonly encountered by engineers require decision making under conditions of uncertainty. Integrating the uncertainties in geometry, strength, loads and its consequences by reliability theory ensures **safety of a structure**.

Safety of structure is one of the core civil engineering concepts. As a beginner several books on Structural Reliability for civil undergraduate and postgraduate program require mathematical expertise to comprehend. This mathematical complexity often creates an aversion towards the subject. Hence, the book answers the common question of "Where do I begin".

During my period of working as a professor of civil engineering also showed me that these complexities can be solved in a systematic manner. This urged me to design a book in such a fashion that it would guide them through the steps of problem solving which is the central theme of all engineering. It also intends to meet the need that exists for an elementary level textbook on reliability based analysis and design for civil engineers.

It is important to emphasize that MATLAB programs were developed by the authors to solve the problems and almost every student should be equipped with such a tool for evaluation of results.

The understanding of the basic principles requires practice by hand computation. Several numerical examples are given in each chapter to help the reader understand the technique involved in solving the problem and develop the skill. The authors have consulted many sources and drawn materials from various authors/experts in this field. ***The authors profoundly express their indebtedness & acknowledge them all.***

The chapters in the book are well planned, so that a beginner in reliability engineering will understand the subject. The book contains ten chapters; introduction outlines two sources of uncertainties which are classified as aleatory and epistemic. The design philosophies, partial safety factors, load and resistance factors are presented. Chapter 1 provides an understanding of important ideas of statistics and probability to illustrate their relevance with applications. Chapter 2 outlines different types of random variables and the probabilistic characteristics of those variables are fitted with theoretical distribution for the observed data. Chapter 3 examines the models using mathematical probability density functions and cumulative distribution functions. Statistical uncertainty arises from the choice of distribution resulting in parameter estimation. Chapter 4 integrates random basic variables and consequences of structural safety or probability of failure of structure using first order second moment method (FOSM). Chapter 5 deals with approximate method to evaluate the safety or failure using Point Estimate method with limitation on the number of basic random variables. Chapter 6 gives a formal presentation on Taylor series expansion using first and second order analyses to evaluate the structural safety. Chapter 7 discusses on Hasofer-Lind and emphasizes on standard method of assessing the probability of safety through reliability index at the component level, circumventing the lack of invariance which are present in other methods. Chapter 8 gives a formal discussion on Monte Carlo simulation demonstrating the sample-to-sample, or sampling variation that can be expected in reality. For this reason, performing simulation is one way of generating a sampling variation that is inherent in statistical problems. Chapter 9 discusses civil engineering structures that are series (determinate), parallel (redundant) and a combination called hybrid systems. Here system safety or failure is discussed. Chapter 10 deals with the rudiments of reliability based design (Load Resistance Factor Design/LRFD) by FOSM, Hasofer-Lind, PEM and Taylor series method. A general expression is used for comparing LRFD and Allowable Stress Design (ASD) for assessing the economy and safety.

All the chapters have sufficient number of solved problems which the reader would definitely appreciate.

Devaraj V
Ravindra R

Acknowledgements

During proofs review and edit phase, very prompt response and necessary help came from doctoral students. Mr. C.B. Yogish, Assistant Director, Karnataka Rural Infrastructure Development Limited, Bangalore; Mr. N. Dharmesh and Mr. Y.B. Madhu Sudhana, Assistant Professors, at East Point College of Engineering and Technology, Bangalore. Special thanks go to my colleagues, Dr. N. Munirudrappa (Mentor) and Dr. L. Govindaraju, Associate Professor at Bangalore University, for all the help they have extended while writing this book.

We offer our sincere thanks to all our colleagues those who have extended helping hand from time-to-time. Even though maximum care has been taken to present the text free from errors, some errors might have crept in inadvertently. We would be grateful to readers and instructors, to send us feedback for improvement.

professor.devaraj@gmail.com **Devaraj V**

ravimurd@gmail.com **Ravindra R**

Notations

1	CoV	Coefficient of variation
2	FoS	Factor of safety
3	R_n	Nominal resistance
4	S_{ni}	Nominal working or service stress
5	γ	Load factor
6	ϕ	Resistance factor
7	M_e	Model error
8	MR	Moment Resistance
9	S	Load effect
10	R	Resistance
11	k_s	Percentile of the load
12	k_R	Percentile of the resistance
13	S_d	Design load
14	S_m	Mean load
15	R_d	Design resistance
16	R_m	Mean resistance
17	W_g	Dead load
18	W_L	Live load
19	P_S	Probability of survival
20	f_S	Probability density function for load

21	f_R	Probability density function for resistance
22	F_S	Cumulative density function for load
23	F_R	Cumulative density function for resistance
24	P_f	Probability of failures
25	μ_x	Mean value, mean value of X
26	σ_x	Standard deviation
27	PMF	Probability Mass Function
28	PDF	Probability Density Function
29	CDF	Cumulative Density Function
30	r_1	Coefficient of skewness
31	r_2	Coefficient of kurtosis
32	ρ	Correlation coefficient
33	λ	Average number of events
34	t	Time
35	ν	Mean rate of occurrences/Average number of occurrences
36	ξ	Parameters of log normal distribution (position factor)
37	λ	Parameters of log normal distribution (scaling factor)
38	*Fact* (*k*-1)	Factorial of (k-1) or (k-1)!
39	Var[X]	Variance of X
40	β	Reliability index
41	X_+	Upper limit
42	X_-	Lower limit
43	(MS)	Margin of safety
44	$E[X]$	Expectation of X
45	H	Hessian matrix
46	N_f	Number of trials
47	R_{Ui}	Uniform random number

48	LCG	Linear Congruential Generator
49	S	Sample space
50	E	Occurrence of an event
51	$B(.)$	Beta function
52	$\Gamma(.)$	Gamma function
53	AFOSM	Advanced First Order Second Moment
54	$\Phi(\ )$	CDF of standard normal variate
55	MCS	Monte Carlo Simulation
56	IT	Inverse Transform method
57	p_{ss}	Probability of survival of system
58	p_{si}	Probability of survival of component
59	p_{fs}	Probability of failure of system
60	p_{fi}	Probability of failure of component
61	COV	Covariance

Contents

Introduction

RELIABILITY METHODS IN CIVIL ENGINEERING

One must look back to the field of structural reliability, to such pioneers as Alfred Freudenthal (1890–1975). Freudenthal's work was followed by a generation of researchers in structural engineering, namely Gumbel, A.H.S. Ang, C.A. Cornell, O. Ditlevesen, A.M. Hasofer, N. Lind, and R. Rackwiz. In the last 30 years, practical probabilistic and reliability methods have been developed to help engineers tackle the analysis, quantify, monitor and assess structural risks, also undertake sensitivity analysis of inherent uncertainties and make rational decisions about the performance of structures over their working life.

The reliability theory is nothing but mathematical engineering judgement, that is, an engineering experience gathered for long and transformed into computational model. Three fundamental concepts are to be borne in mind; they are (i) mathematical statistics and theory of probability; (ii) structural systems or components; and (iii) time (deterioration), by all those who want to master the structural reliability problems.

The structure is analogous to a water filter; the input consists of actions/loads that act on the structure. System (structure) is modelled by considering the elasticity (stiffness), mass (invariant), and damping. Response quantities, namely, displacement, rotation, deformations (stresses, strains) are the output quantities. It is shown in Figs. 1(a) and 1(b).

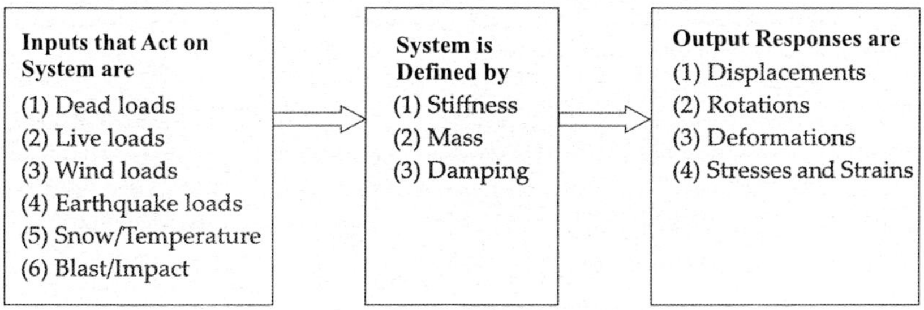

Fig. 1a Input, system and output analysis

PERFORMANCE FUNCTIONS OF (SYSTEM) STRUCTURES

The safety of a structure is assessed by the performance of a whole or a part of it and may be described with reference to a set of limit states with separate acceptable states

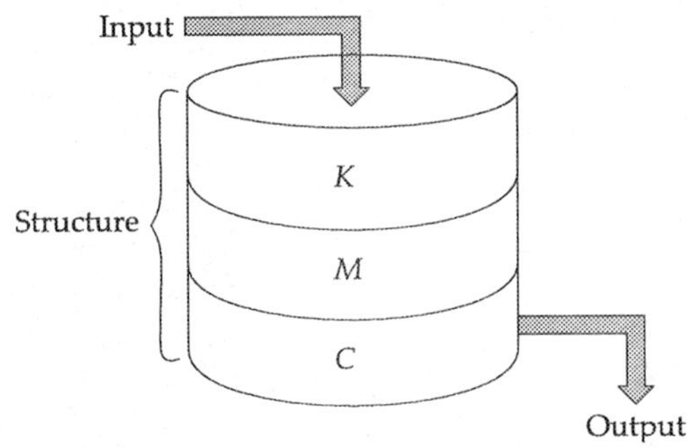

Fig. 1b Structural analysis

of a structure from unacceptable states. The boundary between acceptable (safe) and unacceptable (failure) states may be distinct. Thus, verification of a structure with respect to a particular limit state may be carried out through a mathematical model describing the limit state in terms of a function (called the limit state function) whose values depend on all design parameters.

The limit state equation is given by $g(X) = 0$ where X represents the vector of design parameters (called basic variable vectors) that are relevant to the problem. Basic variables comprise actions and influences, material properties, geometrical data and factors related to the models used for constructing the limit state function. The basic variables exhibit variation in time and space. The different states of limit state of a structure are given in Table 1.

Table 1 Types of limit states

Sl. No.	Limit state	Description	Examples
1.	Ultimate (safety)	Collapse of all or part of structure	Tipping or sliding, rupture, progressive collapse, plastic mechanisms, instability, corrosion, fatigue, deterioration and fire
2.	Damage	Damage to structure	Excessive or premature cracking, deformation or permanent inelastic deformation
3.	Serviceability	Disruption of normal use	Excessive deflection, vibrations and local damage

An engineering structure response depends on many uncertainty factors such as geometry, strength, loads, stiffness, mass, and damping. These uncertainties are shown in Tables 2 to 4. The response is considered satisfactory when the design requirements imposed on the structural behavior are met within an acceptable degree of certainty. Each of these requirements is termed limit state or constraint.

The limit state equation is expressed as

$g(X) = g(R) - g(S) = 0$, where $g(R)$ and $g(S)$ denotes the function of resistance and loads respectively.

The margin of safety,

MS = Strength or resistance effects – Load effects or combination of loads

Hence, $MS = g(R, S) = R - S$

where *R* and *S* denote the resistance and load effect respectively. There are uncertainties due to geometrical size, sectional dimension, material properties, and loads. There are three possible limit states and are as follows:

(i) $\text{MS} > 0$; strength is greater than load on the system or structure, hence it is reliable or safe.

(ii) $\text{MS} < 0$; strength is lower than the load on the system or structure, hence fails.

(iii) $\text{MS} = 0$; strength is equal to load; hence the system or structure is in limit state.

Table 2 Load uncertainties

Sl. No.	Load uncertainties	Examples
1.	Operational loads	Dead load of structures and live load of equipment
2.	Environmental loads	Wind, seismic ground shaking and temperature loads
3.	Accidental loads	Impact, fire and explosions

Uncertainties about resistances have to do with site conditions, static and dynamic soil properties, and how the structure behaves when subjected to load.

Table 3 Resistance uncertainties

Sl. No.	Resistance uncertainties	Examples
1.	Site conditions	Location and character of stratification Thickness of important strata Existence of soft inclusions or gas pockets
2.	Soil properties	Strength deformation in static loading Consolidation parameters Cyclic strength parameters
3.	Structural performance	Strength of critical components or members Fatigue behaviour of structural materials
4.	Soil-structure interaction	Structural and foundation behaviour

Table 4 Some uncertainties in dimensional properties

Sl. No.	Dimensional property	Mean/Specified	CoV	Source
1.	Flange thickness	0.97	0.028	
2.	Web thickness	1.00	0.032	
3.	I–Sectional area	1.00	0.006	Pham & Bridge (1985)
4.	Area of reinforcement steel	0.99	0.025	Mirza & McGregor (1978)

Fundamentals of Structural Reliability Theory. Four basic performances are to be satisfied for a system (structure) to be reliable. The structure shall satisfy the following conditions, namely: (1) strength or capacity to resist different loads and their combination of loads, which relate to maximum load carrying capacity; (2) serviceability under service loads and under normal use; (3) durability with normal maintenance; and (4) structural integrity (instability) during and after the application of loads.

Structural reliability is synthesis of strength, serviceability, durability and structural integrity for structural safety.

It must be emphasized at the outset that reliability approaches neither remove uncertainty nor alleviate the need for judgements in dealing with the real world problems. They provide a way of quantifying those uncertainties and handling them consistently. The evaluation of risk and reliability is closely connected to the field of probability and statistics.

The uncertainties in structural and mechanical engineering are largely deductive: starting from reasonably well-known conditions, models are employed to deduce the behaviour of a reasonably well-specified universe. The uncertainties in geotechnical engineering are largely inductive: starting from limited observations, judgement, knowledge of geology and statistical reasoning are employed to infer the behaviour of a poorly-defined universe.

Table 5 Material properties

Sl. No.	Material property	Mean value	CoV	Source
1.	Modulus of elasticity	1.00	0.06	Ellingwood *et al.* 1980
2.	Shear modulus	1.00	0.06	
3.	Poisson's ratio	1.00	0.03	
4.	Yield strength	1.05	0.10	
5.	Tensile strength	1.10	0.11	AISI 1978

Uncertainty and Randomness

The structural analysis is performed with reference to (1) existing structures and (2) structures being designed.

The loads that are acting and that which will act on a structure are never known exactly. They can be estimated from records of loads on similar constructions, but two such records will never be identical. The uncertainty *can be reduced, but never be eliminated.*

A quantity that is subjected to any of the uncertainties is called a *random variable.* A quantity that does vary in time is called a random process or stochastic process. In opposition to random quantities, a quantity whose random uncertainty is negligible is called *deterministic.*

The probabilistic analysis methods, to assess the structural reliability of a structure subjected to a specific limit and if the constructed facility (structure) is unable to perform as required, then specific limit is called *limit-state.*

A cantilever beam, as shown in Figs. 2(a) and 2(b), with its elastic modulus; E, fluctuates over the length of the beam. The fluctuation of the elastic modulus can be split into two parts, the mean and the fluctuating part, in order to reduce the bias and to better facilitate analysis.

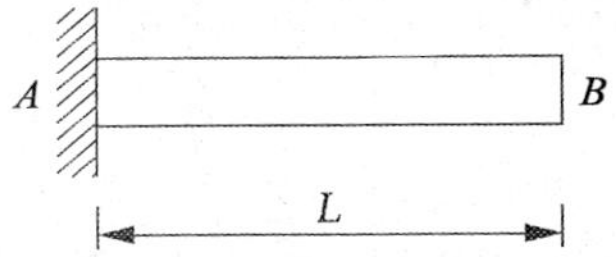

(a) Cantilever beam

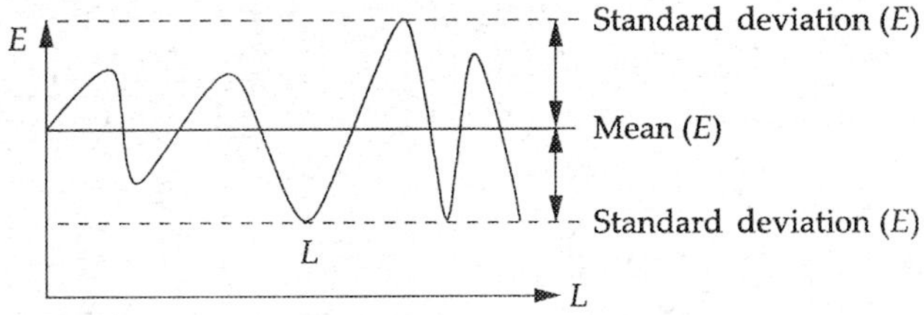

(b) Young's modulus vs span

Fig. 2

$$\text{Random variable } (E) = \text{Mean } (E) \pm \text{Standard deviation } (E)$$

How to Measure Uncertainty. Uncertainty is measured or quantified by coefficient of variation (CoV) of a random quantity.

$$\text{Coefficient of variation (CoV)} = \text{Standard deviation/Mean value}$$

Engineering is the **art** of moulding *materials* we do **not really understand**, into *shapes we cannot* **really analyze**, so as to withstand *forces* we **cannot really assess**, in such a way that the *public does not really suspect.* **The public has faith that structures will be designed and built to perform as intended.**

It is clear that the external actions, strength of the structure and the system (constructed facility) are uncertain and it is shown in Fig. 3.

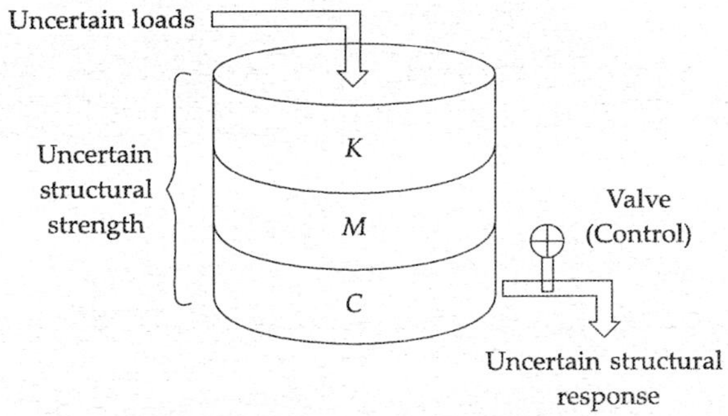

Fig. 3 Uncertain loads, strength and response of structures

In dealing with real world problems, uncertainties are unavoidable. As engineers, it is important that we recognise the presence of all major sources of uncertainty in engineering. The sources of uncertainties may be classified into two broad types, namely: (i) aleatory and (ii) epistemic.

Aleatory Uncertainties. These are associated with inaccuracies in prediction and estimation of reality. The presence of uncertainty in engineering, therefore, is clearly unavoidable; the available data are often incomplete or insufficient and invariably contain variability. Uncertainty is associated with imperfect models of the real world, because of insufficient or imperfect knowledge of reality. The aleatory (data based) uncertainty is associated with the inherent variability of basic information, which is part of the real world problem (within our ability to observe and describe). Much of the aleatory uncertainty that civil engineers must deal with are inherent in nature and therefore, cannot be reduced or modified. Many phenomena or processes of concern to engineers or that engineer must contend with, contain randomness, i.e., the expected outcomes are unpredictable (to some degree). Such phenomena are characterized by field measurement or experimental data that contain significant *variability* that represents the natural randomness of an underlying phenomenon. The observed measurements are different from one experiment to another, even if conducted or measured under apparently identical conditions. The variability inherent in such data or information is *statistical* in nature, and realisation of a specific value (or range of values) involves *probability*. The inherent variability in the observed or measured data can be expressed graphically in the form of *histogram* or *relative frequency diagram*. Further, if two variables are involved, the joint variability may be portrayed as *scatter diagram*.

Epistemic Uncertainty. Uncertainty associated with imperfect models of the real world because of insufficient or imperfect knowledge of reality. On the other hand, epistemic (or knowledge based) uncertainty is associated with imperfect knowledge of the real world and may be reduced through the application of better prediction models and/or improved experiments. In engineering, we have to rely on idealised models of real world in our analysis and predictions for the purpose of making decisions or for planning and developing criteria for the design of an engineering system. These idealized models, which may be mathematical or simulation models (mathematical formulae, equations, numerical algorithms, computer programs) or even laboratory models, are imperfect representations of the real world. The response or results of analysis, estimations or predictions obtained on the basis of such models are inaccurate (with some unknown degree of error) and thus also contain uncertainty. Such uncertainties are therefore, knowledge based and are of the epistemic type. It is reasonable to limit our consideration to the inaccuracy in computing or estimating the *central* or *mean* or *median value* and also in estimating the parameters of the probability *distribution*.

The two types of uncertainty may be combined and analysed as a total uncertainty or treated separately. In either case, the principles of probability and statistics are applied equally.

Uncertainty analysis is characterised by two types that are commonly used to represent aleatory and epistemic uncertainties. The centre of probability density function (PDF) indicates the most probable point and the tail region of the PDF indicates less probable events, i.e. the region of interest. If the information for PDF is not available, the interval of upper and lower bounds of random variables may be appropriate to represent these kinds of uncertainty and this is shown in Fig. 4.

Range of Values Known. The uncertainty may be evaluated by knowing the lower and upper limits of a variable. In engineering, the information may often have to be expressed by the range of possible values of the random variable, the mean value of the variable and the underlying uncertainty may be evaluated by prescribing suitable distribution within the range (Yucemen, Tang, and Ang, 1976).

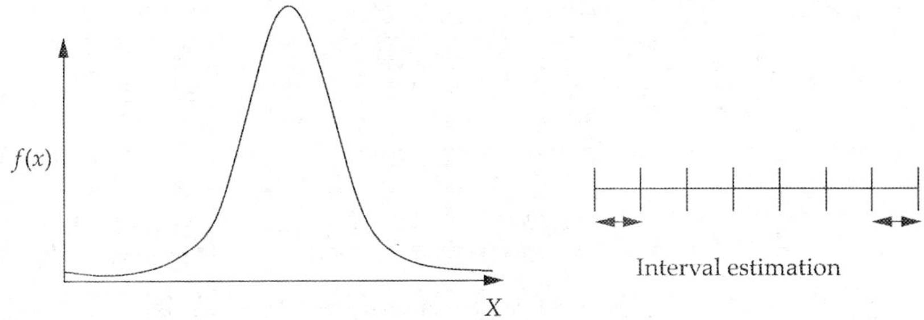

Fig. 4 Uncertainty representation

Framework for Reliability Analysis. The main steps in a reliability analysis of structural component are the following:

1. Identify the basic variables and develop appropriate probabilistic models
2. Define the limit state function for the particular design situation considered
3. Compute reliability index and failure probability
4. Perform sensitivity analysis
5. Specify appropriate time reference period; for time dependent reliability, degradation, life assessment, creep and fatigue.

Structural Reliability. What are the uncertainties an engineer faces in designing structures? These can be divided into two groups: uncertainties in the loads to design (loading conditions or combinations) and uncertainties in the resistance of a structure. As a first approximation, uncertainties in the loading condition have to do with operational loads (dead and live loads on the structure) environmental loads (seismic loads, wind and temperature) and accidental loads (blast, impact, fire and explosion).

Structural reliability is concerned with the calculation and prediction of the probability of limit state violations at any stage during a structure's life. The probability of the occurrence of an event such as a limit state violation is a numerical measure of the chance of occurring. Once the probability of failure is determined, the next goal is to choose a design alternative that improves the structural reliability and minimizes the risk of failure.

The data of each basic elementary variable are modelled as random variables either as independent or dependent with their probability density function or at least with their marginal probability distribution function along with correlation between each basic variable. The basic variables require knowledge of mean and standard deviation, i.e., statistical properties for the random variables and is shown in Fig. 5.

Probabilistic Modelling. Probabilistic methods are convenient tools to describe or model physical phenomena that are too complex to treat with present levels of scientific knowledge. Probabilistic design explicitly incorporates given statistical data for the basic random variables by considering an appropriate joint probability density function or at least marginal density function for each variable, by considering the random variables

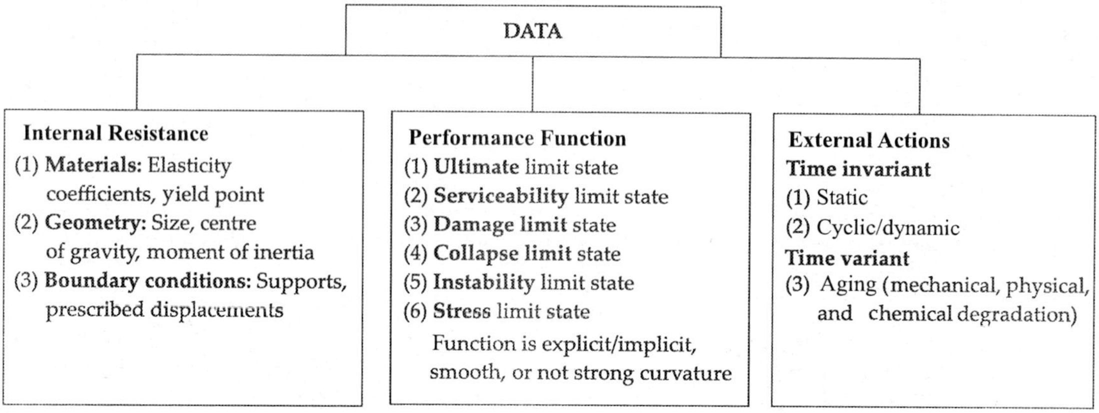

Fig. 5 Data on external actions, internal strength, and performance function on a part of a system

as correlated or uncorrelated. The mechanical models are used to evaluate the internal strength (R) and external actions (S) and combining them with a performance function which relates to the limit states of violation or safety. The mechanical model may be an equivalent reality and analysed based on the behaviour of system as linear, elastoplastic and plastic or by considering the non-linearity in geometry or material or both. The scenario is presented in Fig. 6. The different methods of probabilistic analysis are shown in Fig. 7.

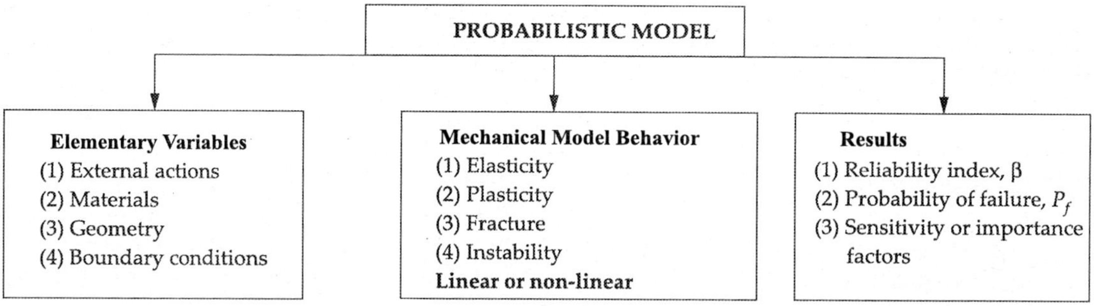

Fig. 6 Probabilistic model, considering elementary basic variable, with a mechanical model, and results

Design Philosophy. Three design philosophies are (i) allowable stress design; (ii) plastic design (iii) load resistance factor design (LRFD).

Allowable Stress Design (ASD). The general expression in this type of design is

$$\frac{R_n}{\text{FoS}} \geq \sum_{i=1}^{m} S_{ni}$$

where,

R_n = Nominal resistance of the structural member expressed in units of stress
S_{ni} = Nominal working or service stress computed working load

FoS = Factor of safety
i = type of load
m = number of load type

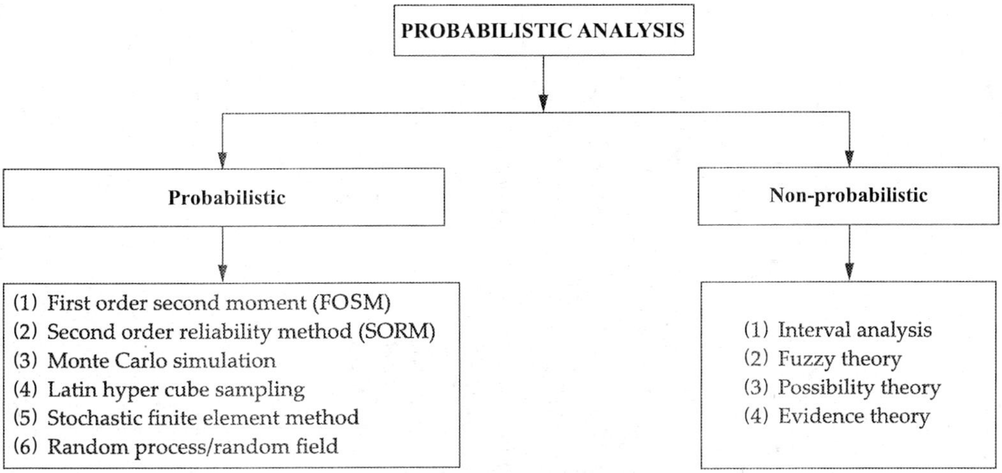

Fig. 7 Probabilistic analysis chart

This uses the first order elastic analysis, the stresses are computed and do not exceed their allowable stresses. The highlights of this philosophy are:

(i) The effects of second order moments are inelastic and considered indirectly.
(ii) In the allowable stress design, the factor of safety is applied only to the resistance term and safety is evaluated at the service load.
(iii) The ASD is characterized by the un-factored "working" loads in conjunction with the single factor of safety applied to the resistance.
(iv) The greater variability and unpredictability of the live load and other loads in comparison with the dead load, a uniform reliability is not possible with ASD.

Plastic Design (PD). The general expression in this type of design is

$$R_n \geq \gamma \sum_{i=1}^{m} S_{ni}$$

where γ = Load factor

The highlights of this philosophy are:

(i) It is ensured that factored load combinations do not exceed the maximum plastic strength of the structure or component.
(ii) Safety is incorporated only in the load term and is evaluated at the ultimate (plastic strength) limit state.
(iii) Applying a factor of safety to the load term is more appropriate than ASD because uncertainty associated with load is higher than associated with resistance.

(iv) Uniform reliability cannot be fully achieved with PD because only a single factor of safety (load factor) is applied to all loads.

(v) Plastic design is superior to ASD in the sense it considers moment redistribution in beams and frames in a more direct way.

(vi) The regions of structure in which significant plasticity (plastic hinges) must satisfy the limitation on the flange and web proportions for plastic design section.

(vii) In addition, sufficient lateral and torsional restraint must be provided to prevent the member from buckling laterally.

(viii) Although material non-linearity has been considered in the first order plastic hinge analysis, the effects of the spread of yielding and member stability are not accounted in the design.

(ix) Thus, like ASD, any insignificant nonlinear effects have to be accounted for approximating in the member strength equations. The member capacity checks are required when plastic analysis is used so as to prevent instability type of failure.

Load Resistance Factor Design (LRFD). It is ensured that the factored load effects do not exceed the factored nominal resistance of the structural members or component. Here there are two safety factors: one applied to the loads, the other to the resistance.

$$\phi R_n \geq \sum_{i=1}^{m} \gamma_i S_{ni}$$

where, ϕ = Resistance factor, γ = Load factor

Sources of Uncertainty. The *first* source is the *inherent randomness* in all physical observations, that is, repeated measurements of the same physical quantity do not yield the same value, due to numerous fluctuations in the environment, test procedure, instruments and, the observer. This is referred to as inherent uncertainty. This type of uncertainty is addressed by collecting large number of observations. This provides good information about the variability of the measured quantity. Number of observations is limited by money and time. This may arise from random and systematic errors in the measurement of these physical quantities.

The *second* source of uncertainty is the *statistical uncertainty*. In this case, one does not have precise information about the variability of the physical quantity of interest due to limited data. The information on variability will vary and depend on the number of sample data collected to predict the reality (population). This is due to reliance on limited information and finite samples.

The *third* source of uncertainty is the *model uncertainty*. In this system analysis of models are only approximate representations of the actual or reality. Computation models capture the essential characteristics of system behaviour through idealised mathematical relationship or numerical procedures. This is related to predictive accuracy of calculation models used.

Model Error for Bending Resistance. To illustrate the model error for bending resistance consider four beam specimens tested, which have failed by bending or flexure. Bending resistance is a measure of our ability (or inability) to predict the strength of a beam (steel-concrete composite beam) in exact form. The experimental results were compared with analytical and numerical methods. The model error (M_e) is the ratio of experimental moment resistance to analytical or numerical moment resistance.

$$\text{Model error, } Me = \frac{MR^{\text{experimental}}}{MR^{\text{analytical}}}$$

Table 6 Experimental, analytical and numerical results of beam test

Beam	Experimental	Analytical	Numerical	Experimental/ Analytical	Experimental/Numerical
A1	193.12	163.24	164.97	1.18	1.17
A2	181.99	163.16	164.97	1.12	1.10
B1	185.93	163.24	164.97	1.14	1.13
B2	185.06	163.16	164.97	1.13	1.12

The model error mean value is called bias and Covariance (COV) is a measure of random error of model.

If $Me > 1$; then resistance evaluated by the model is smaller than the actual resistance.
If $Me < 1$; then resistance evaluated by the model is larger than the actual resistance.
If $Me = 1$; then resistance evaluated by the model is equal to the actual resistance.

Classification of the Levels of Probabilistic Approach. The classification in the levels of probabilistic approach in structural design, analysis and evaluation is presented as follows:

Level 0 corresponds to a *purely deterministic analysis*. The loads and resistances are deterministic values based on the experience and global safety coefficient is chosen. Experience feedback alone justifies the design and the qualitative evaluation of the reliability.

Level 1 is called *semi-probabilistic*. The structural analysis is based on a member safety rather than structural system analysis. The design philosophy is that a conservatively high load effect value (*loads are overestimated,* mean plus some value of standard deviation) must not exceed a conservatively low resistance (*strengths are underestimated,* mean minus some number of standard deviations). It is convenient to define a load factor γ, and a resistance factor ϕ, by the following equations:

$$\gamma = 1 + (k_S \times \text{CoV}_S)$$
$$\phi = 1 - (k_R \times \text{CoV}_R)$$

In which CoV_S, CoV_R and k_S, k_R are the coefficients of variation and percentile of the load and resistance, respectively.

Design Values. The design values of resistance and loads are obtained by dividing and multiplying the resistance and load variables with a factor of safety respectively. The partial safety coefficient is greater than unity for loads and less than unity for strength.

In principle, there is a partial factor associated with each basic variable, which is related to the modelling uncertainties (loading and resistance respectively). It ensures an acceptable range of safety and load combinations, at the same time, resulting in desirable safety and economy of construction.

Factored values are given by

$$S_d = S_m + (k_S \times \sigma_S)$$

$$\Rightarrow \quad S_d = \gamma S_m$$

$$R_d = R_m - (k_R \times \sigma_R)$$

$$\Rightarrow \quad R_d = \phi R_m$$

where, S_d = Design load

S_m = Mean load

R_d = Design resistance and

R_m = Mean resistance

The loads and resistances are introduced by representative values, derived from a more or less rich statistical content. Partial coefficients are also introduced into the design rules. Level 1 is the most widely used method.

Characteristic Values. These are the main representative values of the variables. Characteristic values are further classified as nominal values X_n and values with statistical content. An engineer typically uses characteristic value as average value depending on the direction necessary to approach safety.

Nominal Values. A nominal value is a value assigned to a quantity by the code of practice for an engineer.

Partial Safety Factors and Code Formats. Present codes treat load and resistance variables as deterministic. The characteristic (representative or nominal) value is strictly defined as the value of random variable which has a prescribed probability of not being exceeded (on the unfavourable side) during a reference period.

Deterministic safety checking code format for a beam subjected to flexure is given as

$$\gamma_{sd}(W_g + W_L) \le \frac{1}{\gamma_{R_d}}(f_{y_d} \times Z_d)$$

where the design values of basic variables representing loading, resistance and geometrical variables respectively, which can be obtained from characteristic/representative values and associated partial safety factors, and these factors are related to modelling uncertainties (loading and resistance functions respectively).

Level 2 corresponds to a *partial application of probabilistic methods.* It requires the definition of a reference period and the first order second moment (FOSM) or the second order reliability method (SORM) approximations are allowed. Level 2 is also the basis for the partial coefficient calibration procedure defined in level 1. The basic premise of this method is that for practical applications it would be too complicated to associate and deal with full probability distribution for each random variable. The approach suggested by Cornell was the use of two quantities for each variable. If one envisages the probability density function as a beam cross section, then the mean lies on the centroid and standard deviation is the radius of gyration. The variance, being the square of the standard deviation, is the moment of inertia (also called the second moment). The probability density function is always a shape enclosing a unit area. This approach of using the mean and variance is therefore, referred to as the second moment approximation, and is usually designated as level 2 format. Such format is also referred to as the first order format. The real advantage of the second moment approach is that it introduces consideration of variability *without requiring full probabilistic analysis.* The load and resistance variables do not actually follow normal distribution and then the reliability computed will not be the true reliability. The extension to multiple load or resistances causes a problem known as *lack of invariance.*

Level 3 is the *fully probabilistic method*. It requires knowledge of the joint probability densities of all the basic variables, which is inaccessible in practice in most situations. In order to simplify the problem, certain reasonable assumptions are introduced, namely, the resistance does not vary over time and it restricts the time aspect of the problem on the load side. Time varying loads and critical load combinations involve random variables, rather than random processes. Since both load and resistance are random variables, the evaluation of that probability generally involves a convolution integral. This convolution may be thought of as the product is integrated over all possible values.

$$P_f = \int_0^{\infty} f_S(x) F_R(x)\, dx \quad (R, S \text{ are independent})$$

$$P_f = \int_0^{\infty} [1 - f_S(x)]\, F_R(x)\, dx \quad (R, S \text{ are independent})$$

where, P_f denotes the probability of failure

F_R = cumulative density function for resistance

F_S = cumulative density function for load

f_R = probability density function for resistance

f_S = probability density function for load.

Level 4 is the *fully stochastic method*. It requires the probability of safe performance, that is the probability that each component of the load effect vector is less than the corresponding component of the resistance vector, throughout the space of the structure and during the full design lifetime. This formulation is often termed stochastic format. While this approach is simple in concept, the complexity of dealing with multidimensional stochastic variables makes it intractable in application.

$$P_S = P\left[S(x,t) < R(x,t)\right]$$

where, x is the geometrical position of structure; t is the economic life of structure; P_f denotes the probability of safety.

Structures exist in an uncertain world and the most consistent way to consider the effect of these uncertainties is the theory of probability. Since structural system is too complex to warrant a complete stochastic analysis in most cases, approximate probabilistic models are incorporated into codal formats. Code formulating bodies must have adequate representation by those who have basic understanding of probability concepts, but structural designers may use the code without such knowledge. *The load and resistance factor approach is an attractive format at this point of time.*

Considering the probabilistic approach in structural design, analysis and evaluation, the process is presented in Fig. 8.

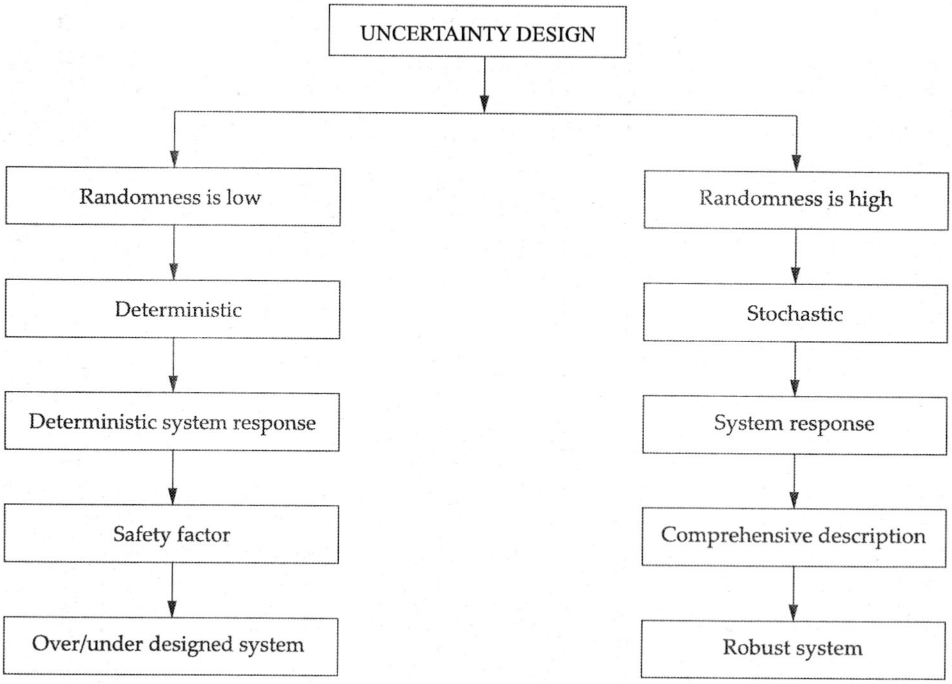

Fig. 8 Uncertainty design process

Concluding Summary

The uncertainties associated with randomness are inherent variables, called aleatory uncertainty, whereas those underlying imperfect models of reality which are knowledge based, called epistemic uncertainty.

The predictive model if improved, the epistemic uncertainty can be reduced; on the other hand, because the inherent variability is part of the nature, the aleatory uncertainty may not be reducible.

In particular, the amount of data affects the definition of variability (i.e., aleatory uncertainty) of the information; however, epistemic uncertainties associated with imperfect models of reality generally require judgemental assessments irrespective of the amount of data.

Probability and statistics is the conceptual and theoretical base for modelling and analysing uncertainty.

1
CHAPTER

Statistics and Probability

1.1 FUNDAMENTALS OF STATISTICS

The results of experiments in engineering are never same in spite of being conducted under similar conditions. For example, the cube strength of concrete of a particular mix, wind loads or live loads on the structure, traffic on the road, strength of beam or column or any experiments to assess the physical properties of a material have outcomes whose results are not the same. Certainly uncertainty exists in the results.

Our daily experience teaches that some data are generated in accordance to the known and precise laws, while other data seem to occur in a purely haphazard way. Data generated in accordance to known and precise laws are called deterministic data. The attributes of deterministic data is that the same data will be obtained, within the precision of the measurements under repeated experiments in well defined conditions.

Consider an illustration on deterministic data. Assume that a construction firm is designated to build a concrete bridge. The firm uses the same formula and same type of ingredients. All cubes are tested for 28 days and the modulus of elasticity at that time is a random variable. If the measured elastic modulus is same, then one can believe it is deterministic. The numerical value of modulus of elasticity for each batch is called *outcome*, or *observed value of the random variable*.

The outcome of events is not deterministic but random or probabilistic. The degree of uncertainty varies from one event to another event. When the results are probabilistic the engineer has to take a decision on estimating the numerical value of the data set. The numerical value of the data set needs to be processed so that it presents useful information of the sample. Statistics is used for this purpose. Statistical methods make scientific judgements in the face of uncertainty and variations. When dealing with a random data set, as one realisation (or one instance) of a set or a set (or ensemble) consisting of a possibly infinite number of realisations of a generating process, is so called random process (or stochastic process).

1.2 DATA ANALYSIS

A general overview of the data in terms of frequencies with which a certain interval of values occurs, both in tabular and in graphical form, is usually advisable as a preliminary step before proceeding to the statistical analysis.

There are two types of data namely, discrete and continuous. The number of persons working, equipment used in the construction firm is discrete (integers) and contains no fractions. The concrete strength is thc rcal value which contains fractions.

A continuous variable represents an estimate of probability density function from histogram. For this purpose, we establish a certain number of equal length intervals of random variable and compute the frequency of occurrence in each of these intervals. In practice, one determines the lowest and highest sample values and divides the range into equal length called class width.

The number of sample values (observations) in a class interval is called frequency, and grouped into a tabular form is called frequency table; the graphical form is called histogram. The summation of probability density estimate of each interval sums up to one.

If one wants the best estimate of the random variable, the mean value, i.e., the quantitative measure of the data centre is commonly used. Most data analyses have the following three descriptive measures—central tendency measures, dispersion measures and percentiles measures.

Central Tendency Measures. The different parameters which describe the central tendency measures are average or mean value, mean value and mode.

The sample mean, $\overline{x}$ of a set of observations (ungrouped data) is given by

$$\overline{x} = \frac{1}{n}\sum_{i=1}^{n} x_i \qquad \ldots(1.1)$$

where n = number of observations, x_i = sample point.

Median is the middle value of the observations when the values of x_i are arranged in a ranked order. If n is an odd number, the median is the value with a rank of $\frac{n+1}{2}$. If n is an even number, then the median equals the average of $\frac{n}{2}$ and $\frac{n+1}{2}$.

Mode value is the most frequently observed data. Mode can be determined with the aid of a histogram.

Dispersion Measures. Dispersion measures describe the level of scatter in the data about central tendency point. Range, variance, standard deviation and coefficient of variation (CoV) are the general measures of dispersion measures.

Range, R. It is given by the difference of the largest and the smallest values of n values of the observation. This descriptive parameter is seldom used as it gives no clear picture of the distribution pattern between the minimum and maximum values.

Variance, s^2. The variance of n observations in a sample (ungrouped data) that are given equal weight is given by

$$s^2 = \frac{1}{n-1}\sum_{i=1}^{n}\left(x_i - \overline{x}\right)^2 \qquad \ldots(1.2)$$

The sum of deviations is always zero, i.e. $\sum_{i=1}^{n}(x_i - \overline{x}) = 0$. The term ($n$ – 1) is used instead of n as there are always (n – 1) independent deviations $(x_i - \overline{x})$. As their sum is always zero, the value of any one particular deviation is always equal to the negative of the sum of other (n – 1) deviations.

Standard deviation, s. It is the positive square root of the variance. It is given by the following expression for an ungrouped data as

$$s = \sqrt{\frac{1}{n-1}\sum_{i=1}^{n}(x_i - \overline{x})^2} \qquad \ldots(1.3)$$

The relative variation between several sets of data is measured by coefficient of variation and is given by the ratio of standard deviation to the mean of the sample.

$$\text{Coefficient of variation, CoV} = \frac{s}{\overline{x}} \qquad \ldots(1.4)$$

Quartile and Other Percentiles. As the median divides a set of data into two equal halves, other division points can also be considered. When an ordered data set is divided into quarters the resulting division points are also called quartiles. The first quartile is the value that has one fourth or 25% of observations below its value.

If the data is grouped the mean, $\overline{x}$ is given by

$$\overline{x} = \frac{\sum_{i=1}^{n} x_i f_i}{\sum_{i=1}^{n} f_i} \qquad \ldots(1.5)$$

and variance, s^2 of a grouped data is given by

$$s^2 = \frac{\left[n\sum_{i=1}^{n} x_i^2 f_i - \left(\sum_{i=1}^{n} x_i f_i\right)^2\right]}{n(n-1)} \qquad \ldots(1.6)$$

where x_i is the data, for the ith class and f_i is the corresponding class frequency.

The first moment of a probability distribution about the origin denotes the mean, while the first central moment about the mean is zero. The second central moment $E(x_i - \overline{x})^2$ denotes the variance. The third central moment $E(x_i - \overline{x})^3$ is a measure of the degree of skewness or asymmetry. The fourth central moment $E(x_i - \overline{x})^4$ provides a measure of the peakedness is called Kurtosis of a distribution.

Since the units of the third central moment are the cube units of the variable, to provide an absolute measure of skewness, Pearson proposed that its value be divided by the cube of standard deviation to yield the dimensionless coefficient of skewness,

$$r_1 = \frac{E\left[(x_i - \overline{x})^3\right]}{\left(\sigma[x_i]^3\right)} \qquad \ldots(1.7)$$

If r_1 is positive, the long tail of the distribution is on the right side of the mean; if it is negative, the long tail is on the left side of the mean as shown in Fig. 1.1.

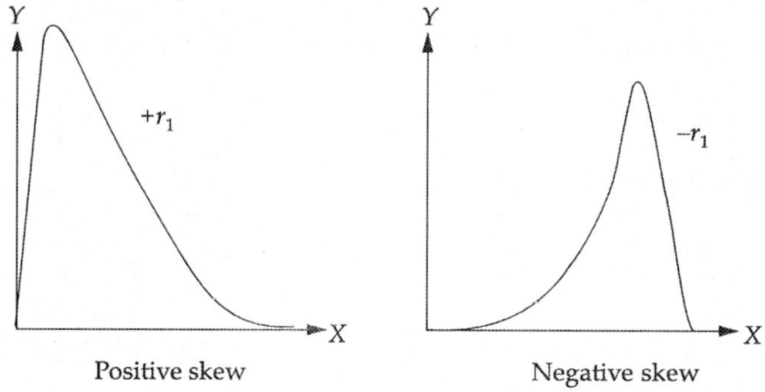

Fig. 1.1 Skewness of distribution

Pearson also proposed the dimensionless coefficient of Kurtosis as a measure of peakedness,

$$r_2 = \frac{E\left[(x_i - \bar{x})^4\right]}{\left(\sigma[x_i]^4\right)} \quad \text{...(1.8)}$$

For distribution that are flat, $r_2 < 3.0$

$$r_2 \geq r_1^2 + 1 \quad \text{...(1.9)}$$

Example 1.1 The compressive strength in N/mm^2 of 20 concrete cubes from a building project is as follows:

17.24	19.73	17.60	19.82	21.42	13.60	13.96	13.87	15.65	13.96
16.18	17.24	18.76	20.07	22.31	14.98	15.64	15.75	16.27	15.58

Find the range, mean, variance, standard deviation and the skewness coefficient.

Solution:

Sl. No.	Cube strength N/mm^2 (x_i)	$(x_i - \bar{x})^2$	$(x_i - \bar{x})^3$
1	17.24	0.0676	0.0176
2	19.73	7.5625	20.79
3	17.60	0.3844	0.24
4	19.82	8.0656	22.906
5	21.42	19.713	87.53

Contd.

6	13.60	11.424	– 38.61
7	13.96	9.120	– 27.54
8	13.87	9.672	– 30.08
9	15.65	1.768	– 2.353
10	13.96	9.120	– 27.54
11	16.18	0.640	– 0.51
12	17.24	0.067	0.018
13	18.76	3.168	5.64
14	20.07	9.548	29.50
15	22.31	28.408	151.42
16	14.98	4.000	– 8.00
17	15.64	1.795	– 2.41
18	15.75	1.512	– 1.86
19	16.27	0.504	– 0.36
20	15.58	1.960	– 2.744
Σ =	**339.63**	**128.505**	**175.47**

Range of the compressive strength from the given data is 13.60 N/mm^2 to 22.31 N/mm^2

$$\text{Mean, } \bar{x} = \frac{\Sigma x_i}{n} = \frac{339.63}{20} = 16.98\ N/mm^2$$

$$\text{Variance, } s^2 = \frac{\Sigma(x_i - \bar{x})^2}{n-1} = \frac{128.63}{19}\ 6.77\ N/mm^2$$

$$\text{Standard deviation, } s = \sqrt{\text{Variance}} = \sqrt{6.77} = 2.601\ N/mm^2$$

$$\text{Coefficient of skewness, } r_1 = \frac{\Sigma(x_i - \bar{x})^3}{n \times s^3} = \frac{175.47}{(20 \times 17.615)} = 0.4980$$

Example 1.2 The field data of soil samples collected from various depths are given below.

Depth (m)	2	3	4	5	6	7
Shear strength (kN/m^2)	14.8	20.3	32.2	39.0	42.0	56.2

Obtain the co-variance, correlation coefficient between the depth and soil shear strength.

Solution:

Sl. No.	Depth (m), x_i	Shear strength (kN/m^2), y_i	$(x_i - \overline{x})(y_i - \overline{y})$	$(x_i - \overline{x})^2$	$(y_i - \overline{y})^2$
1	2	14.8	48.2	6.25	371.72
2	3	20.3	20.67	2.25	189.89
3	4	32.2	0.94	0.25	3.53
4	5	39.0	2.46	0.25	24.21
5	6	42.0	11.88	2.25	62.73
6	7	56.2	55.3	6.25	489.29
Σ	**27**	**204.5**	**139.45**	**17.5**	**1141.37**

Mean depth, $\overline{x} = \dfrac{\Sigma x_i}{n} = \dfrac{27}{6} = 4.5$ m

Mean soil shear strength, $\overline{y} = \dfrac{\Sigma y_i}{n} = \dfrac{204.5}{6} = 34.08$ kN/m^2

Variance in depth, $s_x^2 = \dfrac{\Sigma(x_i - \overline{x})^2}{n-1} = \dfrac{17.5}{(6-1)} = 3.5$ m

Variance in soil shear strength, $s_y^2 = \dfrac{\Sigma(y_i - \overline{y})^2}{n-1} = \dfrac{1141.37}{(6-1)} = 228.27$ kN/m^2

Standard deviation in depth, $s_x = \sqrt{3.5} = 1.87$ m

Standard deviation in soil shear strength, $s_y = \sqrt{228.27} = 15.10$ kN/m^2

Sample co-variance, $s_{xy} = \dfrac{\Sigma(x_i - \overline{x})(y_i - \overline{y})}{n-1} = \dfrac{139.45}{5} = 27.89$

Sample correlation coefficient, $r_{xy} = \dfrac{s_{xy}}{s_x \times s_y} = \dfrac{27.89}{(1.87 \times 15.1)} = 0.9877.$

Example 1.3 Three hundred and three tensile pieces of a certain new brittle lacquer (used for experimental stress analysis) gave the tensile strength which is as given below.

Sample number	Strength interval (kN/m^2)	Class midpoint	Frequency	Relative frequency	Cumulative frequency
1	200–230	215	7	0.023	0.023
2	230–260	245	30	0.099	0.122
3	260–290	275	50	0.165	0.287
4	290–320	305	77	0.254	0.541
5	320–350	335	53	0.175	0.716
6	350–380	365	40	0.132	0.848
7	380–410	395	35	0.116	0.964
8	410–440	425	6	0.020	0.984
9	440–470	455	3	0.010	0.994
10	470–500	485	2	0.006	1.000
		Σ Total	**303**		

(a) Calculate the sample coefficient of variation

(b) The coefficient of skewness is defined as the third central moment divided by the cube of the standard deviation. Calculate the coefficient of skewness.

(c) The coefficient of kurtosis is defined as the fourth central moment divided by the fourth power of the standard deviation. Calculate the coefficient of kurtosis.

(d) Draw the histogram, and compute the mean tensile strength, variance and standard deviation.

Solution: When the number of observations in each increment is divided by 303 (i.e., total number of observations, *N*), one obtains the relative frequency of occurrences of the random variable in the bar graph as shown in Fig. 1.2. Frequency polygon is obtained by joining the mid points of the vertical bars.

The number of class intervals is obtained by Sturge's formula $N_c = 1 + 3.3 \log_{10} (303) = 9.18$

where, N_c is the number of class intervals.

The class interval is chosen as ten.

$$\text{Sample mean} = \bar{x} = \sum \frac{f_x}{N} = 97155/303 = 320.64 \text{ kN/m}^2$$

Sl. No.	Class midpoint (x_i)	Frequency (f_i)	$(x_i \times f_i)$	$(x_i^2 \times f_i)$	$(x_i^3 \times f_i) \times 10^9$	$(x_i^4 \times f_i) \times 10^{11}$
1	215	7	1505	323575	0.0696	0.1496
2	245	30	7350	1800750	0.4412	1.0809
3	275	50	13750	3781250	1.0398	2.8596
4	305	77	23485	7162925	2.1847	6.6633

Contd.

Sl. No.	Class midpoint (x_i)	Frequency (f_i)	$(x_i \times f_i)$	$(x_i^2 \times f_i)$	$(x_i^3 \times f_i) \times 10^9$	$(x_i^4 \times f_i) \times 10^{11}$
5	335	53	17755	5947925	1.9926	6.6751
6	365	40	14600	5329000	1.9451	7.0996
7	395	35	13825	5460875	2.1570	8.5203
8	425	6	2550	1083750	0.4606	1.9575
9	455	3	1365	621075	0.2826	1.2858
10	485	2	970	470450	0.2282	1.1066
	Σ = Total	303	**97155**	**31.98×10^6**	**10.80**	**37.39**

$$\text{Sample mean} = \overline{x} = \sum \frac{f_x}{N} = 97155/303 = 320.64 \text{ kN/m}^2$$

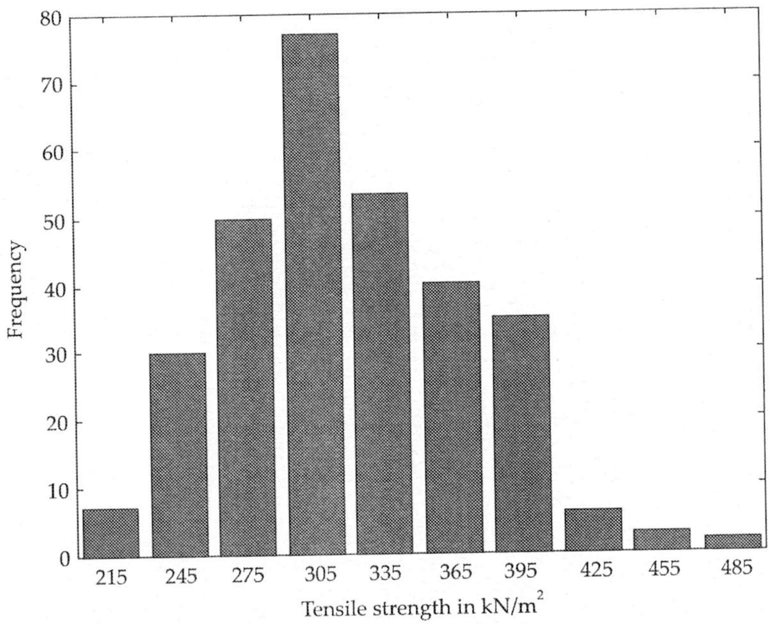

Fig. 1.2 Histogram

The sample variance is a measure of the data's dispersion. Physically, the sample variance can be visualised as a statistical counterpart of area moment of inertia. The sample variance is given by Eq. 1.6.

The sample standard deviation is defined as the square of the sample variance.

Sample standard deviation $s = \sigma[x_i] = 52.32 \text{ (kN/m}^2)$

The sample coefficient of variation is defined as the sample standard deviation divided by the sample mean.

The sample coefficient of variation = $(\text{CoV}) = \dfrac{s}{\bar{x}} = \left(\dfrac{52.32}{320.64}\right) = 16.13\%.$

Third central moment, skewness = $E\,(x_i - \bar{x})^3 = (x_i^3 \times f_i) = 10.80 \times 10^9\ (\text{kN/m}^2)^3$

Fourth moment, kurtosis = $E\,(x_i - \bar{x})^4 = (x_i^4 \times f_i) = 37.39 \times 10^{11}\ (\text{kN/m}^2)^4$

The coefficient of skewness is $r_1 = \dfrac{E[(x_i - \bar{x})^3]}{(\sigma\,[x_i]^3)} = 10.80 \times 10^9/(52.32)^3 = 75408.54$

The coefficient of kurtosis is $r_2 = \dfrac{E[(x_i - \bar{x})^4]}{(\sigma\,[x_i]^4)} = 37.39 \times 10^{11}/(52.32)^4 = 498981.59$

The cumulative frequency diagram is shown in Fig. 1.3.

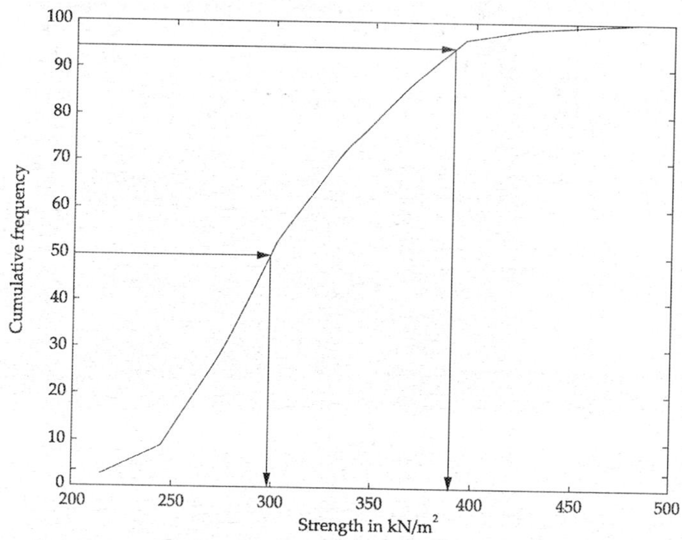

Fig. 1.3 Cumulative frequency

1.3 PROBABILITY AND EVENTS

1.3.1 Introduction

The word probability is used frequently in a sense that a certain event has a good chance of occurring. Mathematically, it is a numerical index that can vary from zero which implies an impossible event to unity which defines absolute certainty and it is shown in Fig. 1.4 in a probability scale. Why is probability necessary? A set of experiments is performed to measure the diameters of a group of rods. Even though the rods are supposed to have the same dimensions, they do not have same dimensions because of manufacturing imperfections and measurement errors.

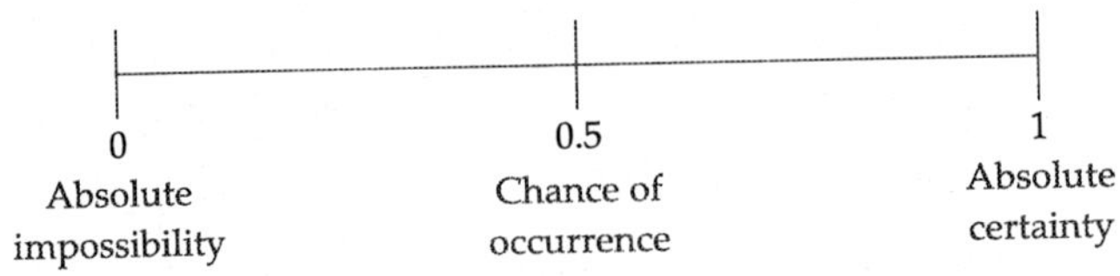

Fig. 1.4 Probability scale

The diameters of rods for the design are to be specified, there is no single value, but possibly many values. These possible values are called *events*. A way to organise this information is needed to accommodate all possibilities in a mathematical way so that computation is possible.

1.4 SETS

A set is a collection of data that share certain characteristics. We will examine the connection between events and probability. A *sample space* is the set of all possibilities. The number of bolts in a box is discrete data and size (diameters) of bolts is a continuous data. The group of possible diameters is a sample space, $S = \{d_1, d_2, d_3, d_n\}$, S is called sample space and d_1 is called realisation or sample point and one of the sample points is called realisation of that sample space. A subset of the sample space is an *event* within the sample space.

1.4.1 Basic Events

Several events are needed for a complete modelling framework.

1. An impossible event, denoted by { } or ϕ, is the event that contains all sample points in the sample space, the impossible or null event is an empty set.
2. A certain or universal event, denoted by S or ω, is the event that contains all the sample points. The certain event is the sample space itself.
3. A complementary event is denoted by the event symbol with an over bar. For example, event is represented as E, and the complementary event as $\overline{E}$, meaning that it contains all sample points in S that are not in E.

An useful visual representation of sets, spaces and their manipulations is provided by the *Venn diagram* as shown in Fig. 1.5.

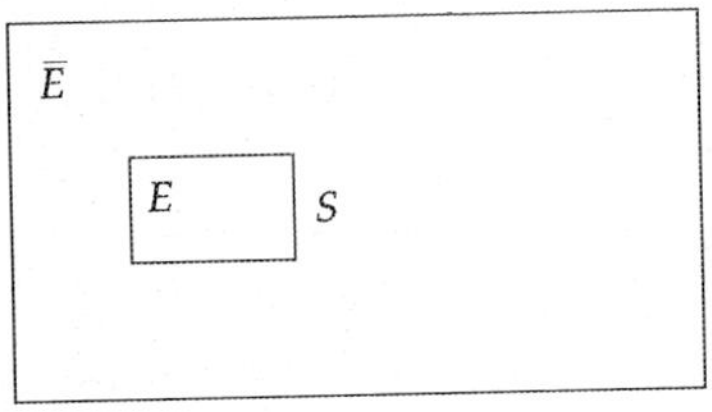

Fig. 1.5 Venn diagram

The probability of success and failure are defined as follows:

$$P(s) = \frac{\text{Number of successes}}{\text{Number of possible outcomes}} \quad \text{... (1.10)}$$

$$P(f) = \frac{\text{Number of failures}}{\text{Number of possible outcomes}} \quad \text{... (1.11)}$$

The sum of the probability is unity.

$$P(s) + P(f) = 1.0 \quad \text{... (1.12)}$$

Example 1.4 A fair coin is tossed consecutively three times and the total number of heads is counted. Find all the possible outcomes. Find the probability of three heads.

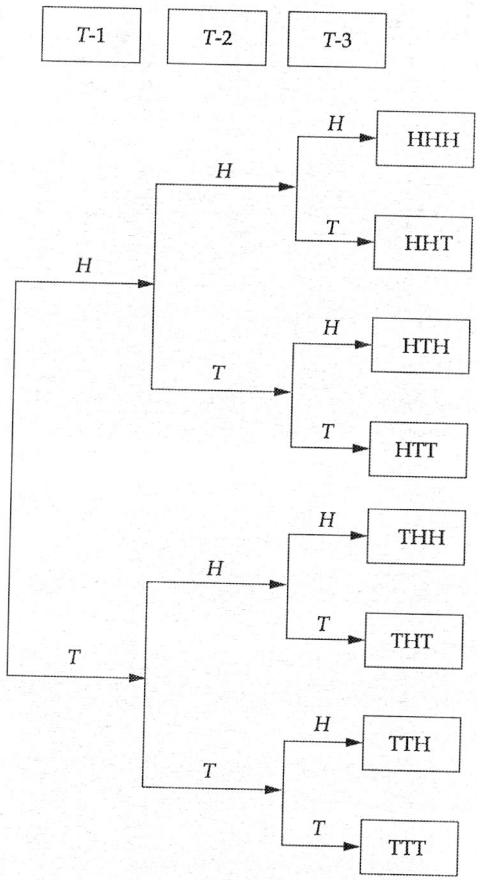

S = {HHH, HHT, HTH, THH, HTT, THT, TTH, TTT}

Fig. 1.6 Tree diagram

The certain or universal event is the set of all possible outcomes, i.e., the total number of heads counted.

$$S = \{0, 1, 2, 3\}$$

If A is the event that the total number of heads is greater than 3, then A is an impossible event, written as

$$A = \{\,\} \quad \text{or} \quad A = \phi.$$

Note that $\{0\}$ is not an impossible event, but the event that no heads appeared.

Let B be the event that the total number of heads is 2, that is $B = \{2\}$.

The complementary event of B is then $\overline{B} = \{0,1,3\}$

It should be noted that the impossible event is a complementary event of the universal event and vice versa, $\overline{S} = \phi$, $\overline{\phi} = S$

$$P(H=3) = \frac{\text{Number of successes}}{\text{Number of possible outcomes}} = \frac{1}{8} = 0.125.$$

Example 1.5 The frequency of vibration of structures varies from 0 Hz to 50 Hz. Find the probability of structural vibration in the interval 10 Hz to 20 Hz.

Solution: It may be assumed that the probability of structural vibration in a given interval is proportional and the sample space is shown in intervals of 1 Hz.

The frequency is in the range $\omega_1 < \omega < \omega_2$, where ω_1 is the lowest frequency and ω_2 is the highest frequency is an example of continuous sample space. The particular value $\omega_1 = 30.0$ is a sample point whose occurrence is a realisation point. All outcomes are equally likely and because there are an uncountable (infinite) number of events. All outcomes are equally likely.

$$P_{(\text{Event})} = \frac{\text{Favourable length}}{\text{Total length}}$$

$$P_{(10\text{ Hz to }20\text{ Hz})} = \frac{20-10}{50} = 0.2$$

In most engineering applications, evaluation of success and failure probabilities cannot be found from a simple knowledge of geometry, design or mathematical specification. It is necessary to obtain experimental evidence to determine the probabilities.

Suppose a point is chosen at random in a region in the plane the outcome of the experiment is said to be uniformly distributed on (or in) the region. As a special case, if a number x is chosen at random from an interval $[a, b]$ and a second number y is chosen (independently from the x choice) at random from the interval $[c, d]$, then the point (x, y) is uniformly distributed in a 2-dimensional rectangle. We choose as the mathematical model a sample space (universe) consisting of all points in the region, where

$$P_{(\text{Event})} = \frac{\text{Favourable area}}{\text{Total area}}$$

1.5 PRACTICAL ENGINEERING CONCEPTS

1.5.1 Objective Probability

Probability can be broadly divided into two categories, i.e., namely, objective probability and subjective probability (Fig. 1.7). The *classical* or *mathematical* or *a priori probability* because if we keep using orderly examples such as fair coins, unbiased dice and standard decks of cards, we can state the answer in advance (*a priori*) without tossing a coin, rolling a dice, or drawing a card. We do not have to perform experiments to make out probability statements about fair coins, standard decks and unbiased dice. Instead, we make *statements based on logical reasoning before any experiments takes place*.

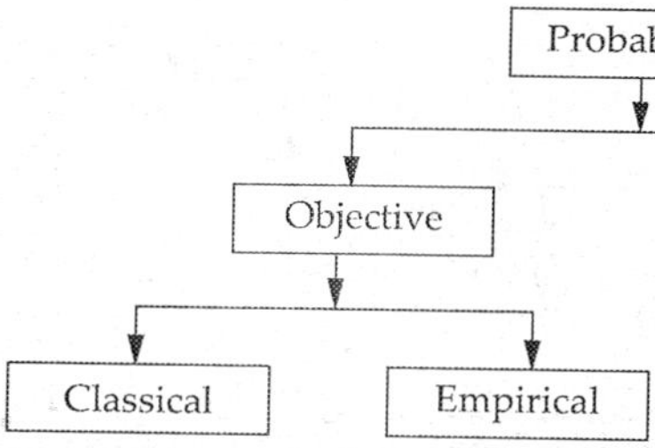

Fig. 1.7 Classification of probability

Real-life situations, data is random and disorderly. Hence probabilities are often expressed in a *statistical* or *empirical* or *frequentist approach*. Suppose we begin asking ourselves complex question such as, "What is the probability that I live up to 85"? or "What is the depth of water striking at a particular location"? This method uses the relative frequencies of past occurrences as probabilities.

We determine something which has happened in the past and use that information to predict the probability that it will happen again in the future. In statistical language, we would say that the relative frequency becomes stable as the number of tosses becomes large. This improved accuracy is not free; although more tosses produce a more accurate probability, we must bear in mind the time and cost of additional observations.

$$P_{(\text{Event occurring})} = \lim_{n \to \infty} \left(\frac{f}{n} \right).$$

1.5.2 Subjective Probability

Subjective probabilities are based on beliefs of the person making the probability assignment. In fact, subjective probability can be defined as the probability assigned to an event by an individual, based on evidence available. This evidence may be in the form of relative frequency of past occurrences, or it may be just an educated guess.

An engineer is deciding whether to allow the construction of a nuclear power plant on a site where there is some evidence of a geological fault. He must ask himself, "What is the probability of major nuclear accident at this location"? The fact that there is no relative frequency of occurrence evidence of previous accidents at this location does not excuse him from making a decision. He must use his best judgement in trying to determine the subjective probabilities of a nuclear accident.

Two geologists facing the problem of assessing the water depth at a site would come up with quite different subjective probabilities for the same event.

In most engineering applications, evaluation of success and failure probabilities cannot be found, from simple knowledge of geometry, design or mathematical specifications. Hence, it is necessary to obtain experimental data or evidence to determine the probabilities. For example, small components like pumps, electronic components etc., repeated testing of a large number of components is possible in mechanical, electrical, and electronics engineering. Whereas in civil engineering, large and expensive testing of dams, towers, power stations, chemical engineering plants, it is not possible or practical to test for destruction. Under these circumstances the system reliability evaluation is assessed by subdividing with the individual events to the probability of overall system.

1.6 VENN DIAGRAMS

The rules that will be discussed for combining probabilities are not always easily understood without the help of some pictorial representation of the concept. In set theory applications are normally made by the use of Venn diagrams. A Venn diagram is normally drawn as a rectangle which represents the total probability space *S*, shown in Fig. 1.5. The area *S* encloses or represents the entire space being considered. There may be two or more events within this space for which the probabilities must be combined. This consideration is restricted to two events *A* and *B*. If event *A* is totally enclosed by event *B*, then event *A* is a subset of event *B* (Fig. 1.8a). Mutually exclusive events do not overlap at all (Fig. 1.8b).

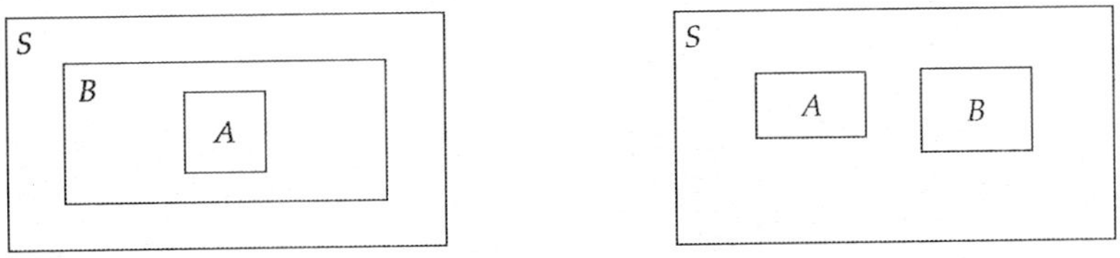

(a) Subset of event (b) Mutually exclusive event

Fig. 1.8

1.7 RULES FOR COMBINING PROBABILITIES

Rule 1 Independent Events

Two events are said to be independent if the occurrence of one event does not affect the probability of other event. No event is affected by the events preceding or following it.

Examples of Independent Event

The vehicles taking a left or right turn are independent (Fig. 1.9), wind or earthquake load acting on the structure and its strength. A fair dice and a fair coin are tossed. The two events are assumed to be independent, i.e., the outcome of one does not affect the

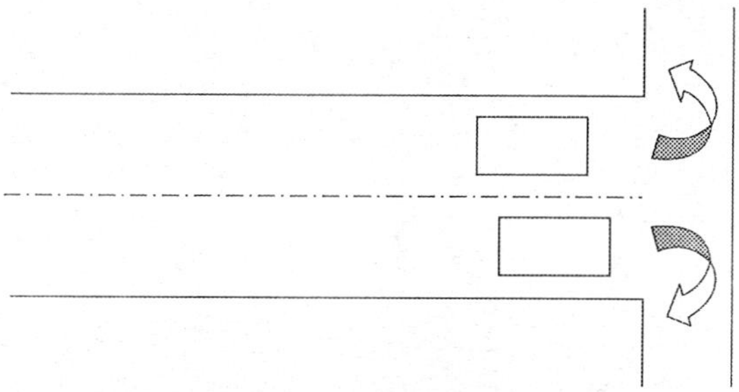

Fig. 1.9 Vehicles taking turn at a *T* junction

outcome of the other. Therefore, the fair dice showing any dots are not affected by the outcome of coin head.

Rule 2 Mutually Exclusive Events

Two events are said to be mutually exclusive (or disjoint) if they cannot occur at the same time.

Examples of Mutually Exclusive Events

(i) When tossing a coin the events are head or tail. These are mutually exclusive because both *cannot occur simultaneously.*

(ii) A vehicle on a road can take left or right turn, the *occurrence of one event excludes* the happening of the other.

(iii) A structural tension member can fail by yielding (ductile) or fracture (brittle).

(iv) A structure can fail or be safe; the occurrence of one event excludes the happening of the other.

(v) All outcomes are equally likely but we can't use Eq. 1.10 because there are uncountable (infinite) number of events. All outcomes are equally likely.

(vi) $P_{(\text{Event})} = \dfrac{\text{Favourable length}}{\text{Total length}}$

(vii) $P_{(\text{Event})} = \dfrac{\text{Favourable area}}{\text{Total area}}$

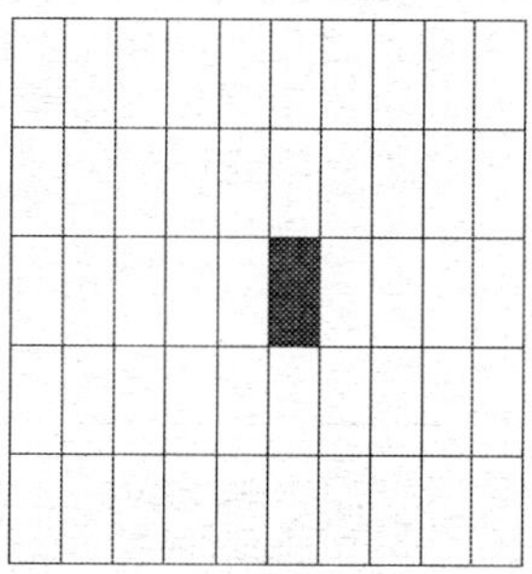

Fig. 1.10 Area of probability

The entire sample space is represented by a rectangle and events are represented by parts of a rectangle. If two events are mutually exclusive, their parts of the rectangle will not overlap, as shown in Fig. 1.10. The rectangle is divided into 50 equal, non-overlapping parts. We shall let the rectangle have an area of 1 (because the probability of certain event is 1). The probability of an event is the area of its parts of the rectangle. There the rectangle is divided into 50 equal, non-overlapping parts. The area of any square is 0.02 (1/50).

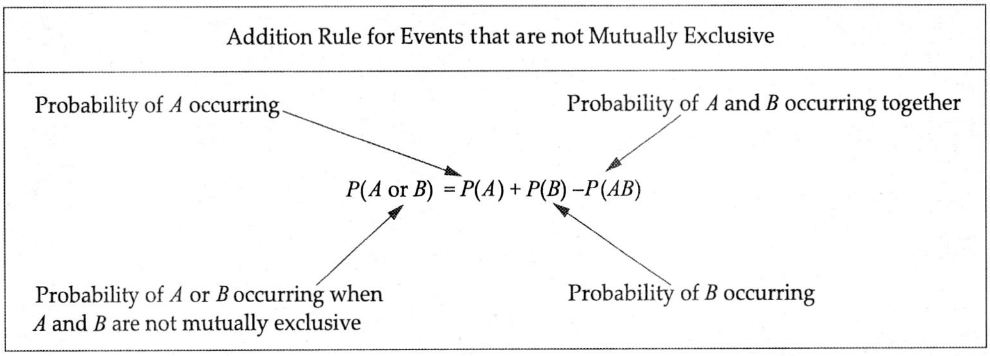

Rule 3 Complementary Events

The two outcomes of an event are said to be complementary, when one outcome does not occur, the other must. This is represented by Venn diagram shown in Fig. 1.11.

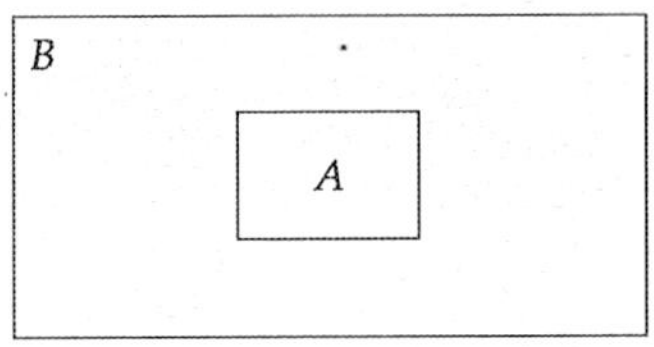

Fig. 1.11 Complementary event

The two outcomes A and B have probabilities $P(A)$ and $P(B)$, then
$P(A) + P(B) = 1$ or $P(B) = P(A^c)$ where, $P(A^c)$ is the probability of A NOT occurring.

Example 1.6 A structure can be in safe event (S) or failure event (F). State the outcomes are complementary.

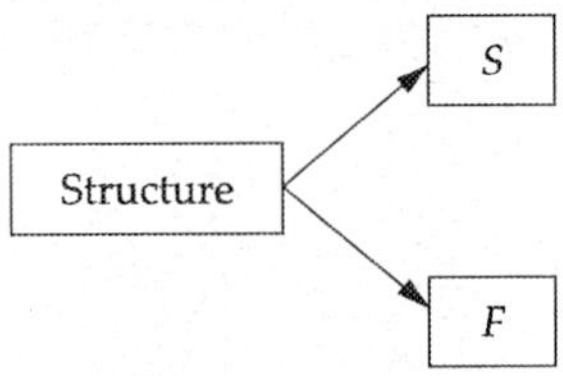

$$P(S) + P(F) = 1.0 \quad \text{or} \quad P(F) = P(S^c)$$

It is evident that complementary events are also mutually exclusive but the converse in not necessarily true, i.e., two mutually exclusive events are not necessarily complementary.

1.8 PROBABILITIES UNDER CONDITIONS OF STATISTICAL INDEPENDENCE

When two events occur, the outcome of the first event may or may not have an effect on the outcome of the second event, i.e., the events may be either dependent or independent. The occurrence of one event *has no effect* on the probability of the occurrence of any other event and it is called *statistical independence*. There are three types of probabilities under statistical independence:

(i) Marginal
(ii) Joint
(iii) Conditional

Conditional Probability of Statistically Independent Event

Probability of Event B given that A has occurred → $P(B|A) = P(B)$ ← Probability of Event B

$$P(B|A) = P(B)$$

Rule 4 Conditional Events: Multiplication Rule

Conditional events are events which occur conditionally on the occurrences of another event or events.

Consider two events A and B and also consider the probability of A occurring under the condition that event B has occurred. This is described mathematically as $P(A|B)$ in which the vertical bar is interpreted as GIVEN and the complete probability expression is interpreted as the conditional probability of A occurring GIVEN that B has occurred.

$$P(A|B) = \frac{\text{Number of ways } A \text{ and } B \text{ can occur}}{\text{Number of ways } B \text{ can occur}}$$

Joint Probability for Statistically Dependent Events

Joint probability of A and B occurring together or in succession happened → $P(BA)$

Probability of event B given that event A has happened → $P(B|A)$

Probability of event A happening → $P(A)$

$$P(BA) = P(B|A) \times P(A)$$

The number of ways *A* and *B* can occur is the hatched area shown in Fig. 1.12 and is represented mathematically as $(A \cap B)$, the probability that of which can be expressed as

$$P(A \mid B) = \frac{P(A \cap B)}{P(B)} \quad \ldots(1.13)$$

$$P(B \mid A) = \frac{P(A \cap B)}{P(A)} \quad \ldots(1.14)$$

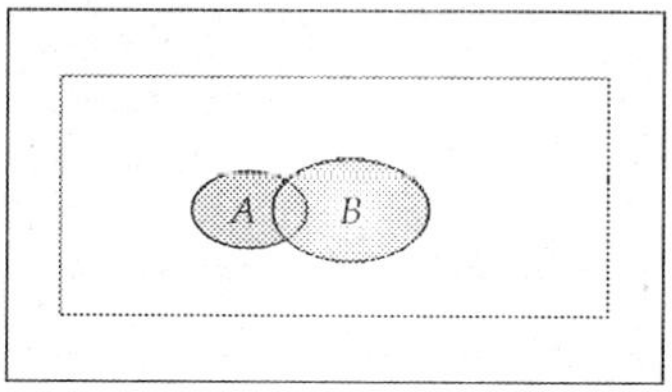

Fig. 1.12 Intersection of events

Example 1.7 A 50 kilometre road length, shown in Fig. 1.13, has uniform traffic volume and accidents are equally likely to occur anywhere on the highway. What is the probability of accident in stretch *A* on the condition that an accident in stretch *B* has occurred?

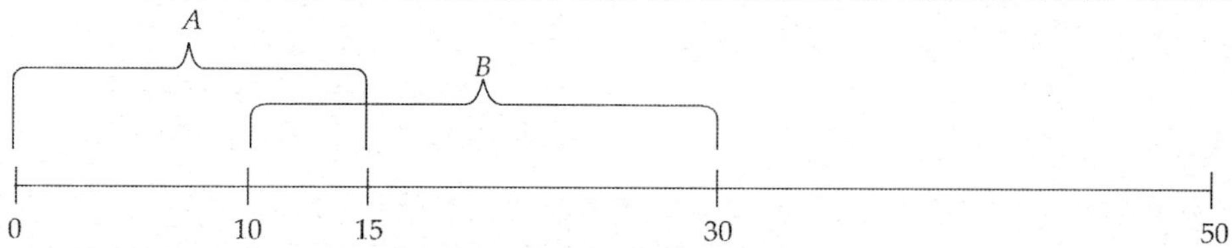

Fig. 1.13 Intersection of events on a highway

Solution: The accidents are equally likely anywhere along the highway. It may be assumed that the probability of an accident in a given interval of the highway is proportional to the distance of the interval. (i) *A* = an accident in kilometres 0 to 15 (ii) *B* = an accident in kilometres 10 to 30.

$$P_{(\text{Event})} = \frac{\text{Favourable length}}{\text{Total length}}$$

$$\text{Marginal probability of event } P(A) = \frac{15}{50} = 0.3$$

$$\text{Marginal probability of event } P(B) = \frac{(30-10)}{50} = \frac{20}{50} = 0.4$$

$$P(A \cap B) = \frac{5}{50} = 0.1$$

$$P(A \mid B) = \frac{P(A \cap B)}{P(B)} = \frac{0.1}{0.4} = 0.25$$

Rule 5 Simultaneous Occurrences of Events

The simultaneous occurrences of events A and B are the occurrences of both A and B. Mathematically, it is known as intersection of the two events and is represented as

$$P(A \cap B), (A \text{ and } B) \text{ or } (AB)$$

In this rule there are two cases to consider, the first is when the two events are independent and the second is when they are dependent.

(a) Events are independent

If two events are independent, then the probability of occurrence of each event is not influenced by the probability of occurrence of the other. In this case

$$P(A \mid B) = P(A) \text{ and } P(B \mid A) = P(B)$$

or

$$P(A \cap B) = P(A) \times P(B)$$

If there n independent events, the principle can be extended to give

$$P(A_1 \cap A_2 \cap A_3 \cap \ldots \ldots A_i \cap \ldots. A_n) = \prod_{i=1}^{n} P(A_i) \qquad \ldots(1.15)$$

Example 1.8 A truss is loaded as shown in Fig. 1.14. The probability of failure of individual members of the trusses A and B are 0.05 and 0.04 respectively. Assuming the failure of truss members is statistically independent find the probability of failure of the truss.

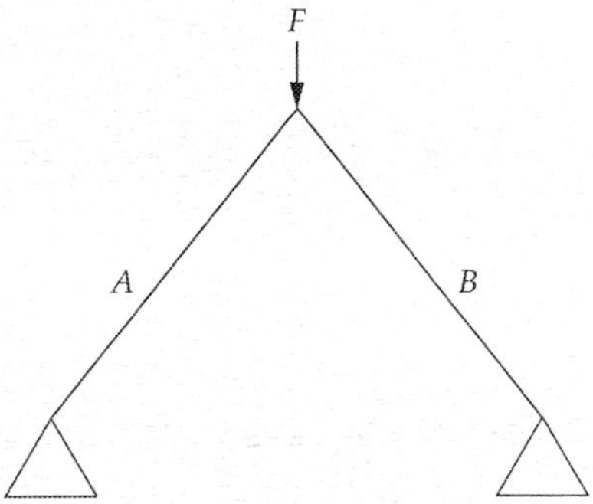

Fig. 1.14 Statistically independent events of trusses

$$P(A \text{ failure} \cap B \text{ failure}) = P(A \text{ failure})\, P(B \text{ failure}) = 0.05 \times 0.04 = 0.002$$

(b) Events are dependent

If two events are not independent, the probability of occurrence of one event is influenced by the probability of occurrence of the other. In this case Eqs 1.13 and 1.14 can still be used but not simplified as in the case of independence.

$$P(A \cap B) = P(B \mid A) \times P(A)$$

$$P(A \cap B) = P(A \mid B) \times P(B) \qquad \ldots(1.16)$$

Example 1.9 A manufacturer orders for ready mix concrete from three different suppliers. From the past records, the manufacturer finds that the three suppliers S_1,

S_2, and S_3 have the following percentage of defective concrete, i.e., 3%, 5% and 4% respectively. The concrete supplied by suppliers is 500 cum, 200 cum, and 350 cum from supplier 1, 2 and 3 respectively. What is the probability that the concrete is defective?

Solution: Total quantity of concrete is 1050 cum; the probability that the concrete supplied by supplier 1 at random is

$$P(S_1) = 500/1050 = 10/21.$$

The probability that the concrete supplied by supplier 2 at random is

$$P(S_2) = 200/1050 = 4/21$$

The probability that the concrete supplied by supplier 3 at random is

$$P(S_3) = 350/1050 = 7/21$$

The total probabilities of defective concrete supplied by the three companies are
The probability of defective $P(D) = 0.014 + 0.00952 + 0.0133 = 0.0368$
The defective concrete supplied by all the three companies is 3.68%

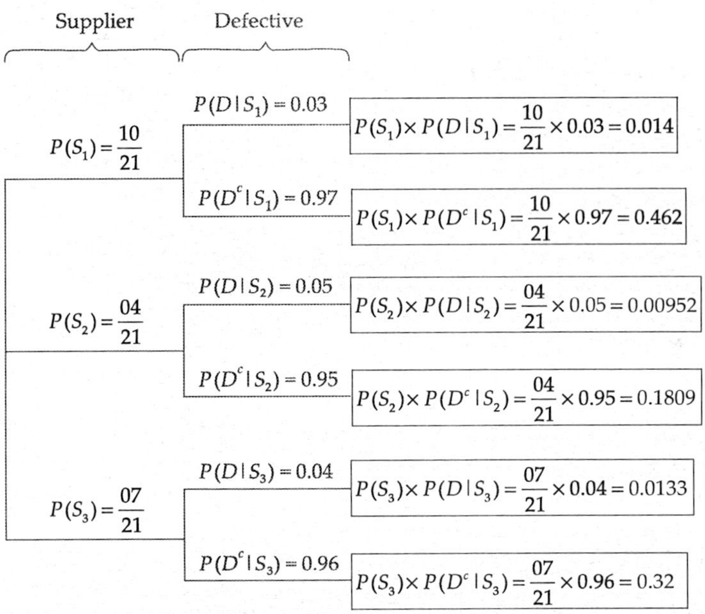

Fig. 1.15 Dependent events

Rule 6 Occurrence of at Least of Two Events

The occurrence of at least one of two events A and B is the occurrence of A or B or both.

In terms of Venn diagram, this occurrence is the hatched area shown in Fig. 1.12. Mathematically, it is known as the union of the two events and is expressed as

$$(A \cup B), (A \text{ or } B) \text{ or } (A + B)$$

In this rule *there are **three** cases to consider*; the events are independent but mutually exclusive, the events are independent and mutually exclusive and finally the events are not independent.

(a) Events are independent but not mutually exclusive
There are two ways in which the appropriate equation can be deduced using an analytical method and using the Venn diagram.

$$P(A \cup B) = P(A \text{ or } B \text{ or both } A \text{ and } B) = P(A) + P(B) - P(A) \times P(B) \quad ...(1.17)$$

(b) Events are independent and mutually exclusive.

$$P(A \cup B) = P(A \text{ or } B \text{ or both } A \text{ and } B) = P(A) + P(B) \quad ...(1.18)$$

(c) Events are not independent

$$P(A \cup B) = P(A) + P(B) - P(A \cap B) \quad ...(1.19)$$

$$P(A \cup B) = P(A) + P(B) - P(A \mid B) \times P(B) \quad ...(1.20)$$

$$P(A \cup B) = P(A) + P(B) - P(B \mid A) \times P(A) \quad ...(1.21)$$

Example 1.10 The probability that single pump will fail to operate over a period of 1 year is 5 per cent. Pump failures are assumed to be independent events. The pumps are in series. Find the probability of failure of pump.

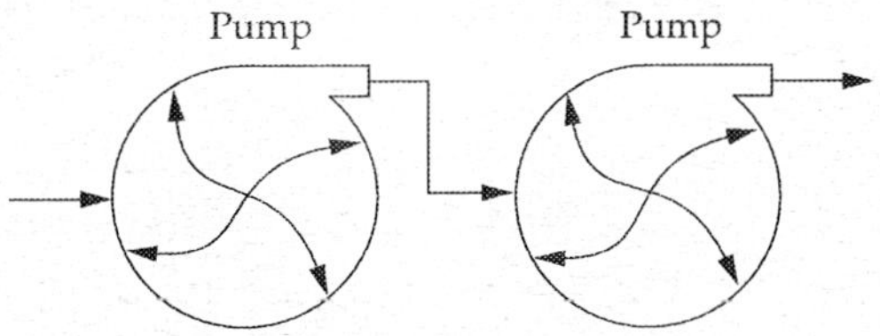

Fig. 1.16 Pumps in series

Solution:

$$P(P_1 \cup P_2) = P(P_1) + P(P_2) - (P_1 \cap P_2)$$

$$P(P_1 \cup P_2) = P(P_1) + P(P_2) - P(P_1) \times P(P_2)$$

$$P(P_1 \cup P_2) = 0.05 + 0.05 - 0.05 \times 0.05 = 0.0975$$

Example 1.11 The foundation of a tall building may fail either by bearing capacity, or by excessive settlement. Let B and S represent the failure modes. The $P(B) = 0.001$ and $P(S) = 0.008$ and $P(B \mid S)$ equals the probability of failure in bearing capacity given that it has excessive settlement = 0.1, determine the probability failure of the foundation.

Solution:

$$P(F) = P(B \cup S) = P(B) + P(S) - P(B \mid S) \times P(S)$$

$$P(F) = 0.001 + 0.008 - 0.1 \times 0.008 = 0.0082$$

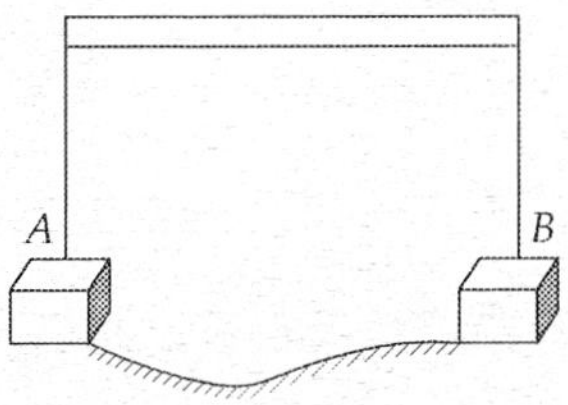

Fig. 1.17 Settlement of a frame

Rule 7 Applications of Conditional Probability (Total Probability)

The concept of conditional probability was introduced in Rule 4 when it was considered that an occurrence of event A was dependent on another event B.

This principle can be extended to consider occurrence of an event B which is dependent on a number of mutually exclusive events A_i.

$$P(A \cap B) = P(A \mid B) \times P(B)$$

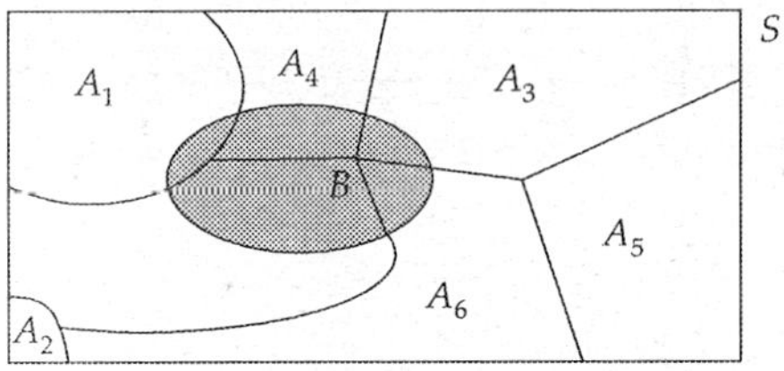

Fig. 1.18 Total probability of mutually exclusive event

$$P(B) = P(A_1 \cap B) + P(A_2 \cap B) + \ldots\ldots + P(A_n \cap B)$$

$$P(B) = \sum_{i=1}^{n} P(B \mid A_i) \times P(A_i) \qquad \ldots(1.22)$$

Example 1.12 A factory production line is manufacturing bolts at two plants. Plant 1 meets 70% of the requirement and Plant 2 meets 30%. It is known from previous experience that Plant 1 produces 90% standard and Plant 2 only 80%. Evaluate how many will be meeting the standard if the 1000 bolts are purchased by a customer.

A = the event that the bolt meets the standard

B_1 = the event that bolt meets the standard, which is produced by Plant 1

B_2 = the event that bolt meets the standard, which is produced by Plant 2

$$P(A \mid B_1) = 0.90;\ P(A \mid B_2) = 0.8;\ P(B_1) = 0.7;\ P(B_2) = 0.3$$

$$P(A) = P(A \mid B_1) \times P(B_1) + P(A \mid B_2) \times P(B_2)$$

$$P(A) = 0.9 \times 0.7 + 0.8 \times 0.3 = 0.87$$

For every purchase of 1000 bolts by the customer 870 bolts meet the standard.

Rule 8: De Morgan's Rule

In set theory the De Morgan's rule relates sets and their complements. *The complements of unions and intersections are equal to the intersections and unions of the respective complements.*

$$\overline{E_1 \cup E_2} = \overline{E_1} \cap \overline{E_2} \qquad \ldots(1.23)$$

$$\overline{E_1 \cup E_2 \cup \ldots E_n} = \overline{E_1} \cap \overline{E_2} \cap \ldots \overline{E_n} \qquad \ldots(1.24)$$

$$\overline{\overline{E_1} \cup \overline{E_2} \cup \ldots \cup \overline{E_n}} = E_1 \cap E_2 \cap E_3 \cap \ldots E_n \qquad \ldots(1.25)$$

$$\overline{E_1 \cap E_2 \cap E_3 \cap \ldots E_n} = \overline{E_1} \cup \overline{E_2} \cup \overline{E_3} \cup \ldots \overline{E_n} \qquad \ldots(1.26)$$

This is the duality relation. De Morgan's theorem can be proved by Fig. 1.19.

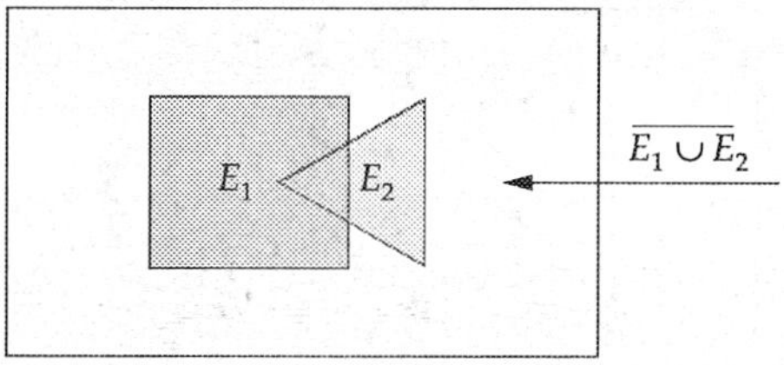

Fig. 1.19 De Morgan's theorem

Example 1.13 A pumping station consists of two pumps, as shown in Fig. 1.16. Clearly the pumps will fail if either pump breaks down; thus P_1 = failure of pump 1, and P_2 = failure of pump 2, then what is the reliability of pump.

Failure of pump = $P(P_1 \cup P_2)$

Apply De Morgan's rule

No failure of pump or reliability of pump

$$\overline{P_1 \cup P_2} = \overline{P_1} \cap \overline{P_2}$$
$$\overline{P_1 \cup P_2} = \overline{P_1}\ \overline{P_2}$$

Reliability = $(1 - P_1) \times (1 - P_2) = (1 - 0.05) \times (1 - 0.05) = 0.9025$

Rule 9: Baye's Theorem

Baye's theorem offers a powerful statistical method of evaluating new information and revisiting our prior estimates (based upon limited information only) of the probability that things are in one state or another. If correctly used, it makes it unnecessary to gather masses of data over a long period of time in order to make good decisions based on probabilities.

We begin with an analysis with prior probability estimates (hypothesis) for specific events of interest. Then from sources such as a sample, a specific report or a product test we obtain some additional new information (evidence) about the events, with this new information, we update the prior probability values by computing revised probabilities, referred to as posterior probabilities. Baye's theorem provides a means for making these posterior probabilities.

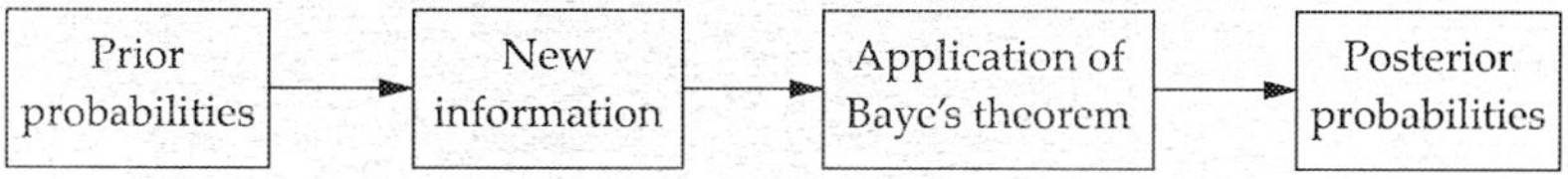

Fig. 1.20 Baye's theorem

$$P(H \mid E) = \frac{P(E \mid H) \times P(H)}{P(E \mid H_1) \times P(H_1) + P(E \mid H_2) \times P(H_2) + P(E \mid H_3) \times P(H_3)}$$

Prior $P(H)$
Posterior probability = $P(H \mid E)$
Likelihood of event = $P(E \mid H)$
Normalising $P(E) = P(E \mid H_1) \times P(H_1) + P(E \mid H_2) \times P(H_2) + P(E \mid H_3) \times P(H_3)$

Example 1.14 A factory production line is manufacturing bolts using three machines *A*, *B* and C. The total output of machine *A* is 25%, machine *B* for 35% and machine C for the 40%. It is known from previous experience with the machine that 5% of the output from machine *A* is defective, 4% from machine *B* and 2% from machine C. A bolt is chosen at random from the production line and found to be defective. What is the probability that came from (a) machine *A* (b) machine *B* (c) machine *C*?

Solution: Let D = {bolt is defective}, *A* = {bolt is produced from machine *A*}, *B* = {bolt is produced from machine *B*}, C = {bolt is produced from machine C}. We know that $P(A) = 0.25$, $P(B) = 0.35$ and $P(C) = 0.4$. Also $P(D \mid A) = 0.05$, $P(D \mid B) = 0.04$, $P(D \mid C) = 0.02$. A statement of Baye's theorem for three events *A*, *B* and *C* is

$$P(A \mid D) = \frac{P(D \mid A) \times P(A)}{P(D \mid A) \times P(A) + P(D \mid B) \times P(B) + P(D \mid C) \times P(C)}$$

$$P(A \mid D) = \frac{0.05 \times 0.25}{0.05 \times 0.25 + 0.04 \times 0.35 + 0.02 \times 0.4} = 0.3623$$

$$P(B \mid D) = \frac{P(D \mid B) \times P(B)}{P(D \mid A) \times P(A) + P(D \mid B) \times P(B) + P(D \mid C) \times P(C)}$$

$$P(B \mid D) = \frac{0.04 \times 0.35}{0.05 \times 0.25 + 0.04 \times 0.35 + 0.02 \times 0.4} = 0.4057$$

$$P(C \mid D) = \frac{P(D \mid C) \times P(C)}{P(D \mid A) \times P(A) + P(D \mid B) \times P(B) + P(D \mid C) \times P(C)}$$

$$P(C \mid D) = \frac{0.02 \times 0.4}{0.05 \times 0.25 + 0.04 \times 0.35 + 0.02 \times 0.4} = 0.2318$$

Example 1.15 A college has three hundred teaching faculty. During the past year 48 of the teaching faculty earned a pay rise. At the beginning of the year, the college offered a special training seminar which was attended by 138 teaching faculty. Twenty seven of the teaching faculty who earned a rise attended the training seminar.

(a) What is the probability that a teaching faculty who attended the seminar earned a pay rise?

(b) If a teaching faculty is selected at random, what is the probability that the person attended the seminar and earned a pay rise?

Solution:

(a) The probability of teaching faculty attended the seminar and earned a pay rise

$$P(B \mid A) = \frac{27}{138} = 0.1956$$

(b) The probability of selecting a teaching faculty at random and attended the seminar and earned a pay rise

$$P(A \cap B) = P(A) \times P(B \mid A) = \frac{138}{300} \times \frac{27}{138} = \frac{27}{300} = 0.09$$

Sl. No.	**Event**	**Prior** $P(A_i)$	**Conditional** $P(B \mid A_i)$	**Joint probabilities** $P_i(A \text{ and } B)$	**Posterior probabilities** $P(A_i \mid B)$
1	A_1	0.46	0.1956	0.0899	0.5620
2	A_2	0.54	0.1296	0.0699	0.4375
			Σ sum	**0.15996**	**1.0000**

The number of persons who got pay rise are 48, the number of persons who earned a pay rise and attended the seminar is 27, the rest 21 earned a pay rise and did not attend the seminar.

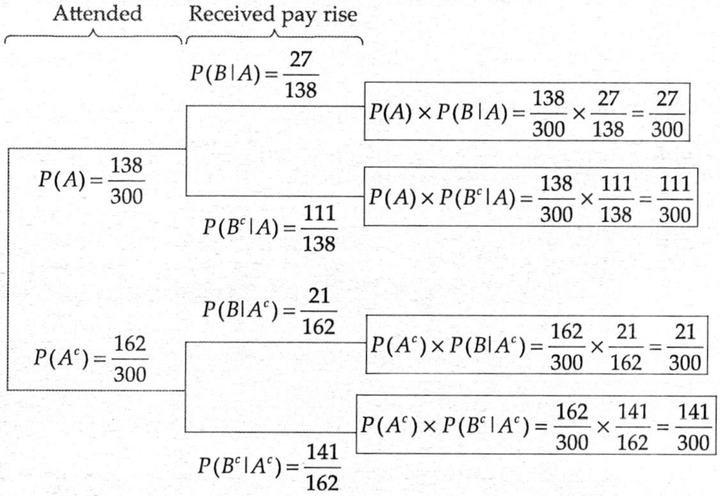

Fig. 1.21 Pay rise who attended the seminar

UNSOLVED PROBLEMS

1. The following test data for random concentrated load and a random uniformly distributed load which is acting on structure is given below.

Sample number	P_1(kN)	P_2(kN/m)	Sample number	P_1(kN)	P_2(kN/m)
1	22.7293	12.0545	11	24.8643	13.8641
2	34.0272	10.2886	12	33.4490	11.0183
3	31.3139	11.4124	13	29.9795	13.7911
4	26.4656	13.1344	14	31.3139	13.0651
5	39.2925	14.0246	15	26.2877	10.9891
6	23.6189	10.4054	16	26.3766	13.0177
7	28.6451	13.4117	17	26.0653	13.1198
8	29.8461	12.5799	18	29.9795	11.0475
9	26.9994	11.0767	19	26.2877	10.9891
10	26.3766	12.1420	20	24.3306	10.2544

(a) Plot the frequency histogram for each random load.
(b) Plot the cumulative frequency histogram for each random load.
(c) Plot the three-dimensional frequency histogram for the two random loads.
(d) Plot the three-dimensional cumulative frequency histogram for the two random loads.
(e) Calculate the sample mean, sample variance, and sample coefficient of variation for each random load.
(f) Calculate the sample covariance and sample correlation coefficient for the two random loads.

2. The shrinkage limit of a particular clay, the moisture content in percentage of dry weight was determined on 20 specimens with the following results.

15.2	16.7	15.8	14.6	18.1	17.2	18.0	15.9	16.1	16.9
14.8	17.6	18.2	16.9	17.3	16.5	15.6	16.7	15.8	18.2

Compute the mean, range, mean deviation, standard deviation and coefficient of variation.

3. Three hundred and three tensile pieces used for experimental stress analysis gave the following tensile strengths at the period of 7 days.

Sl. No.	Strength (kN/m^2)	Frequency	Sl. No.	Strength (kN/m^2)	Frequency
1	200–230	7	6	350–380	40
2	230–260	30	7	380–410	35
3	260–290	50	8	410–440	6
4	290–320	77	9	440–470	3
5	320–350	53	10	470–500	2

(a) Compute the frequency table.
(b) Draw the histogram and frequency polygon.
(c) Draw the cumulative frequency diagram.
(d) Calculate the mean, median and mode.
(e) Compute the standard deviation and coefficient of variation.

4. Five hundred and forty one passenger car spot speeds were found to be listed and it is given below.

Sl. No.	Speed interval (km/hr)	Frequency	Sl. No.	Speed interval (km/hr)	Frequency
1	8–16	0	7	56–64	31
2	16–24	1	8	64–72	16
3	24–32	34	9	72–80	3
4	32–40	146	10	80–88	2
5	40–48	178	11	88–96	0
6	48–56	130	12	–	–

(a) Compute the frequency table.
(b) Draw the histogram and frequency polygon.
(c) Draw the cumulative frequency diagram.
(d) Calculate the mean, median and mode.
(e) Compute the standard deviation, coefficient of variation and correlation.
(f) If the 85th and 98th percentiles are employed in speed regulation and design, respectively, compute the speed limit and the design speed for the locality.

5. On a construction site it was required to make concrete with a specified minimum compressive strength of 17 MPa. The minimum was understood to be a value exceeding by not less than 95% of test results. The values of strength at 28-day cube strength, in MPa of 50 test cubes are given below.

1	2	3	4	5	6	7	8	9	10
22.3	18.2	24.8	23.5	19.8	27.0	25.9	29.2	24.1	24.3
21.7	23.4	23.3	28.3	23.4	20.0	26.1	15.2	20.0	22.8
32.0	25.6	20.8	23.2	26.9	22.8	22.1	21.2	23.4	18.1
23.4	23.0	23.6	25.4	22.2	17.8	28.0	28.8	22.1	20.1
20.8	23.0	28.2	23.0	30.4	25.1	24.8	19.0	18.2	25.1

Calculate the mean strength, range, mean deviation, standard deviation, and coefficient of variation. Find whether the specification requirements are satisfied.

6. The following set of test data for two concentrated loads, P_1 and P_2, acting on the structure.

Sample Number	P_1(kN)	P_2(kN)	Sample Number	P_1(kN)	P_2(kN)
1	10.1414	20.3274	11	12.7213	20.0605
2	7.1613	14.1891	12	7.3837	20.1050
3	6.7610	11.7427	13	15.7459	22.5069
4	5.7824	14.7674	14	11.2979	15.9238
5	11.2090	12.3210	15	17.0358	18.2368
6	12.9437	13.7443	16	11.7427	22.7293
7	9.6522	19.9715	17	16.6800	17.4806
8	11.6982	17.6586	18	14.8118	15.1677
9	17.3027	14.6339	19	16.7590	23.6189
10	5.8269	19.9270	20	9.8746	20.5053

(a) Plot the frequency histogram for each random load.
(b) Plot the cumulative frequency histogram for each random load.
(c) Plot the three-dimensional frequency histogram for the two random loads.
(d) Plot the three-dimensional cumulative frequency histogram for the two random loads.
(e) Calculate the sample mean, sample variance and sample coefficient of variation for each random load.
(f) Calculate the sample covariance and sample correlation coefficient for the two random loads.

7. The following test data for random concentrated load and a random uniformly distributed load which is acting on structure is given below.

Sample number	P_1(kN)	P_2(kN/m)	Sample number	P_1(kN)	P_2(kN/m)
1	22.7293	12.0545	11	24.8643	13.8641
2	34.0272	10.2886	12	33.4490	11.0183
3	31.3139	11.4124	13	29.9795	13.7911
4	26.4656	13.1344	14	31.3139	13.0651
5	39.2925	14.0246	15	26.2877	10.9891
6	23.6189	10.4054	16	26.3766	13.0177
7	28.6451	13.4117	17	26.0653	13.1198
8	29.8461	12.5799	18	29.9795	11.0475
9	26.9994	11.0767	19	26.2877	10.9891
10	26.3766	12.1420	20	24.3306	10.2544

(a) Plot the frequency histogram for each random load.
(b) Plot the cumulative frequency histogram for each random load.
(c) Plot the three-dimensional frequency histogram for the two random loads.
(d) Plot the three-dimensional cumulative frequency histogram for the two random loads.
(e) Calculate the sample mean, sample variance and sample coefficient of variation for each random load.
(f) Calculate the sample covariance and sample correlation coefficient for the two random loads.

8. Three construction techniques are commonly used to transfer column load into the ground; footings, pile, and caissons. The location and type of structure dictate the technique to be used. Over a period of years it is noted that footings are used 40% of the time and other two share the remaining 60% equally. The probability of failure noted for each is 0.02 for the footings, 0.01 for the piles, and 0.005 for the caissons. If a structure shows intolerable distress, what is the probability that it has been placed on (a) footing, (b) a pile, or (c) caissons?
9. A boring is contemplated at the site of a structure. If it is performed, it will reveal whether the underlying soil is sand, silt, or clay. If the underlying soil is sand or silt, the column loadings of the structures can be transferred into the ground by using isolated footings. On the other hand, if the underlying soil is clay, caisssons or piles will need to be considered. Caissons will be used only if the underlying bedrock is

within 15 m of the ground surface; otherwise, piles are required. Construct a tree diagram for the various possibilities.

10. The probability that a bridge on a secondary highway will perform adequately for 20 years is $\frac{1}{3}$. The probability that the highway approaches will do the same is $\frac{1}{4}$. If both act independently, compute the probability that: (a) both will last more than 20 years; (b) at least one will be performing satisfactorily in 20 years; (c) neither will survive for 20 years; (d) only the bridge will survive for 20 years.
11. Two cables are used to lift load w normally only cable A will be carrying the load, cable B participate in carrying the load. But if cable A breaks, then B will have to carry the full load until A is replaced. The probability that cable A will break is 0.02. The probability that B will fail if it has to carry the load by itself is 0.3. (a) What is the probability that both cables break; (b) if the load remains lifted, what is the probability that none of the cable failed?
12. The reliability of one single path of a pump is 0.4. To obtain system reliability of 0.8, compute the number of parallel paths.

2

CHAPTER

Random Variables

2.1 INTRODUCTION

Many random phenomena in engineering are associated with numerical outcomes of some physical quantity. The possible outcome of a random phenomenon can be identified numerically or artificially. *A random variable is a numerical variable whose specific value cannot be predicted with certainty before an experiment.* The outcomes or events are identified by values of a function and such a function is called a random variable. Hence, a random variable is a function of sample space of an experiment such that there is a numerical value of the random variable corresponding to each outcome of an experiment. Different simple events in engineering like occurrence of earthquakes, strength of concrete, traffic count, runoff in hydrological studies, etc., may have some associated values relating to its certainty/uncertainty. The value assumed by a random variable depends on the outcome of the experiment and has probability associated with each occurrence of event in the sample space. In short, random variable is a tool or device to identify events in numerical terms. The probability measures associated with a numerical value of a random variable is described by probability distribution or probability law, i.e. Probability Mass Function (PMF)/Probability Density Function (PDF) and Cumulative Density Function (CDF).

2.2 TYPES OF RANDOM VARIABLE

Random variables can be discrete or continuous. Random variable can assume either integer values or fractional values. A variable which assumes integer values is called discrete variable and which assumes fractional value is called continuous variable.

2.2.1 Characteristic of a Random Variable

In science and engineering, we are interested in numerical outcomes. There are two types of outcome in random experiment, namely, numerical and descriptive. In a dice experiment we obtain the numerical (outcomes as 1, 2, 3, 4, 5, and 6) while tossing a coin the outcome turnout as head or tail, hence the sample space in non-numerical (outcome as head or tail) or descriptive. Instead of dealing with descriptive values, it is easy to assign numerical value say, '1' to head and '0' to tail. The interpretation is easy and attractive from mathematical point of view. This rule of *mapping* from the original sample space (numerical or descriptive) to a *numerical* (real value) sample space, subjected to certain constraints is a *random variable*. Hence, a random variable is a real valued function which is *single valued and not multi-valued*.

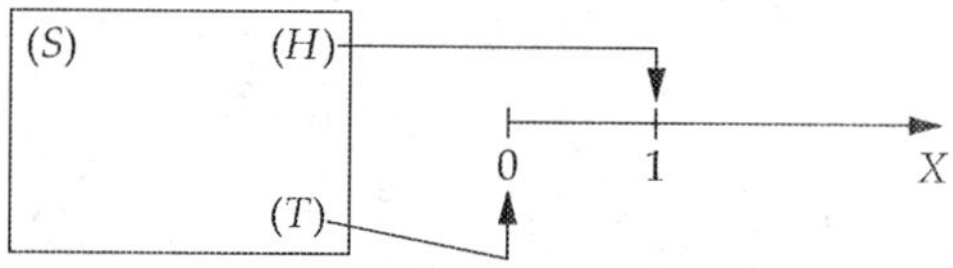

Fig. 2.1 Mapping of $S = (H, T)$ on to real line

2.2.2 Discrete Variable

The PMF of a discrete variable, X is given by the function $p_X(x) = P(X = x)$. The standard convention is to denote a random variable by an upper case letter, while the lower case letter is used to represent the outcome or realisation of the corresponding random variable. The statement $(X = x)$ means that the random variable X assumes a value x. If a and b are any given real numbers then $(a \le X \le b)$ will represent the statement that the random variable X assumes a value greater than a and less than or equal to b. It means probability of the variable X taking the value of x. Some examples of discrete random variables are:

(i) The number of earthquakes occurring in a specified time period
(ii) Marks obtained by a student in examination
(iii) Number of floods occurring in a specified time period.

Three conditions need to be satisfied for a valid PMF and are as follows:

(i) $0 \le p_X(x) \le 1$ ...(2.1)

(ii) $\sum_{i=0}^{n} p_X(x_i) = 1$ ...(2.2)

(iii) $P[a \le X \le b] = \sum_{x \ge a}^{x \le b} p_X(x_i)$ ...(2.3)

The CDF is given by $F_X(x)$ and is defined by $F_X(x) = P(X \le x)$. The CDF of discrete variable is given by $F_X(x) = \sum_{i=0}^{i} p_X(x_i)$.

Example 2.1 Check whether the following functions serve as probability distribution

(a) $p_X(x) = (x - 2)/2$ for $x = 1$
(b) $p_X(x) = x^2/25$ for $x = 0, 1, 2, 3, 4$

Solution:

(a) When $x = 1$,

$p_X(1) = -0.5$ and less than zero and violates Eq. (2.1).

Hence, $p_X(x) = (x - 2)/2$ is not a valid function for PMF.

(b) When $x = 0, 1, 2, 3, 4$ for the function $p_X(x) = x^2/25$, we have

$$p_X(0) = 0,\ p_X(1) = 1/25,\ p_X(2) = 4/25,\ p_X(3) = 9/25 \text{ and } p_X(4) = 16/25.$$

The value of the function, that is, the probabilistic values for each value of x from 0 to 4 are equal to or greater than zero and Eq. (2.1) is satisfied. But the summation of all the probabilistic value of all events is

$$\sum_{i=0}^{4} p_X(x_i) = 0 + \frac{1}{25} + \frac{4}{25} + \frac{9}{25} + \frac{16}{25}$$

$$= \frac{30}{25} \text{ and greater than one, which violates Eq. (2.2).}$$

Hence, $p_X(x) = x^2/25$ for $x = 0, 1, 2, 3, 4$ is not a valid function to serve as PMF.

Example 2.2 A contractor is planning to purchase 3 bulldozers for a new project in a remote area. The possible condition of bulldozers after 6 months is either good (G), or bad (B). What are the possible statuses of the bulldozers?

Solution: A random experiment consists of independent toss of a fair single coin (bulldozer) and $H(G)$ denotes head and $T(B)$ denotes tail. If three coins (3 bulldozers) are tossed the possible statuses (sample space) is shown below in the probability table of random variable X.

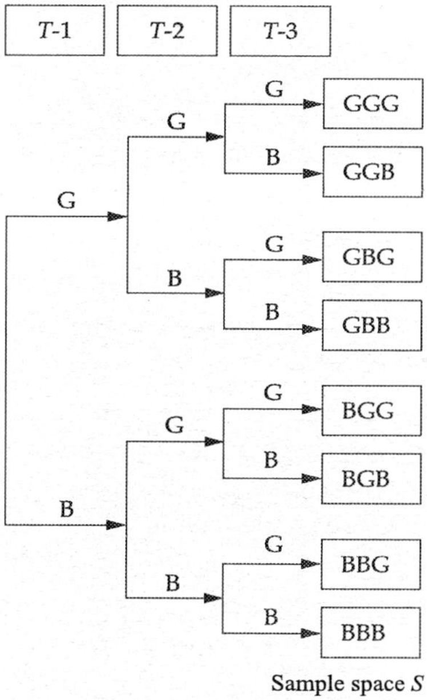

Fig. 2.2 Tree diagram

S = (GGG, GBB, BGB, BBG, GGB, GBG, BGG, BBB) the probability of each outcome S is 1/8.

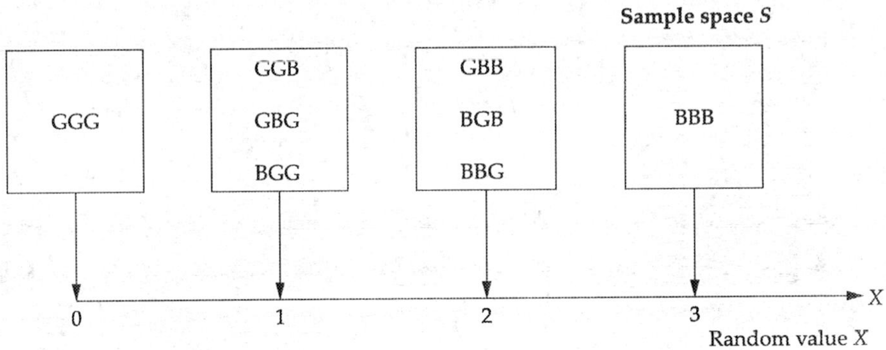

Fig. 2.3 Mapping of sample space on random value

Probability table of random variable *X*

X	0	1	2	3
$p_X(x)$	1/8	3/8	3/8	1/8

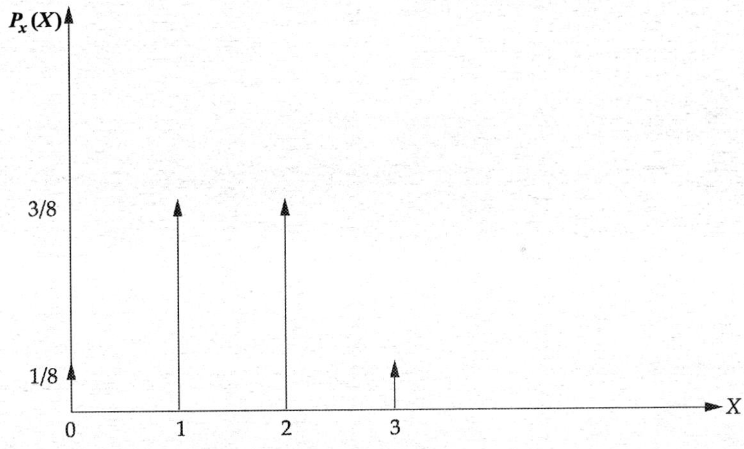

Fig. 2.4 Probability mass function

The Cumulative Distribution Function (CDF) of X can easily be written from the above table

$$F_X(x) = \begin{cases} 0 & \text{if } x<0 \\ (1/8) & \text{if } 0 \le x < 1 \\ (4/8) & \text{if } 1 \le x < 2 \\ (7/8) & \text{if } 2 \le x < 3 \\ 1 & \text{if } x \ge 3 \end{cases}$$

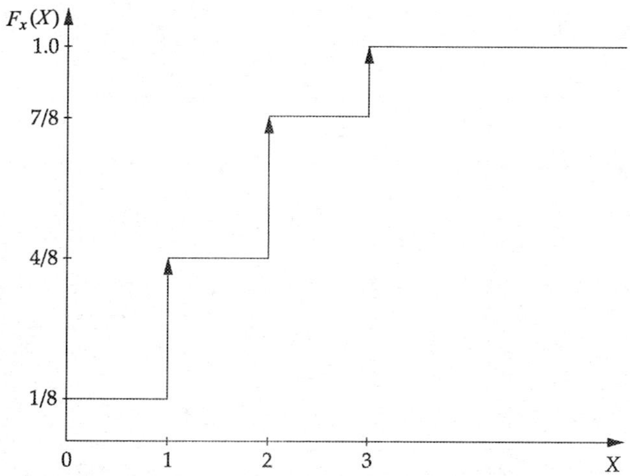

Fig. 2.5 Cumulative distribution function

2.3 TWO-DIMENSIONAL RANDOM VARIABLES

A two-dimensional random variable is mapping of the points in the sample space to ordered pairs $[x, y]$. Usually, when dealing with a pair of random variables, the sample space naturally partitions itself so that it can be viewed as a combination of two simpler sample spaces. For example, if the experiment was to observe the density and weight of soil. The range of soil density could fall in some set, which we will call sample space S_1, while the range of the weight would fall within sample space S_2. The overall sample space of the experiment could be viewed as $S = (S_1 \times S_2)$. The outcome of the experiment, the pair of random variables (X, Y) is merely a mapping of the outcome S to a pair of numerical values $(x(S), y(S))$. It would be natural to choose $x(S)$ to be the density (kN/m^3) while $y(S)$ is the weight of soil in (kN). The density functions of $f_X(x)$ and $f_Y(y)$ do partially characterize the experiment; they do not completely describe the situation. We accomplish this through the joint cumulative distribution function and the joint density function.

2.4 JOINT PROBABILITY MASS FUNCTION

When the random variables are discrete, it is convenient to work with probability mass function straightforward and it is possible to extend the concept to a pair of random variables X, Y given by $P_{X,Y}(x, y) = P(\{X = x\} \cap \{Y = y\})$. In particular, suppose the random variable X takes on values from the set $\{x_1, x_2, \ldots x_M\}$ and the random variable Y takes on values from the set $\{y_1, y_2, y_3 \cdots y_N\}$. Here either M and/or N could be potentially infinite or both could be finite.

Properties of joint probability mass function:

(1) $0 \le P_{X,Y}(x_m, y_n) \le 1$ ...(2.4)

(2) $\sum_{m=1}^{M}\sum_{n=1}^{N} P_{X,Y}(x_m, y_n) = 1$ ...(2.5)

(3) $\sum_{n=1}^{N} P_{X,Y}(x,y) = P_X(x_m)$ ...(2.6)

(4) $\sum_{m=1}^{M} P_{X,Y}(x,y) = P_Y(y_n)$ ...(2.7)

Further, the joint PDF or the joint CDF of a pair of discrete random variables can be related to the joint PMF through the use of delta function or step function by

(5) $f_{X,Y}(x,y) = \sum_{m=1}^{M}\sum_{n=1}^{N} P_{X,Y}(x_m, y_n)\delta(x-x_m)\delta(y-y_n)$ ...(2.8)

(6) $f_{X,Y}(x,y)\sum_{m}^{M}\sum_{n}^{N} P_{X,Y}(x_m, y_n)\delta(x-x_m)\delta(y-y_n)$ ...(2.9)

The PMF of random variable X given for the value of Y, then, by the definition of conditional probability is:

(7) $P(X=x|Y=y) = \frac{P(X=x, Y=y)}{P(Y=y)} = \frac{P_{X,Y}(x,y)}{P_Y(y)}$ ...(2.10)

We refer to this as the conditional PMF of X given Y. By way of notation we write

$$P_{X|y}(x|y) = \frac{P_{X,Y}(x,y)}{P_Y(y)} \qquad \text{...(2.11)}$$

2.5 MATHEMATICAL EXPECTATION

The mean of a random variable is also called its expected value. The notation $E(X)$ is used to denote the expected value. Thus, the mean, μ_X and expected value $E(X)$ are the same quantity and can be used interchangeably.

The mean of a discrete random variable X or the population mean is expressed as

$$E(X) = \Sigma(\text{value}) \times (\text{probability}) = \sum_{\text{all}\cdot x_i} x_i\, p_X(x_i) \qquad \text{...(2.12)}$$

Here, the sum extends over all the distinct values of X.

The variance, Var(X) is given by the expression

$$\text{Var}(X) = \sigma_X^2 = \sum_{\text{all}\cdot x_i} (x_i - \mu)^2 \times p_X(x_i) \qquad \text{...(2.13)}$$

i.e., summation of the product of squares of deviation from mean and the corresponding probability.

In general the rth moment about the mean, μ is given by $E(X-\mu)^r$ and the expectation of $E(X^r)$ is given by $\Sigma(x_i-\mu)^r \times p_X(x_i)$.

Suppose x is a discrete random variable which takes the values $x_1, x_2, x_3, \ldots\ldots x_n$ with respect to probabilities $p_X(x_1), p_X(x_2), p_X(x_3), \ldots\ldots p_X(x_n)$ then it is well known that from Eq. (2.2), $\Sigma p_X(x_i)$ equals unity or

$$p_X(x_1)+p_X(x_2)+p_X(x_3)+\ldots\ldots+p_X(x_n)=1.00$$

The mathematical expectation of random variable, X or the expected value of X can be obtained by treating probabilities $p_X(x_1), p_X(x_2), p_X(x_3), \ldots\ldots p_X(x_n)$ as loads acting at a distance $x_1, x_2, x_3, \ldots\ldots x_n$ from a point. The resultant of the force system is then given by

$$\mu=\frac{p_X(x_1)x_1+p_X(x_2)x_2+\ldots\ldots+p_X(x_n)x_n}{p_X(x_1)+p_X(x_2)+\ldots\ldots+p_X(x_n)} \quad \ldots(2.14)$$

or

$$E(X) = \mu_X=\frac{\sum_{all\cdot x_i} p_X(x_i)\times x_i}{\sum_{all\cdot x_i} p_X(x_i)} \quad \ldots(2.15)$$

Since $\sum_{all\cdot x_i} p_X(x_i) = 1,$

$$E(X) = \mu_X \quad \ldots(2.16)$$

As stated earlier that variance of X is given by $E(X-\mu_X)^2$

Therefore

$$\text{Var}(X),\ \sigma_x^2 = E(X-\mu_X)^2 \quad \ldots(2.17)$$

or

$$\sigma_x^2 = E(X^2+\mu_X^2-2X\mu_X)$$

$$= E(X^2)+\mu_X^2-2\mu_X E(X)$$

$$= E(X^2)+\mu_X^2-2\mu_X^2$$

$$= E(X^2)-\mu_X^2$$

Hence,

$$E(X^2)=\sigma_X^2+\mu_X^2 \quad \ldots(2.18)$$

The third moment of X about the mean gives the skewness

i.e.

$$\text{Skewness} = E(X-\mu_X)^3 \quad \ldots(2.19)$$

The skewness coefficient, r_1 is given by $\dfrac{E(X-\mu_X)^3}{\sigma_X^3}$ ...(2.20)

The fourth moment of X about the mean gives the Kurtosis

i.e.

$$\text{Kurtosis} = E(X-\mu_X)^4 \quad \ldots(2.21)$$

The coefficient of kurtosis, r_2 is given by $\dfrac{E(X-\mu_X)^4}{\sigma_X^4}$ ...(2.22)

Example 2.3 Two dices rolled wherein the random variable varied from 2 to 12 corresponding to the sum of dots of two dices. Find the mean, variance, standard deviation, coefficient of variation, coefficient of skewness and coefficient of kurtosis.

Solution:

Random variable, x_i	PMF, $p_X(x_i)$	$x_i \times p_X(x_i)$	$(x_i - \mu_X) = (x_i - 7)$	$(x_i - \mu_X)^2 = (x_i - 7)^2$	$(x_i - \mu_X)^2 \times p_X(x_i)$	$(x_i - \mu_X)^3 \times p_X(x_i)$	$(x_i - \mu_X)^4 \times p_X(x_i)$
2	$\frac{1}{36}$	$\frac{2}{36}$	–5	25	$\frac{25}{36}$	$\frac{-125}{36}$	$\frac{625}{36}$
3	$\frac{2}{36}$	$\frac{6}{36}$	–4	16	$\frac{32}{36}$	$\frac{-128}{36}$	$\frac{512}{36}$
4	$\frac{3}{36}$	$\frac{12}{36}$	–3	9	$\frac{27}{36}$	$\frac{-81}{36}$	$\frac{243}{36}$
5	$\frac{4}{36}$	$\frac{20}{36}$	–2	4	$\frac{16}{36}$	$\frac{-32}{36}$	$\frac{64}{36}$
6	$\frac{5}{36}$	$\frac{30}{36}$	–1	1	$\frac{5}{36}$	$\frac{-5}{36}$	$\frac{5}{36}$
7	$\frac{6}{36}$	$\frac{42}{36}$	0	0	0	0	0
8	$\frac{5}{36}$	$\frac{40}{36}$	1	1	$\frac{5}{36}$	$\frac{5}{36}$	$\frac{5}{36}$
9	$\frac{4}{36}$	$\frac{36}{36}$	2	4	$\frac{16}{36}$	$\frac{32}{36}$	$\frac{64}{36}$
10	$\frac{3}{36}$	$\frac{30}{36}$	3	9	$\frac{27}{36}$	$\frac{81}{36}$	$\frac{243}{36}$
11	$\frac{2}{36}$	$\frac{22}{36}$	4	16	$\frac{32}{36}$	$\frac{128}{36}$	$\frac{512}{36}$
12	$\frac{1}{36}$	$\frac{12}{36}$	5	25	$\frac{25}{36}$	$\frac{125}{36}$	$\frac{625}{36}$
$\Sigma =$ 77	**1.00**	$\mathbf{\frac{252}{36}}$	0		$\mathbf{\frac{210}{36}}$	**0**	$\mathbf{\frac{2898}{36}}$

Mean of the discrete random variable, $X = \mu_X$, is given

$$= E(X) = \sum_{\text{all}\cdot x_i} x_i \times p_X(x_i) = \frac{252}{36}$$

$$= 7$$

$$\text{Var (X)} = \sigma_X^2 = \sum_{\text{all}\cdot x_i} (x-\mu)^2 \, p_X(x_i) = \frac{210}{36}$$

$$= 5.833$$

Standard deviation of variable, $\sigma_X = \sqrt{\text{Var(X)}}$

$$= 2.415$$

Coefficient of variation, $(\text{CoV})_X = \dfrac{\sigma_X}{\mu_X}$

$$= \frac{2.415}{7}$$

$$= 0.345$$

Coefficient of skewness, $r_1 = \dfrac{E(X-\mu_X)^3}{\sigma_X^3}$

$$= \frac{\sum_{\text{all}\cdot x} (x-\mu_X)^3 \times p_X(x)}{\sigma_X^3}$$

$= 0$, hence, Symmetrical Distribution

Coefficient of kurtosis, $r_2 = \dfrac{E(X-\mu_X)^4}{\sigma_X^4}$

$$= \frac{\sum_{\text{all}\cdot x_i} (x-\mu_X)^4 \times p_X(x_i)}{\sigma_X^4} = \left(\left(\frac{2898}{36}\right) \Big/ \left(\frac{210}{36}\right)^2 \right)$$

$$= 2.36$$

2.6 ALGEBRA OF EXPECTATION

The expectations possess certain properties. For a random variable, expectation of a constant is a constant.

Hence,

$$E(a) = ax \qquad \ldots (2.23)$$

$$E(aX) = a \times E(X)$$

$$= a\mu_X \qquad \ldots (2.24)$$

where μ_X is the mean of the variable X

Supposing $$Y = a + bX$$

Then
$$\begin{aligned} E[Y] &= E[a + bX] \\ &= a + bE[X] \\ &= a + b\mu_X \end{aligned} \quad ...(2.25)$$

Similarly, if $Y = a_1X_1 + a_2 X_2 + a_3 X_3 + \ldots\ldots$

$$E[Y] = E\,[a_1 X_1 + a_2 X_2 + a_3 X_3 + \ldots\ldots]$$

$$\mu_Y = a_1 E[X_1] + a_2E[X_2] + a_3 E[X_3] + \ldots\ldots$$

$$= a_1 \mu_{X_1} + a_2 \mu_{X_2} + a_3 \mu_{X_3} + \ldots\ldots \quad ...(2.26)$$

For any random variables X, Y.... Z,

$$E\,(X + Y + \ldots\ldots + Z) = E(X) + E(Y) + \ldots\ldots\, E(Z) \quad ...(2.27)$$

If the variables are independent

$$E(X \times Y \ldots\ldots\ldots Z) = E(X) \times E(Y) \times \ldots. \times E(Z) \quad ...(2.28)$$

Properties of variances:

(i) Var $(a) = 0$ ...(2.29)

(ii) Var $(aX) = a^2$ Var (X) ...(2.30)

(iii) Var $(a + bX) = b^2$ Var (X) ...(2.31)

2.6.1 Covariance

If X and Y are random variables with their mean μ_X and μ_Y respectively, the covariance (COV) between X and Y is given by

$$\begin{aligned} \text{COV}\,(X, Y) &= E\,[(X - \mu_X)\,(Y - \mu_Y)] \\ &= E\,(XY - \mu_X Y - \mu_Y X + \mu_X \mu_Y) \\ &= E\,(XY) - \mu_X E\,(Y) - \mu_Y E\,(X) + E\,(\mu_X \mu_Y) \end{aligned}$$

or
$$\text{COV}\,(X, Y) = E\,(XY) - E\,(X) \times E\,(Y) \quad ...(2.32)$$

However, if X, Y are independent then

$$E\,(XY) = E\,(X) \times E\,(Y)$$

Hence, COV (X, Y) = 0

If a function $U = a_1X_1 + a_2X_2 + \ldots\ldots + a_n X_n$ where $X_1, X_2 \ldots. X_n$ are random variables and $a_1, a_2, \ldots. a_n$ are constants then

$$\begin{aligned} E\,(U) &= E\,(a_1 X_1 + a_2X_2 + \ldots\ldots + a_n X_n) \\ &= E\,(a_1X_1) + E(a_2X_2) + \ldots\ldots.. + E(a_n X_n) \\ &= a_1E\,(X_1) + a_2E(X_2) + \ldots\ldots.. + a_n E(X_n) \end{aligned}$$

$$U - E(U) = a_1[X_1 - E\,(X_1)] + a_2[X_2 - E\,(X_2)] + \ldots\ldots.. + a_n[X_n - E\,(X_n)]$$

Squaring both sides

$$[U - E(U)]^2 = a_1^2 [X_1 - E(X_1)]^2 + a_2^2 [X_2 - E(X_2)]^2 + \dots$$

$$\dots + a_n^2 [X_n - E(X_n)]^2 + 2a_1a_2 [X_1 - E(X_1)] [X_2 - E(X_2)]$$

$$\text{Var}(U) = [a_1^2 \text{Var}(X_1) + a_2^2 \text{Var}(X_2) + \dots + a_n^2 \text{Var}(X_n)$$

$$+ 2a_1a_2 \text{COV}(X_1X_2) + 2a_{n-1} a_n \text{COV}(X_{n-1}X_n)] \quad \dots(2.33)$$

Corollary. If only $a_1 = a_2 = 1$ and $a_3 = a_4 = \dots = a_n = 0$

$$\text{Var}(X_1 + X_2) = \text{Var}(X_1) + \text{Var}(X_2) + 2a_1a_2 \text{COV}(X_1, X_2) \quad \dots(2.34)$$

$$\text{Var}(X_1 - X_2) = \text{Var}(X_1) + \text{Var}(X_2) - 2a_1a_2 \text{COV}(X_1, X_2) \quad \dots(2.35)$$

Example 2.4 The study duration and grade point average (GPA) of students graduating with Bachelor degree from an engineering school were studied. With X defined as the number of years it takes to graduate and Y as the GPA, it was observed that X could be 4, 5 or 6 years and Y could be 2, 3 or 4. The following Table shows the number of students for each combination of X and Y.

(1) Compute the marginal PMF of X and Y

(2) If only a GPA of 3 is under consideration ($Y = 3$), compute the conditional PMF of X.

(3) Determine the COV(X, Y) and the corresponding correlation coefficient between X and Y.

Y (GPA)	**X (years)**		
	4	5	6
2	5	15	60
3	50	80	20
4	20	40	10

Solution:

Sl. No.	**X**	**Y**	**Frequency**	**Relative frequency**
1	4	2	5	0.017
2	5	2	15	0.050
3	6	2	60	0.200
4	4	3	50	0.167
5	5	3	80	0.267
6	6	3	20	0.067
7	4	4	20	0.067
8	5	4	40	0.133
9	6	4	10	0.033

The Marginal PMF of X is

$$P_X(4) = 0.017 + 0.167 + 0.067 = 0.251$$
$$P_X(5) = 0.05 + 0.267 + 0.133 = 0.450$$
$$P_X(6) = 0.200 + 0.067 + 0.033 = 0.3$$

The marginal PMF of Y is

$$P_Y(2) = 0.017 + 0.05 + 0.200 = 0.267$$
$$P_Y(3) = 0.167 + 0.267 + 0.067 = 0.501$$
$$P_Y(4) = 0.067 + 0.133 + 0.033 = 0.233$$

The conditional PMF of X given Y

$$p_{X|Y}(6|3) = \frac{p_{X,Y}(6,3)}{p_{Y(3)}} = \frac{0.067}{0.501} = 0.134$$

$$p_{X|Y}(5|3) = \frac{p_{X,Y}(5,3)}{p_{Y(3)}} = \frac{0.267}{0.501} = 0.533$$

$$p_{X|Y}(4|3) = \frac{p_{X,Y}(4,3)}{p_{Y(3)}} = \frac{0.167}{0.501} = 0.333$$

The COV (X, Y) and the correlation coefficient $\rho_{X,Y}$ can be estimated by using the equation

$$\rho_{X,Y} = \frac{\text{COV}(X, Y)}{\sigma_X \, \sigma_Y}$$

$$E(X) = [4 \times 0.251 + 5 \times 0.450 + 6 \times 0.3] = 5.054$$
$$\text{Var}(X) = [(4 - 5.054)^2 \times 0.251 + (5 - 5.054)^2 \times 0.450 + (6 - 5.054)^2 \times 0.3] = 0.5486$$
$$\sigma_X = 0.7406$$

Similarly,

$$E(Y) = [2 \times 0.267 + 3 \times 0.501 + 4 \times 0.233] = 2.969$$
$$\text{Var}(X) = [(2 - 2.969)^2 \times 0.267 + (3 - 2.969)^2 \times 0.501 + (4 - 2.969)^2 \times 0.233] = 0.4989$$
$$\sigma_Y = 0.7063$$
$$E(XY) = [4 \times 2 \times 0.017 + 5 \times 2 \times 0.05 + 6 \times 2 \times 0.20 + 4 \times 3 \times 0.167 + 5 \times 3 \times 0.267 + 6 \times 3 \times 0.067 + 4 \times 4 \times 0.067 + 5 \times 4 \times 0.133 + 6 \times 4 \times 0.033] = 14.775$$
$$\text{COV}(X, Y) = E(XY) - E(X)\,E(Y) = [14.775 - (5.054 \times 2.969)] = -0.2303$$

Correlation coefficient, $\rho_{X,Y} = \dfrac{-0.2303}{(0.7407 \times (0.7063)} = -0.44$

Example 2.5 The joint probability mass function of a bi-variate discrete random variable is given by the following Table.

$X \rightarrow$	**1**	**2**	**3**	**Total**
$Y \downarrow$				
1	0.1	0.1	0.2	0.4
2	0.20	0.3	0.1	0.6
Total	**0.3**	**0.4**	**0.3**	**1.0**

Find (i) the marginal probability mass function of X and Y; (ii) the conditional distribution of X given $Y = 1$; and (iii) $P(X + Y < 4)$.

Solution:

Case (i) From the definition of probability mass function (X, Y),

$$p_X(x_i) = \sum_{y_j} p_{XY}(1, y_j) = \sum_{y_j}^{2} p_{XY}(1, y_j)$$

When $x = 1$, $p_X(x=1) = \sum_{y_j} p_{XY}(1, y_j)$

$= p_{XY}(1, 1) + p_{XY}(1, 2) = 0.1 + 0.2 = 0.3$

When $x = 2$, $p_X(x=2) = \sum_{y_j=1}^{2} p_{XY}(2, y_j) = p_{XY}(2, 1) + p_{XY}(2, 2) = 0.1 + 0.3 = 0.4$

When $x = 3$,

$$p_X(x=3) = \sum_{y_j=1}^{2} p_{XY}(3, y_j) = p_{XY}(3, 1) + p_{XY}(3, 2) = 0.2 + 0.1 = 0.3$$

The marginal probability mass function X is

$$\begin{aligned} P_X(x) &= 0.3, \text{ when } x = 1 \\ &= 0.4, \text{ when } x = 2 \\ &= 0.3, \text{ when } x = 3. \end{aligned}$$

The marginal probability mass function of Y is given by $p_Y(y_j) = \sum_{x_i} p_{XY}(x_i\, y_i)$

When $y = 1$

$$p_Y(y=1) = \sum_{x_i}^{3} p_{XY}(x_i, 1) = p_{XY}(1, 1) + p_{XY}(2, 1) + p_{XY}(3, 1) = 0.1 + 0.1 + 0.2 = 0.4$$

When $y = 2$

$$p_Y(y=2) = \sum_{x_i}^{3} p_{XY}(x_i, 2) = p_{XY}(1, 2) + p_{XY}(2, 2) + p_{XY}(3, 2) = 0.2 + 0.3 + 0.1 = 0.6$$

The marginal probability mass function Y is

$$P_Y(y) = 0.4, \text{ when } y = 1$$
$$= 0.6, \text{ when } y = 2$$

Case (ii) The conditional distribution of X given $Y = 1$

$$P(X = x | Y = 1) = \frac{P(X = x \cap Y = 1)}{P(Y = 1)}$$

When $x = 1$, $P(X = x | Y = 1) = \dfrac{P(X = x \cap Y = 1)}{P(Y = 1)} = \dfrac{p_{XY}(1, 1)}{p_Y(1)} = \dfrac{0.1}{0.4} = 0.25$

When $x = 2$, $P(X = 2 | Y = 1) = \dfrac{P(X = x \cap Y = 1)}{P(Y = 1)} = \dfrac{p_{XY}(2, 1)}{p_Y(1)} = \dfrac{0.1}{0.4} = 0.25$

When $x = 3$, $P(X = 3 | Y = 1) = \dfrac{P(X = x \cap Y = 1)}{P(Y = 1)} = \dfrac{p_{XY}(3, 1)}{p_Y(1)} = \dfrac{0.2}{0.4} = 0.5$

Case (iii)

$$P(X + Y < 4) = p\,\{(x, y)/x + y < 4 \text{ where } x = 1, 2, 3; y = 1, 2\}$$
$$= p\,\{(1, 1), (1, 2), (2, 1)\} = p_{XY}(1, 1) + p_{XY}(1, 2) + p_{XY}(2, 1)$$
$$= 0.1 + 0.1 + 0.2 = 0.4$$

Example 2.6 Find the mean and variance of Y, if the random variables X_1 and X_2 are defined as

$$Y = a_1X_1 + a_2X_2$$

Solution:

$$Y = a_1X_1 + a_2X_2$$

Mean of Y is $E[Y] = E\,[a_1X_1 + a_2X_2]$

$$E[Y] = a_1\,E[X_1] + a_2\,E[X_2]$$

Variance of Y is $\text{Var}\,(Y) = \sigma_Y^2 = E[Y^2] - [E(Y)]^2$

$$\sigma_Y^2 = E[(a_1X_1 + a_2X_2)^2] - E[(a_1X_1 + a_2X_2)]^2$$

$$\sigma_Y^2 = E[(a_1^2X_1^2 + a_2^2X_2^2 + 2a_1X_1a_2X_2] - \{a_1\,E[X_1] + a_2\,E[X_2]\}^2$$

$$\sigma_Y^2 = a_1^2\,E[X_1^2] + a_2^2\,E[X_2^2] + 2a_1a_2E[X_1\,X_2] - a_1^2\,E[X_1]^2 - a_2^2\,E[X_2]^2 - 2a_1a_2\,E[X_1]\,E[X_2]$$

Rearranging, we obtain

$$\sigma_Y^2 = a_1^2\,E[X_1^2] - a_1^2\,E[X_1]^2 + a_2^2E[X_2^2] - a_2^2\,E[X_2]^2 + 2a_1a_2E[X_1\,X_2] - 2a_1a_2E[X_1]\,E[X_2]$$

$$\sigma_Y^2 = a_1^2\,\{E[X_1^2] - E[X_1]^2\} + a_2^2\{E[X_2^2] - E[X_2]^2\} + 2a_1a_2\{E[X_1\,X_2] - E[X_1]\,E[X_2]\}$$

$$\sigma_Y^2 = a_1^2\,\sigma_{Y1}^2 + a_2^2\,\sigma_{Y2}^2 + 2a_1a_2\text{COV}\,(X_1, X_2)$$

Example 2.7 The total cost C to manufacture a concrete panel in precast unit is $C = 1.5X + 2Y$ where X is the cost of material, Y is the cost of labour and constants 1.5 and 2 are overhead cost multipliers for material and labour. The cost X and Y are assumed to be uncorrelated random variables with mean values $\mu_X = 5000$/panel and $\mu_Y = 2500$/panel respectively with standard deviation $\sigma_X = 500$/panel, $\sigma_Y = 250$/panel. Find the mean, variance and total cost of manufacture.

Solution: Given cost, $C = 1.5X + 2Y$

$$\begin{aligned}\text{Mean cost, } \mu_C &= E(C) = 1.5E(X) + 2E(Y)\\ &= 1.5\,\mu_X + 2\,\mu_Y\\ &= 1.5(5000) + 2(2500)\\ &= 12500\end{aligned}$$

$$\begin{aligned}\text{Variance of the cost Var}(C) &= \text{Var}(1.5X + 2Y)\\ &= (1.5)^2\,\sigma_X^2 + (2)^2\,\sigma_Y^2 + 2(1.5)(2)\,\text{COV}(X, Y)\end{aligned}$$

X and Y being uncorrelated random variables, i.e., independent COV$(X, Y) = 0$

$$\text{Var}(C) = (1.5^2)(500^2) + (2^2)(250^2)$$
$$\text{Var}(C) = 812500$$

$$\begin{aligned}\text{Standard deviation of the cost } \sigma_C &= \sqrt{\text{Var}(C)}\\ &= \sqrt{812500} = 901.39\end{aligned}$$

$$\begin{aligned}\text{The total cost of manufacture } C &= \mu_C \pm \sigma_C\\ &= 12500 \pm 901.39\\ &= 13401.39 \text{ or } 11598.61\end{aligned}$$

2.7 CONTINUOUS VARIABLE

A random variable X can take all possible values between certain limits or in an interval which may be finite or infinite. Some examples of continuous random variables are (i) the strength of concrete; (ii) density of soil; (iii) the length of room; (iv) the wind load; and (v) the angle between two sides. A random variable which can take both discrete and continuous values is called a mixed random variable.

Since a continuous random variable will typically have a zero probability of taking on a specific value, we avoid talking about such probabilities. The CDF can also be used to measure the probability that a random variable takes on a value in certain interval. The probability measures of a continuous random variable is defined by probability density function (PDF), and is given by the following equation defined between intervals a and b

$$P(a \le X \le b) = \int_a^b f_X(x)\,dx \qquad \ldots(2.36)$$

The properties of a cumulative density function (CDF) of variable X if it assumes value between ($-\infty$ and $+\infty$) is given by

$$F_X(x) = P(X \le x) = \int_{-\infty}^{x} f_X(x)\,dx \qquad \text{...(2.37)}$$

If the variable, X assumes value from (0 to ∞), the CDF is given by

$$F_X(x) = P(X \le x) = \int_{0}^{x} f_X(x)\,dx \qquad \text{...(2.38)}$$

The derivative of the CDF, $F_X(x)$ gives the PDF of the variable, X.

Hence,
$$f_X(x) = \frac{dF_X(x)}{dx} \qquad \text{...(2.39)}$$

The following rules are must for satisfying a CDF, $F_X(x)$

(i) $F_X(-\infty) = 0$ ...(2.40)

(ii) $F_X(+\infty) = 1$ ...(2.41)

(iii) $\int_{-\infty}^{\infty} f_X(x)\,dx = 1$ ...(2.42)

(iv) $f_X(x) \ge 0$ ...(2.43)

(v) $F_X(x) \ge 0$ ...(2.44)

The condition $P(a \le X \le b) = \int_{a}^{b} f_X(x)\,dx$ can also be rewritten as

$$P(a \le X \le b) = \int_{-\infty}^{b} f_X(x)\,dx - \int_{-\infty}^{a} f_X(x)\,dx \qquad \text{...(2.45)}$$

2.7.1 Expectation of Continuous Random Variable

The continuous random variable X with PDF $f_X(x)$, the expected value of function $g(X)$, of the random variable is given by

$$E[g(X)] = \int_{-\infty}^{+\infty} g(x) f_X(x)\,dx \qquad \text{...(2.46)}$$

In general the nth moment of a random variable X is defined as

$$E\left[(X)^n\right] = \int_{-\infty}^{+\infty} (x)^n f_X(x)\,dx \qquad \text{...(2.47)}$$

If the random variable is continuous whose probability density function is $f_X(x)$, then

The mean value μ, is $E(X) = \int_{-\infty}^{+\infty} x \times f_X(x)\,dx$ ...(2.48)

Variance of X, σ_X^2 is given by $E(X - \mu_X)^2 = \int_{-\infty}^{+\infty} (x - \mu_X)^2 f_X(x)\,dx$ ...(2.49)

The third moment of X about the mean gives the skewness

i.e., skewness = $E(X - \mu_X)^3 = \int_{-\infty}^{+\infty} (x-\mu_X)^3 f_X(x)\,dx$ ...(2.50)

The coefficient of skewness, r_1 is given by $\dfrac{E(X-\mu_X)^3}{\sigma_X^3}$ ...(2.51)

The fourth moment of X about the mean gives the kurtosis

i.e., kurtosis = $E(X - \mu_X)^4 = \int_{-\infty}^{+\infty} (x-\mu_X)^4 f_X(x)\,dx$ ...(2.52)

The coefficient of kurtosis, r_2 is given by $\dfrac{E(X-\mu_X)^4}{\sigma_X^4}$ ...(2.53)

Example 2.8 Find the mean, standard deviation, coefficient of skewness, coefficient of kurtosis and coefficient of variation for the PDF

$$f_X(x) = \begin{cases} 0.5x & \text{for}\,(0 \le x \le 2) \\ 0 & \text{otherwise} \end{cases}$$

Solution:

Mean, $\mu_X = \int_{-\infty}^{\infty} x\, f_X(x)\, dx = \int_{-\infty}^{0} x\, f_X(x)\, dx + \int_{0}^{2} x\, f_X(x\, dx) + \int_{2}^{\infty} x\, f_X(x)\, dx$

$$\mu_X = \int_0^2 x\, f_X(x\, dx) = \int_0^2 (x)\,(0.5x)\, dx = \frac{4}{3}$$

$$\text{Var}(X) = \sigma_X^2 = \int_{-\infty}^{\infty} (X-\mu_X)^2 f_X(x)\, dx = \int_0^2 \left(x - \frac{4}{3}\right)^2 (0.5x)\, dx = \frac{2}{9}$$

Standard deviation, $\sigma_X = \sqrt{\left(\frac{2}{9}\right)} = 0.4714$

Skewness = $E(X - \mu_X)^3 = \int_{-\infty}^{+\infty} (x-\mu_X)^3 f_X(x)dx = \int_0^2 \left(x - 4/3\right)^3 0.5\, x\, dx = -0.059$

The coefficient of skewness, r_1 is given by $\dfrac{E(X-\mu_X)^3}{\sigma_X^3} = \dfrac{-0.059}{(0.4714)^3} = -0.563$

Kurtosis = $E(X - \mu_X)^4 = \int_{-\infty}^{+\infty} (x-\mu_X)^4 f_X(x)dx = \int_0^2 \left(x - 4/3\right)^4 0.5\, x\, dx = 0.1185$

The coefficient of kurtosis, r_2 is given by $\dfrac{E(X-\mu_X)^4}{\sigma_X^4} = \dfrac{0.1185}{(0.4714)^4} = 2.4$

$$\text{Coefficient of variation, (CoV)} = \frac{\sigma_X}{\mu_X} = \frac{0.4714}{\left(4/3\right)} = 0.35355$$

Example 2.9 Find value of K so that the following function can serve as a probability density function.

$$f_X(x) = \begin{cases} 0 & \text{for } x \le 0 \\ Kxe^{-4x^2} & \text{for } x > 0 \end{cases}$$

Solution: For a valid PDF Eq. (2.42) has to be satisfied, i.e.

$\int_{-\infty}^{\infty} f_X(x)dx = 1$, using the boundary conditions we have equation rewritten as

$$\int_0^{\infty} K \cdot x \cdot e^{-4x^2}\, dx = 1$$

Using $u = 4x^2$ the above integral can be expressed as

$u = 4x^2$

Now, differentiate w.r.t x

$$\frac{du}{dx} = 8x$$

$$\Rightarrow \qquad du = 8x \times dx$$

$$\therefore \qquad \frac{du}{8x} = dx$$

Now,

$$\int_0^{\infty} K \cdot x \cdot e^{-4x^2} dx = 1$$

$$\Rightarrow \qquad \int_0^{\infty} K \cdot x \cdot e^{-u} \frac{du}{8x} = 1$$

$$\Rightarrow \qquad \frac{K}{8} \int_0^{\infty} e^{-u}\, du = 1$$

$$\text{or} \qquad \frac{K}{8}\left[-e^{-u}\right]_0^{\infty} = 1$$

Hence, $K = 8$ for a valid PDF.

Example 2.10 A random variable X has the following density function.

$$f_X(x) = \begin{cases} Kx^2 & \text{for } -3 \le x \le 3 \\ 0 & \text{elsewhere} \end{cases}$$

Find (a) $P(1 \le x \le 2)$, (b) $P(x \le 2)$, (c) $P(x > 1)$

Solution: The value of K is evaluated first for valid PDF

$\int_{-3}^{3} Kx^2\,dx = 1$. Integrating we get the value of $K = \frac{1}{18}$ for a valid PDF

(a) $P(1 \le x \le 2) = \int_{1}^{2} Kx^2\,dx$

$$= \int_{1}^{2} \frac{1}{18}x^2\,dx = \frac{1}{18}\left[\frac{x^3}{3}\right] = \frac{1}{18} \times \frac{1}{3}[2^3 - 1]$$

$$= \frac{7}{54}$$

(b) $P(x \le 2) \quad = \int_{-3}^{2} \frac{1}{18}x^2\,dx$

$$= \frac{35}{54}$$

(c) $P(x > 1) \quad = \int_{1}^{3} \frac{1}{18}x^2\,dx = \frac{18}{18 \times 3} \times [x^3]_1^3$

$$= \frac{26}{54}$$

Example 2.11 Find the CDF for the following PDF of random variable, x.

(a) $f_X(x) = \begin{cases} 6x - 6x^2 & \text{for } 0 \le x \le 1 \\ 0 & \text{otherwise} \end{cases}$

(b) $f_X(x) = \begin{cases} \frac{x}{4} e^{-\frac{x}{2}}\,dx & \text{for } 0 \le x \le 1 \\ 0 & \text{otherwise} \end{cases}$

Solution: If $f_X(x)$ is a PDF then the CDF is given by $F_X(x) = \int_{-\infty}^{x} f_X(x)\,dx$

(a) $F_X(x) = \int_{-\infty}^{x} (6x - 6x^2)\,dx$

Applying the given boundary conditions

$$= \int_{-\infty}^{0} (6x - 6x^2)\,dx + \int_{0}^{x} (6x - 6x^2)\,dx$$

$$= [0 + (3x^2 - 2x^3)]$$

Hence, CDF is $F_X(x) = (3x^2 - 2x^3)$ for $0 \le x \le 1$

(b) CDF, $F_X(x) = \int_{-\infty}^{x} \frac{x}{4} e^{-\frac{x}{2}} dx$

$$= \int_{-\infty}^{0} \frac{x}{4} e^{-\frac{x}{2}} dx + \int_{0}^{x} \frac{x}{4} e^{-\frac{x}{2}} dx$$

$$= 0 + \int_{0}^{x} \frac{x}{4} e^{-\frac{x}{2}} dx$$

$$= \frac{1}{4} \int_{0}^{x} x e^{-\frac{x}{2}} dx$$

$$= \frac{1}{4} \left[\left(x \frac{e^{-\frac{x}{2}}}{-\left(\frac{1}{2}\right)} \right) - \left(1 \frac{e^{-\frac{x}{2}}}{\left(\frac{1}{4}\right)} \right) \right]_0^x \qquad \left\{ \because \int uv dv = u \int dv - \int v du \right\}$$

$$= \left(1 - e^{-\frac{x}{2}} - \frac{x}{2} e^{-\frac{x}{2}} \right)$$ for $(0 < x < \infty)$ is the CDF.

Example 2.12 A continuous random variable has CDF, $F_X(x)$ as follows

$$F_X(x) = \begin{cases} 0 & \text{for } (x < 1) \\ C(x-1)^4 & \text{for } (1 \le x \le 3) \\ 1 & \text{for } (x > 3) \end{cases}$$

Obtain PDF and evaluate the value of C.

Solution: Using Eq. (2.39), PDF is given by $f_X(x) = \frac{dF_X(x)}{dx}$

$$\text{PDF, } f_X(x) = \begin{cases} 0 & \text{for } (x < 1) \\ 4C(x-1)^3 & \text{for } (1 \le x \le 3) \\ 0 & \text{for } (x > 3) \end{cases}$$

The value of C is evaluated using Eq. (2.42), i.e.,

$$\int_{1}^{3} 4C(x-1)^3 dx = 1$$

The above equation gives the solution for the value of $C = \frac{1}{16}$

Hence, the PDF can be written as

$$\text{PDF,} \quad f_X(x) = \begin{cases} 0 & \text{for } (x < 1) \\ \dfrac{(x-1)^3}{4} & \text{for } (1 \le x \le 3) \\ 0 & \text{for } (x > 3) \end{cases}$$

Example 2.13 Evaluate the value of K for a proper PDF function given as

$$f_X(x) = Ke^{-\lambda x} \quad \text{for } (x \ge 0)$$

Also find the CDF.

Solution: Using Eq. 2.42, we have

$$\int_0^\infty Ke^{-\lambda x}\, dx = 1$$

or

$$K = \lambda$$

Hence, PDF is given by $f_X(x) = \lambda e^{-\lambda x}$

The CDF is given by

$$F_X(x) = P(X \le x) = \int_0^x f_X(x)\, dx$$

$$= \int_0^x \lambda e^{-\lambda x}\, dx$$

$$= \left[1 - \frac{1}{e^{-\lambda x}}\right]$$

or

$$\text{CDF, } F_X(x) = \left[1 - \frac{1}{e^{-\lambda x}}\right] \quad \text{for } (x \ge 0)$$

Example 2.14 PDF of a random variable is given by $f_X(x) = x^2/9$ for $0 \le x \le 3$, find the mean, median, standard deviation, skewness, and coefficients of variation.

$$\text{Mean } (\mu_X) = \int_{-\infty}^{+\infty} x f_X(x)\, dx = \int_0^3 x \frac{x^2}{9}\, dx = \frac{x^4}{36} = 2.25$$

$$\text{Median } (\tilde{\mu}_X) = \int_{-\infty}^{+\infty} x f_X(x)\, dx = \int_0^t \frac{x^2}{9}\, dx = 0.5; \frac{t^3}{27} = 0.5; \text{hence } t = 2.381$$

$$\text{Var}(X) = \int_{-\infty}^{+\infty} (x - \mu)^2 f_X(x)\, dx = \int_0^t (x - 2.25)^2 \frac{x^2}{9}\, dx = 0.3375$$

$$\sigma_X = \sqrt{\text{Var}(X)} = \sqrt{0.3375} = 0.5809$$

$$\text{Skewness } \mu_3 = \int_{-\infty}^{+\infty} (x - \mu)^3 f_X(x)\, dx = \int_0^t (x - 2.25)^3 \frac{x^2}{9}\, dx = -0.168$$

Skewness coefficient $\mu_3 = \frac{\mu^3}{\sigma^3} = -0.857$

Example 2.15 The bearing capacity of soil under a column footing foundation is known to vary between 200 MPa to 400 MPa. The probability density function within this range is given as

$$f_X(x) = \begin{cases} k\left(1 - \frac{x}{400}\right) & \text{for } 200 \text{ MPa} \le x \le 400 \text{ MPa} \\ 0 & \text{elsewhere} \end{cases}$$

If the column is designed to carry a load of 300 MPa. What is the probability of failure of foundation?

Solution:

$$F_X(x) = P(X \le x) = \int_0^x f_X(x)\,dx$$

$$f_X(x) = \int_{200}^{400} k\left(1 - \frac{x}{400}\right) dx = 1.0 = k\left(x - \frac{x^2}{800}\right)\Bigg]_{200}^{400}$$

$$k\left\{\left[400 - \left(\frac{400^2}{800}\right)\right] - \left[200 - \left(\frac{200^2}{800}\right)\right]\right\} = 1.0$$

$$\therefore \qquad k = \frac{1}{50}$$

$$p_s = \int_{200}^{300} k\left(1 - \frac{x}{400}\right) dx = \frac{1}{50}\left[\left(x - \frac{x^2}{800}\right)\right]_{200}^{300} = 0.75$$

The probability of failure of footing is $p_f = 1 - p_s = 1 - 0.75 = 0.25$.

Example 2.16 The probability of operation of a motor continuously without stopping for a period of one year and this random event as success S, $P(S) = 0.98$. The pump stops before the end of one year period as failure F and $P(F) = 0.02$. An engineer using this pump of this kind finds that if pump fails, it will cost ₹ 11.25 lakh owing to the production delays, repair and lost profits $V(F)$ = ₹ 11.25 lakh, if pump does not fail, there are no monetary losses or $V(S)$ = ₹ 0. What is the expected loss or long-term loss. The cost of operating a wastewater treatment system which is a function of its flow $C = 20\,X^{0.8}$, where C is the annual cost of maintenance in lakhs and the x is the flow in million gallons per day (Mgal/day) and it follows a uniform distribution:

$$f_X(x) = 1/5 \quad \text{for} \quad 5 \le x \le 10 \text{ Mgal/day.}$$

Solution: Expected monetary value (EMV) = $V(S)\,P(S) + V(F)\,P(F)$

$$\text{EMV} = 0 \times 0.98 + 11.25 \times 0.02 = 0.225 \text{ lacs}$$

The expectation of flow

$$E(X) = \text{Mean}(\mu_X) = \int_{-\infty}^{+\infty} x\, f_X(x)\,dx = \int_0^3 x\frac{1}{5}\,dx = \frac{x^2}{10} = 0.9 \text{ Mgal/day}$$

$$\text{Var}(X) = E(X^2) = \int_{-\infty}^{+\infty} x^2\, f_X(x)\,dx = \int_5^{10} x^2\frac{1}{5}\,dx = \left[\frac{x^3}{15}\right]_5^{10} = 58.33\left(\frac{\text{Mgal}}{\text{day}}\right)^2$$

$$\sigma_X{}^2 = \text{Var}(X) = E(X^2) - [E(X)]^2 = 58.33 - 0.9^2 = 57.52$$

$$\sigma = 1.4422 \text{ MG/day}$$

$$C = 20\ X^{0.8} = g(x)$$

The expected cost

$$E[g(x)] = \int_{-\infty}^{+\infty} g(x)\, f_X(x)\,dx = \int_5^{10} 20x^{0.8}\frac{1}{5}\,dx = \frac{4x^{1.8}}{1.8} = ₹\ 99.947 \text{ lacs/year}$$

2.8 MULTIPLE RANDOM VARIABLES

Some events are a result of two or more physical processes, and the events may be mapped into two or more dimensions. Such a process requires two or more random variables. An example in this respect can be a case of productivity at the construction site which depend on manpower and manhours. We may use random variable for manpower and another variable for manhours.

When two variables are considered simultaneously, the joint probability mass function (PMF) of discrete random variables, X and Y is

$$p_{X,Y}(x, y) = P(X = x, Y = y) \quad \text{...(2.54)}$$

The joint PMF of discrete random variables can also be expressed as

$$p_{X,Y}(x, y) = P((X = x) \cap (Y = y)) \quad \text{...(2.55)}$$

The joint CDF is given by

$$F_{X,Y}(x, y) = P(X \le x, Y \le y) \quad \text{...(2.56)}$$

Equation (2.56) can also be defined by

$$F_{X,Y}(x, y) = P((X \le x) \cap (Y \le y)) \quad \text{...(2.57)}$$

Hence, the joint CDF discrete variables X and Y,

$$F_{X,Y}(x, y) = \sum_{x_i \le x} \times \sum_{y_i \le y} p_{X,Y}(x_i, y_j) \quad \text{...(2.58)}$$

It is the sum of probabilities associated with all point pairs (x_i, y_j) in the subset $\{(x_i \le x), (y_j \le y)\}$

The conditions that are to be satisfied for a joint distribution are

(i) $p_{X,Y}(x_i, y_j) \geq 0$

(ii) $\sum_{\text{all}\cdot x_i} \cdot \sum_{\text{all}\cdot y_j} p_{X,Y}(x_i, y_j) = 1.00$

(iii) $F_{X,Y}(-\infty, -\infty) = 0$; $F_{X,Y}(\infty, \infty) = 1$

(iv) $F_{X,Y}(-\infty, y) = 0$; $F_{X,Y}(\infty, y) = F_Y(y)$

(v) $F_{X,Y}(x, -\infty) = 0$; $F_{X,Y}(x, \infty) = F_X(x)$

Morever, $F_{X,Y}(x, y)$ is a non-negative, and non-decreasing function of x and y.

If the random variables X and Y are continuous, joint probability density function (PDF) is given by

$$f_{X,Y}(x, y)\, dxdy = P[x \leq X \leq x + dx, y \leq Y \leq y + dy] \quad \text{...(2.59)}$$

Then

$$F_{X,Y}(x, y) = \int_{-\infty}^{x} \int_{-\infty}^{y} f_{X,Y}(u,v)\, dv \cdot du \quad \text{...(2.60)}$$

The joint PDF is also given by

$$f_{X,Y}(x, y) = \frac{\partial^2 F_{X,Y}(x,y)}{\partial x\, \partial y} \quad \text{...(2.61)}$$

The joint PDF is also defined as

$$P(a_1 \leq x \leq a_2, b_1 \leq y \leq b_2) = \int_{a_1}^{a_2} \int_{b_1}^{b_2} f_{X,Y}(x, y) dxdy \quad \text{...(2.62)}$$

Which is the volume under the surface $f(x, y)$

The following conditions are to be satisfied for a proper PDF

(i) $f_{X,Y}(x, y) \geq 0$ for all values of x and y

(ii) $\int_{-\infty}^{\infty} \int_{-\infty}^{\infty} f_{X,Y}(x, y) dxdy = 1.00$

(iii) $\int_{a}^{b} \int_{c}^{d} f_{X,Y}(x,y) dxdy = P(a \leq X \leq b, c \leq Y \leq d)$

$$F_{XY}(x, y) = \int_{-\infty}^{x1} \int_{-\infty}^{y1} f_{XY}(x, y) dxdy \quad \text{...(2.63)}$$

(iv) The correlation between two random variables X and Y is defined as

$$E(XY) = \int_{-\infty}^{+\infty} \int_{-\infty}^{+\infty} xy\, f_{XY}(x,y) dxdy \quad \text{...(2.64)}$$

If two random variables are uncorrelated with each other, if the correlation between X and Y is equal to the product of their means. $E(XY) = E(X)\, E(Y)$.

The random variables are orthogonal to each other, if the correlation between X and Y is equal to zero.

(v) The covariance between random variables X and Y is then

$$\text{COV}(X,Y) = \int_{-\infty}^{+\infty}\int_{-\infty}^{+\infty} (x-\mu_X)(y-\mu_Y)\, f_{XY}(x,y)\,dxdy \quad \text{...(2.65)}$$

$$\text{COV}(X, Y) = E(XY) - \mu_X\mu_Y \quad \text{...(2.66)}$$

When COV(X, Y) is zero, the two random variables X and Y are uncorrelated. When the random variables are uncorrelated the correlation is equal to the product of their means.

(vi) The correlation between two random variables X and Y is defined as

$$\rho = \frac{\text{COV}(X,Y)}{\sigma_X\,\sigma_Y} = \frac{E(X,Y) - \mu_X\mu_Y}{\sigma_X\,\sigma_Y} \quad \text{...(2.67)}$$

The correlation coefficient is a number that varies between –1 and +1. When the correlation coefficient is 1 or –1, the variables are perfectly correlated and or have linear relationship between them. When ρ is +1, they are positively correlated or are directly proportional. When ρ is –1 they are negatively correlated or inversely proportional. If correlation is 0, then the random variables are uncorrelated. *If the two random variables are statistically independent then they are uncorrelated and the covariance is zero but converse is not true.*

(vii) The joint probability density function $f_{XY}(x, y)$ of two random variables X and Y, the marginal probability density function of one of the random variables is the integral of the joint probability density function over other random variable.

The marginal probability density function of X is

$$f_X(x) = \int_{-\infty}^{+\infty} f_{XY}(x,y)\,dy \quad \text{...(2.68)}$$

The marginal probability density function of Y is

$$f_Y(y) = \int_{-\infty}^{+\infty} f_{XY}(x,y)\,dx \quad \text{...(2.69)}$$

(viii) Conditional probability of two cumulative distribution functions of random variables X and given $Y = y$ is

$$F\left(\frac{X}{Y}\right) = \frac{F_{XY}(x,y)}{F_Y(x)},\ F_Y(y) > 0 \quad \text{...(2.70)}$$

Conditional probability of two cumulative distribution function of random variables Y and given $X = x$ is

$$F\left(\frac{y}{x}\right) = \frac{F_{XY}(x, y)}{F_X(x)}, F_X(x) > 0 \qquad \text{...(2.71)}$$

(ix) The conditional probability of two probability density functions of random variables X and given $Y = y$ is

$$f_{X|Y}\left(\frac{x}{y}\right) = \frac{f_{XY}(x, y)}{f_Y(x)}, f_Y(y) > 0 \qquad \text{...(2.72)}$$

The conditional probability of two probability density functions of random variables Y and given $X = x$ is

$$f_{Y|X}\left(\frac{y}{x}\right) = \frac{f_{XY}(x, y)}{f_X(x)}, f_X(x) > 0 \qquad \text{...(2.73)}$$

(x) The $(m, n)^{th}$ joint moment of two random variables X and Y is

$$E(X^m Y^n) = \int_{-\infty}^{+\infty}\int_{-\infty}^{+\infty} x^m y^n f_{XY}(x, y)\, dxdy \qquad \text{...(2.74)}$$

Example 2.17 The joint probability density function of a bi-variate random variable (X, Y)

$$f_X(x) = \begin{cases} k(x+y) & \text{for } (0 \le x \le 2) \\ 0 & \text{for } (0 \le y \le 2) \end{cases}$$

(i) Find k
(ii) Find the marginal probability density function of X and Y
(iii) Are X and Y are independent?
(iv) Find $f_{X|Y}(y | x)$ and $f_{Y|X}(y | x)$
(v) Evaluate $P(0 < y < \frac{1}{2} | X = 1)$.

Solution:

$$\int_{-\infty}^{\infty}\int_{-\infty}^{\infty} f_{X,Y}(x, y)\, dxdy = 1.00 \Rightarrow \int_0^2\int_0^2 k(x+y)\, dxdy = 1.0$$

$$\int_0^2 \left[\frac{x^2}{2} + xy\right]_0^2 dy = 1 \rightarrow k\int_0^2 (2+2y)\, dy = k[2y + y^2]_0^2 = 1; k = \frac{1}{8}$$

(i) The marginal PDF of X is given by

$$f_X(x) = \int_{-\infty}^{+\infty} f_{XY}(x, y) dy = \frac{1}{8}\left[xy + \frac{y^2}{2}\right]_0^2 = \frac{(1+x)}{4}$$

The marginal PDF of Y is given by

$$f_Y(y) = \int_{-\infty}^{+\infty} f_{XY}(x, y)\,dx = \frac{1}{8}\int_0^2 (x+y)\,dx = \frac{(1+y)}{4}$$

(ii) $f_X(x)\, f_Y(y) \neq f_{XY}(x, y)$ hence, X and Y are not indepdent

(iii) Conditional PDF $f_{Y|X}(y/x)$ is given by

$$f_{Y|X}\left(\frac{y}{x}\right) = \frac{f_{XY}(x, y)}{f_X(x)};\ \frac{\frac{1}{8}(x+y)}{\frac{1}{4}(x+1)} = \frac{1\,(x+y)}{2\,(x+1)}\ 0 < x < 2; 0 < y < 2$$

Conditional PDF $f_{X|Y}(x/y)$ is given by

$$f_{X|Y}\left(\frac{y}{x}\right) = \frac{f_{XY}(x, y)}{f_Y(x)};\ \frac{\frac{1}{8}(x+y)}{\frac{1}{4}(y+1)} = \frac{1\,(x+y)}{2\,(y+1)}\ 0 < x < 2; 0 < y < 2$$

(iv) $P(0 < Y < 1/2 \mid X = 1) = \int_0^{1/2} f_{Y|X}(y|x=1)\,dy = \frac{1}{x}\int_0^{1/2} \frac{(1+y)}{2}\,dy = \frac{5}{32}.$

Example 2.18 The joint probability density function of a bi-variate random variable (X, Y)

$$f_{XY}(x, y) = \begin{cases} k(x+y) & \text{for } 0 \le x1 \\ 0 & \text{for } 0 \le y \le 2 \end{cases}$$

Find k and hence ρ_{xy}.

Solution:

$$\int_0^2\int_0^1 k(x+y)\,dxdy = \int_0^2 k\left[\frac{x^2}{2} + xy\right]_0^1 dy = 1 \rightarrow k\int_0^2\left(\frac{1}{2} + y\right)dy = 1$$

$$k\left[\frac{1}{2}y + \frac{y^2}{2}\right]_0^2 = 1; k = \frac{1}{3}$$

Marginal density function of X is

$$f_X(x) = \int_{-\infty}^{+\infty} f_{XY}(x, y)\,dy = \int_0^2 \frac{1}{3}(x+y)\,dy = \frac{1}{3}\left[xy + \frac{y^2}{2}\right]_0^2 = \frac{1}{3}(2x+2)$$

Marginal density function of Y is

$$f_Y(y) = \int_{-\infty}^{+\infty} f_{XY}(x, y)\,dx = \int_0^1 \frac{1}{3}(x+y)\,dx = \frac{1}{3}\left[xy + \frac{x^2}{2}\right]_0^1 = \frac{1}{3}\left(y + \frac{1}{2}\right)$$

$$E(X) = \int_{-\infty}^{+\infty} x\,f_X(x)\,dx = \int_0^1 x\,\frac{1}{3}(2x+2)\,dx = \frac{2}{3}\left[\frac{x^3}{3} + \frac{x^2}{2}\right]_0^1 = \frac{5}{9}$$

$$E(X^2) = \int_0^1 x^2\,\frac{1}{3}(2x+2)\,dx = \frac{2}{3}\left[\frac{x^4}{4} + \frac{x^3}{3}\right]_0^1 = \frac{7}{18}$$

$$E(Y) = \int_{-\infty}^{+\infty} y\,f_Y(y)\,dy = \int_0^2 y\left(\frac{\left(\frac{1}{2}+y\right)}{3}\right)dy = \frac{1}{3}\left[\frac{y^3}{3} + \frac{y^2}{4}\right]_0^2 = \frac{11}{9}$$

$$E(Y^2) = \int_0^2 y^2\left(\frac{\left(\frac{1}{2}+y\right)}{3}\right)dy = \frac{1}{3}\left[\frac{y^3}{6} + \frac{y^4}{4}\right]_0^2 = \frac{16}{9}$$

$$E(XY) = \int_0^{+2}\int_0^1 xy\,f_{XY}(x, y)\,dxdy = \int_0^2\int_0^1 xy\,\frac{1}{3}(x+y)\,dxdy$$

$$= \frac{1}{3}\int_0^2\left[\frac{x^2y^2}{2} + \frac{x^3}{3}y\right]_0^1 dy = \frac{1}{3}\int_0^2\left[\frac{y^2}{2} + \frac{y}{3}\right]_0^1 dy = \frac{2}{3}$$

$$\text{Var}\,(X) = \sigma_X^2 = E(X^2) - [E(X)]^2 = \frac{7}{18} - \left(\frac{5}{9}\right)^2 = \frac{13}{162}$$

$$\text{Var}\,(Y) = \sigma_Y^2 = E(Y^2) - [E(Y)]^2 = \frac{16}{9} - \left(\frac{11}{9}\right)^2 = \frac{23}{81}$$

$$\rho_{XY} = \frac{E(XY) - E(X)\,E(Y)}{\sigma_X\,\sigma_Y} = \frac{\frac{2}{3} - \left(\frac{5}{9}\right)\left(\frac{11}{9}\right)}{\sqrt{\frac{13}{162} \times \frac{23}{81}}} = -0.0817$$

2.9 TRANSFORMATION OF RANDOM VARIABLES

If the input is a random variable with a particular PDF or CDF and it undergoes a transformation which results in another random variable. Here single random variable is transformed into another random variable Y by the transformation T. The transformation T between X and Y may such that Y is either one-to-one or non-one-to-one transformation.

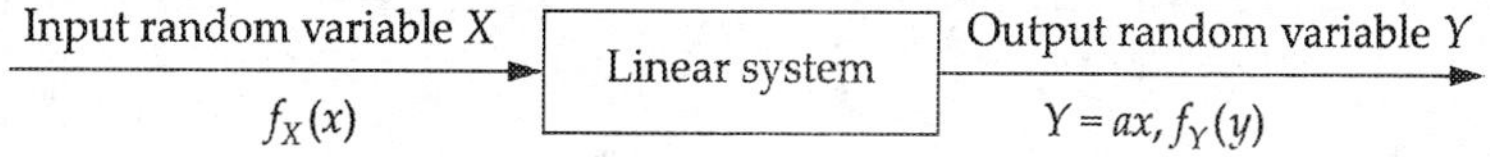

Fig. 2.6 Transformation of $Y = g(X)$

$$f_Y(y) = f_X(x)\left|\frac{dx}{dy}\right| \quad \text{or} \quad f_Y(y) = \frac{f_X(x)}{\dfrac{dy}{dx}} \qquad \ldots(2.75)$$

It is more convenient to compute PDF of the new random variable. For the more general case when $Y = g(X)$ has n finite number of real roots, all the transformation which is non-one-to-one, will be broken up into transformations each of which one-to-one and the results are summated together.

Example 2.19 Suppose X is a random variable with quadratic transformation, $Y = 1/X^2$. Find the density function of Y.

Solution: Here the function has two roots for $Y > 0$, $x_1 = -\dfrac{1}{\sqrt{y}}$; $x_2 = \dfrac{1}{\sqrt{y}}$

$$f_Y(y) = \frac{f_X(x_1)}{\dfrac{dy}{dx_1}} + \frac{f_X(x_2)}{\dfrac{dy}{dx_2}}$$

$$\frac{dy}{dx_1} = \frac{-2}{x^3}; \; \frac{dy}{dx_2} = \frac{-2}{x^3};$$

$$f_Y(y) = \frac{f_X\left(\dfrac{-1}{\sqrt{y}}\right)}{2y^{-3/2}} + \frac{f_X\left(\dfrac{1}{\sqrt{y}}\right)}{2y^{-3/2}}$$

$$f_Y(y) = \frac{y^{3/2}}{2}\left[f_X\left(\frac{-1}{\sqrt{y}}\right) + f_X\left(\frac{1}{\sqrt{y}}\right)\right]$$

2.9.1 Transformation of Two Random Variables

Given two random variables X and Y, suppose we create new random variable U, V, according to some 2×2 transformation. The multivariable calculus states that if a transformation of the form, maps infinitesimal region $A_{x,y}$ to region $A_{u,v}$ then the ratio of the areas of these regions is given by the absolute value of the Jacobian of the transformation.

$$U = g_1(X, Y),$$
$$V = g_2(X, Y)$$

$$\frac{\text{Area of } A_{x,y}}{\text{Area of } A_{u,v}} = J\begin{vmatrix} x & y \\ z & w \end{vmatrix} = \det\begin{bmatrix} \frac{\partial x}{\partial u} & \frac{\partial y}{\partial u} \\ \frac{\partial x}{\partial v} & \frac{\partial y}{\partial v} \end{bmatrix} \quad \text{...(2.76)}$$

$$f_{UV}(u, v) = \frac{f_{XY}(x, y)}{J\begin{vmatrix} u & v \\ x & y \end{vmatrix}} = f_{XY}(x, y)\, J\begin{vmatrix} x & y \\ z & w \end{vmatrix} \quad \text{...(2.77)}$$

Example 2.20 If X and Y are two continuous random variables with joint density function. Find the joint probability distribution function of $u = x^2$ and $v = xy$.

Solution:
$$f_{XY}(x, y) = \begin{cases} 4xy & \text{for } 0 < x < 1, 0 < y < 1 \\ 0 & \text{otherwise} \end{cases}$$

We know that $f_{UV}(u, v) = \dfrac{f_{XY}(x, y)}{J\begin{vmatrix} u & v \\ x & y \end{vmatrix}}$

$$J = \begin{vmatrix} 2x & 0 \\ y & x \end{vmatrix} = 2x^2 = 2u$$

$$f_{UV}(u, v) = \frac{f_{XY}(x, y)}{J\begin{vmatrix} u & v \\ x & y \end{vmatrix}} = \frac{4xy}{2x^2} = \frac{2v}{u}$$

When $x = 0, y = 0$; $u = 0, v = 0$; when $x = 1, y = 1$, and $u = 1, v = 1$.

2.10 CHEBYSHEV'S INEQUALITY

Statement: If X is a random variable with expected value, μ and standard deviation, σ, then the probability that the random value of the variate will lie within the range of $k\sigma$ on each side of the mean, μ *is at least* $\left(1 - \frac{1}{k^2}\right)$. Chebyshev's inequality is expressed by the following equation.

$$P[(\mu - k\sigma) \le x \le (\mu + k\sigma)] \ge \left(1 - \frac{1}{k^2}\right) \quad \text{... (2.78)}$$

In other words by Chebyshev's inequality the probability of getting a value which deviates from mean, μ by *at least* $k\sigma$ is *at most* $\left(\frac{1}{k^2}\right)$

or
$$P[\ (x - \mu) \geq k\sigma] \leq \left(\frac{1}{k^2}\right) \quad \text{...(2.79)}$$

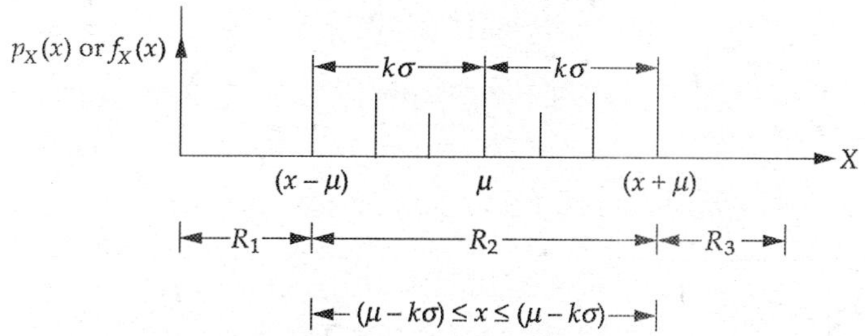

Fig. 2.7 Graphical representation of Chebyshev's inequality

2.10.1 Proof of the Chebyshev's Inequality

Variance of X, is given by

$$\sigma^2 = E(X - \mu)^2$$

$$= \sum_{\text{all}\cdot x_i} (X - \mu)^2\ p_X(x_i)$$

If the random variable distribution is assumed to be divided in three zones R_1, R_2 and R_3 as in Fig. 2.7, then the expression for variance can be written as

$$\sigma^2 = \sum_{R_1}(X - \mu)^2\ p_X(x_i) + \sum_{R_2}(X - \mu)^2\ p_X(x_i) + \sum_{R_3}(X - \mu)^2\ p_X(x_i) \quad \text{...(2.80)}$$

where X in region R_1 is given by $X \leq (\mu - k\sigma)$

or
$$(X - \mu) \geq k\sigma \quad \text{...(2.81)}$$

Variable, X in the region R_2 is given by $(\mu - k\sigma) \leq X \leq (\mu + k\sigma)$ ...(2.82)

Similarly, in the region R_3, the variable X is given by the expression

or
$$X \geq (\mu + k\sigma) \quad \text{or} \quad (X - \mu) \geq k\sigma \quad \text{...(2.83)}$$

If the regions R_1 and R_3 are taken then the variance of the distribution shall be greater than the sum of variance of regions R_1 and R_3. Hence, we have the following expression

$$\sigma^2 \geq \sum_{R_1}(X - \mu)^2 \cdot p_X(x_i) + \sum_{R_3}(X - \mu)^2 \cdot p_X(x_i) \quad \text{...(2.84)}$$

in the regions R_1 and R_3, $(X - \mu) \geq k\sigma$ or $(X - \mu)^2 \geq (k\sigma)^2$. Substituting in Eq. (2.84), we have

$$\sigma^2 \geq \sum_{R_1} (X-\mu)^2 p_X(x_i) + \sum_{R_3} (X-\mu)^2 p_X(x_i) \qquad \text{...(2.85)}$$

or

$$\sigma^2 \geq \sum_{R_1} (k\sigma)^2 p_X(x_i) + \sum_{R_3} (k\sigma)^2 p_X(x_i) \qquad \text{...(2.86)}$$

or

$$\frac{1}{k^2} \geq \sum_{R_1} p_X(x_i) + \sum_{R_3} p_X(x_i) \qquad \text{...(2.87)}$$

The two terms on right hade side of Eq. (2.87) represents the summation of probability of the random variate in regions R_1 and R_3. Hence, Eq. (2.87) can be expressed as

$$\frac{1}{k^2} \geq (1 - \text{Probability of the variate, } X \text{ in region } R_2)$$

or

$$\text{Probability of the variate in region } R_2 \geq \left(1 - \frac{1}{k^2}\right)$$

or

$$P[(\mu - k\sigma) \leq x \leq (\mu + k\sigma)] \geq \left(1 - \frac{1}{k^2}\right) \qquad \text{...(2.88)}$$

This proves the Chebyshev's inequality.

Hence, by Chebyshev's inequality, probabilistic statement can be made without the complete knowledge of PDF/PMF, only the mean and standard deviation are required.

Example 2.21 The mean value for the strength parameter of soil is 25 degrees. Estimate the minimum probability that a random sample of this soil will have the strength parameter between 15 degrees and 35 degrees. Assume the standard deviation of the strength parameter is 2.5 degrees.

Solution: We have, $(\mu - k\sigma) = 15$ and $(\mu + k\sigma) = 35$

or

$$2k\sigma = (35 - 15)$$

Substituting standard deviation as 2.5 degrees, the value of $k = 4$.

Hence, by Eq. (2.88) we have

$$P[15 \leq \alpha \leq 35] \geq \left(1 - \frac{1}{k^2}\right)$$

$$\geq \left(1 - \frac{1}{4^2}\right)$$

$$\geq (0.9375)$$

Hence, the probability is at least 0.9375 that strength parameter of the random sample of soil lies between 15 degrees and 35 degrees without knowing the distribution of strength parameter of the soil.

UNSOLVED PROBLEMS

1. Consider the beam *AB* as shown in Fig. 2.8 loaded with two random loads P_1 and P_2. The loads are assumed statistically independent with means, $\mu_{P1} = 4$ kN, and $\mu_{P2} = 6$ kN and standard deviations, $\sigma_{P1} = 0.4$ kN and $\sigma_{P2} = 0.5$ kN. The shear force Q_B and the bending moment M_B at the support *B* are $Q_B = \frac{1}{27}(13P_1 + 23P_2)$, $M_B = \frac{6}{27}(4P_1 + 5P_2)$. Determine the mean, standard deviation, covariance and correlation coefficient between shear force and bending moment.

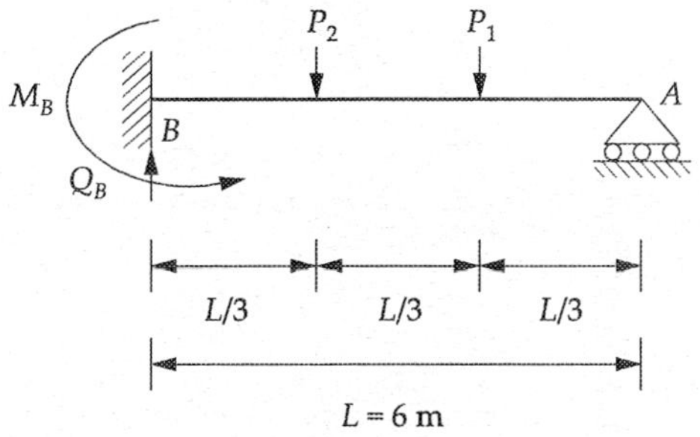

Fig. 2.8

2. A performance function is given by $Y = 3X_1 - 2X_2$ where $\mu_{X_1} = 16.6$; $\mu_{X_2} = 18.8$, $\sigma_{X_1} = 2.45$; $\sigma_{X_2} = 2.83$. The two variables are correlated, and the covariance is equal to 2.0. Obtain the coefficient of variation when X_1 and X_2 are uncorrelated and correlated.
3. Find the CDF for a proper PDF function given as

$$f_X(x) = \begin{cases} 0 & \text{for } x < 1 \\ \frac{1}{2} + \frac{1}{4}(x-3) & \text{for } 1 \le x \le 3 \\ \frac{1}{2} - \frac{1}{4}(x-3) & \text{for } 3 \le x \le 5 \\ 0 & \text{for } x \ge 5 \end{cases}$$

4. A triangular PDF shown in Fig. 2.9 $f_X(x) = \begin{cases} \frac{2}{(b-a)} \frac{(x-a)}{(m-a)} & \text{for } a \le x \le m \\ \frac{2}{(b-a)} \frac{(x-a)}{(m-a)} & \text{for } m \le x \le b \end{cases}$

Show that mean, $\mu_X = \dfrac{(a+m+b)}{3}$, coefficient of variation, $\text{CoV} = \left[\dfrac{1}{2} - \dfrac{(ab+am+bm)}{6\mu_X^2}\right]^{\frac{1}{2}}$

and probability $P(X \geq 0)\left[1 - \dfrac{(x-a)^2}{(b-a)(m-a)}\right]$ for $a < x < m$.

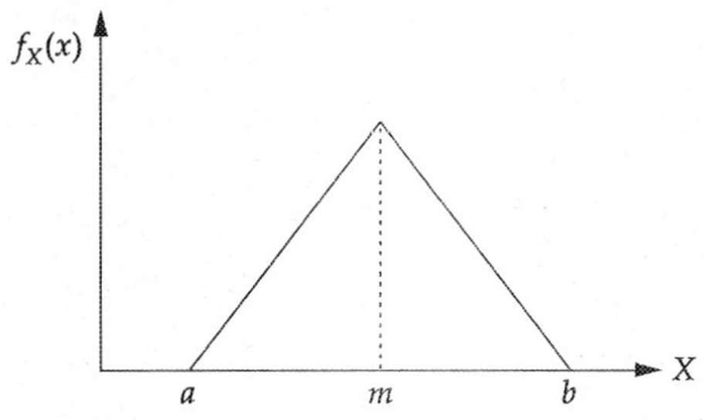

Fig. 2.9

5. A normal N and a tangential T force act on a surface inclined at an angle θ from the horizontal as shown in Fig. 2.10. The force $F_N = N \sin(\theta) + T \cos(\theta)$. If N and T are random variables, show that the expression of $E[F_N]$ and $\text{Var}[F_N]$ are $E[F_N] = [N] \sin(\theta) + [T] \cos(\theta)$; $\text{Var}[F_N] = \text{Var}[N] \sin^2(\theta) + \text{Var}[T] \cos^2(\theta) + 2\rho\, \sigma_N \sigma_T \sin(\theta) \cos(\theta)$.

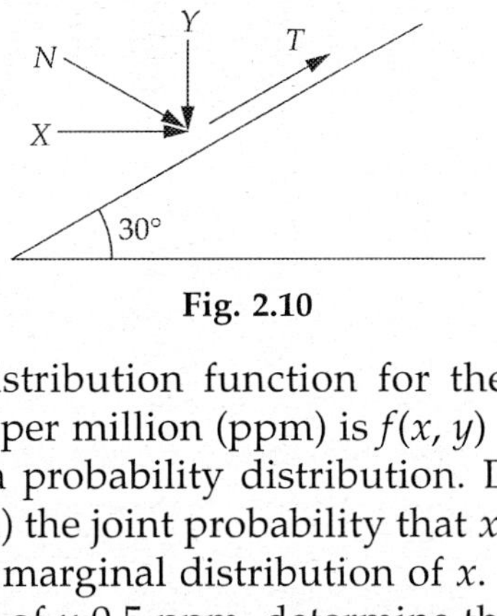

Fig. 2.10

6. The joint probability distribution function for the concentration levels of two pollutants (x, y) in parts per million (ppm) is $f(x, y) = 2 - x - y$, $0 < x < 1$, $0 < y < 1$. (i) Show that $f(x, y)$ is a probability distribution. Determine: (ii) the cumulative distribution function; (iii) the joint probability that $x \leq 1/2$, $y \leq 3/4$; (iv) the marginal distribution of y; (v) the marginal distribution of x. (vi) Are x and y independent? (vii) If the concentration of y 0.5 ppm, determine the probability $x \leq 0.25$.
7. If X_1, X_2, X_3 and X_4 are random variables with $E(X_1) = 2$, $E(X_2) = 4$, $E(X_3) = -1$, $E(X_4) = 0$, $\sigma_{X1}^2 = 4$, $\sigma_{X2}^2 = 8$, $\sigma_{X3}^2 = 6$, $\sigma_{X4}^2 = 1$, $\text{COV}(X_1, X_4) = 1/2$, $\text{COV}(X_2, X_3) = 0$, $\text{COV}(X_2, X_4) = 2$, $\text{COV}(X_3, X_4) = 0$.
Find the mean and variance of $Y = -3X_1 + 2X_2 + 6X_3 - X_4$.

8. The mean and standard deviation of the force in the members of a truss Fig. 2.11 is given as $E[T_1] = 73.2$ N, $E[T_2] = 51.7$ N and $\sigma_{T1}^2 = 8.784, \sigma_{T2}^2 = 6.204$ and the vertical

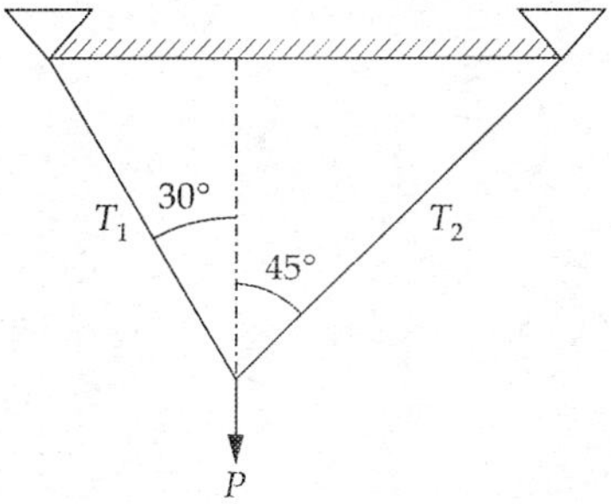

Fig. 2.11

equilibrium $P = T_1 \cos 30° + T_2 \cos 45°$. Estimate the mean value and standard deviation of the load P, given the correlation between T_1, T_2 is zero.

3
CHAPTER

Functions of Random Variables

3.1 TYPES OF RANDOM VARIABLES DISTRIBUTION

A random variable can assume a discrete value or a continuous value. Based on the type of value a random variable assumes, distributions are classified as (a) discrete distribution and (b) continuous distribution. Distributions of discrete and continuous distributions are listed below.

Types of discrete distribution:

(i) Binomial distribution
(ii) Poisson's distribution

Types of continuous distribution:

(i) Uniform distribution
(ii) Exponential distribution
(ii) Normal distribution
(iii) Log-normal distribution
(iv) Gamma distribution
(v) Triangular distribution
(vi) Extreme type I distribution
(vii) Extreme type II distribution
(viii) Extreme type III distribution
(ix) Beta distribution

3.2 BINOMIAL DISTRIBUTION

Engineering studies may involve occurrence of events in a sequence of repeated trials. For example, the chances of vehicles taking left or right turn at a *T* intersection in traffic density studies, specific level of water in a dam every year, chances of performance of construction equipment on daily basis, tossing of coin, passing or failure of student in any class. These events have only two possible outcomes of occurrence or non-occurrence. Problems of this type of repeated trials are modelled by Bernoulli sequence having the following assumptions:

(i) Each trial has only two outcomes—occurrences and non-occurrences.
(ii) The probability of occurrences of any event in each trial is constant.
(iii) The trials are statistically independent.

The coefficients of the binomial expansion can, of course, be computed quite rapidly using tables of factorial values. If, however, appropriate factorial values are not available and n is large, Stirling's formula can be used. This gives the factorial. The error involved is less than 0.1 per cent for n greater than 100 and less than 1 per cent for n greater than 10.

Approximation for the factorial is given by

$$n! \cong e^{-n} n^n \sqrt{2\pi n}.$$

If an event of repeated trials following Bernoulli sequence has a probability of occurrence p, then the probability of exactly x occurrences among n trials is given by binomial PMF as

$$P(X = x) = {}^nC_x \times p^x \times q^{n-x} \quad \text{for } x = 0, 1, 2, 3 \ldots n \qquad \ldots(3.1)$$

where q is probability of non-occurrence = $1 - p$. The term ${}^nC_x$ in the above expression is called Binomial coefficient and is given by

$${}^nC_x = \frac{n!}{x!(n-x)!} \qquad \ldots(3.2)$$

The mean and variance of a random variable following Binomial distribution are

$$\text{Mean, } \mu_X = np \qquad \ldots(3.3)$$

$$\text{Var (X)}, \sigma_X^2 = n \times p \times q \qquad \ldots(3.4)$$

Before concluding, a very useful property of the binomial distribution is observed, i.e., if the parameter p is less than 0.5, the binomial distribution will be skewed to the right, on the other hand, if p exceeds 0.5, the distribution will be skewed to the left. For large values on n, the binomial distribution approaches normal distribution.

Example 3.1 A concrete masonry block manufacturing unit produces 1000 blocks per day and has a probability of failure of 5%. What are the characteristics of non-defective blocks?

Solution: The outcome of the manufacturing unit shall be defective or non-defective, it follows binomial distribution. Probability of failure is 5%, hence probability of success, $p = (1 - q)$ is 95%.

Expected mean, μ_X of non-defective blocks (success) $= n(p) = n(1 - q)$
$= 1000(1 - 0.05)$
$= 950$ blocks

Variance, $\text{Var (X)} = n \times p \times q$

Standard deviation, $\sigma_X = \sqrt{n \times p \times q} = \sqrt{(1000) \times (0.95) \times (0.05)} = 6.8920$

$$\text{CoV} = \frac{\sigma_X}{\mu_X} = \frac{6.8920}{950} = 0.00725$$

Example 3.2 In a precision bombing attack there is 50% chance that any one bomb will hit the target. At least 2 direct hits are required to destroy the target completely. How many bombs must be dropped to give 99% chance to completely destroy the target?

Solution: The outcome of bombing event will have two outcomes, i.e., will hit the target or will not hit the target. Hence, the event follows Bernoulli trial and binomial distribution applicable.

Probability of success, p is 50%, i.e. 0.5, hence probability of failure, $q = (1 - p)$ is 0.5.

Given P(at least 2 bombs to destroy the target completely) = 0.99

$$= P(X \geq 2) = 0.99$$
$$= [1 - P(X < 2)] = 0.99$$
$$= [1 - \{P(X = 0) + P(X = 1)\}] = 0.99$$

Using PMF for binomial distribution, we have, $P(X = x) = {}^nC_x \times p^x \times q^{n-x}$

$$P(X = 0) = {}^nC_0 \times (0.5)^0 \times (0.5)^{n-0} = (1)\,(1)\,(0.5^n) = 0.5^n$$

$$P(X = 1) = {}^nC_1 \times (0.5)^1 \times (0.5)^{n-1} = (n)\,(0.5)\,(0.5^{n-1}) = n \times 0.5^n$$

$$= [1 - \{0.5^n + n \times 0.5^n\}] = 0.99$$

or

$$(1 - 0.99) = 0.5^n (1 + n)$$

Solving $n = 10$

Example 3.3 The probability of a bridge deck getting submerged during floods is 0.05 in any year. What is the probability that the deck getting submerged at least once in next five years?

Solution: The time interval of submergence of the deck is one year, i.e., the return period; hence submergence can be described as an event which follows Bernoulli trial.

Here, $p = 0.05$, $q = (1 - 0.05) = 0.95$ and $n = 5$

Probability that the deck getting submerged at least once in next five years = $P(X \leq 1)$
Expressing $P(X \leq 1) = P(X = 0) + P(X = 1)$

$$= {}^5C_0 \times (0.05^0) \times (0.95^{(5-0)}) + {}^5C_1 (0.05^1) \times (0.95^{(5-1)})$$
$$= 0.977$$

3.3 POISSON DISTRIBUTION

Many random events occur in the continuous dimension of time or space. Some examples could be: cracks may occur at any point in a continuous weld, earthquakes could occur along any point of fault zone, accident at any time on a highway or tornados, hurricanes at any time during the year. The number of occurrences can be considered random variables with Poisson distribution.

However, if the time or space interval is divided into smaller intervals and events will either occur or does not occur such problems could be modelled by binomial distribution. If the time interval approaches zero, the binomial distribution becomes Poisson distribution.

If X_t is the number of occurrences of an event in time t, then the PMF of Poisson distribution is given by

$$P(X_t = x) = \frac{\lambda^x e^{-\lambda}}{x!} \quad \text{for} \quad x = 0, 1, 2, \ldots\ldots \qquad \ldots(3.5)$$

where, $\lambda = vt$ and λ denotes the average number of events in time, t, v denotes the mean rate of occurrences or average number of occurrences of event per unit time or space interval.

The parameters describing Poisson distribution are

$$\text{Mean, } \mu_X = \lambda \qquad \ldots(3.6)$$

$$\text{Variance, } \sigma_X^2 = \lambda \qquad \ldots(3.7)$$

Example 3.4 The seismological record of a place shows that there were 3 earthquakes per year over last 15 years. Assuming the occurrence of earthquakes follows Poisson distribution. What is the probability of (i) no earthquake next year; (ii) exactly 3 earthquakes next year; or (iii) 2 or more earthquakes in next year?

Solution: The PMF of a Poisson distribution is given by

$$P(X_t = x) = \frac{\lambda^x e^{-\lambda}}{x!} \quad \text{for} \quad x = 0, 1, 2, \ldots\ldots$$

Here, λ = 3 denotes the average number of earthquakes per year in time period of 15 years.

(i) Probability of no earthquake next year = $P(X_t = 0) = \dfrac{3^0 e^{-3}}{0!} = 0.0497$

(ii) Probability of exactly 3 earthquakes next year = $P(X_t = 3) = \dfrac{3^3 e^{-3}}{3!} = 0.2240$

(iii) Probability of 2 or more earthquakes next year = $P(X_t \geq 2)$

$$P(X_t \geq 2) = 1 - [P(X_t = 0) + P(X_t = 1)]$$

$$= 1 - \left[\frac{3^0 e^{-3}}{0!} + \frac{3^1 e^{-3}}{1!}\right]$$

$$= 0.80085$$

Example 3.5 The building safety in an earthquake-prone area is under study. The records indicate that there were four strong earthquakes in the area over last 100 years. What is the probability that (i) there will be no strong earthquakes in the area in 50 years or (ii) there will be only two strong earthquakes in 50 years?

Solution: The mean number of earthquakes per unit time (year) = $\nu = \frac{\lambda}{t} = \frac{4}{100}$

Hence, the number of earthquakes over a period of 50 years = $\lambda = \nu t = \frac{4}{100}(50) = 2$

(i) Probability that there will be no strong earthquakes in the area in 50 years is

$$P(X_t = 0) = \frac{2^0 e^{-2}}{0!} = 0.135$$

(ii) Probability that there will be two strong earthquakes in the area in 50 years is

$$P(X_t = 2) = \frac{2^2 e^{-2}}{2!} = 0.271$$

3.4 EXPONENTIAL DISTRIBUTION

This distribution is also known as negative exponential distribution and is related to Poisson process. The time T_1 till the first occurrence of the event in Poisson process follows an exponential distribution. Since no events which follows Poisson distribution occurs in the time t, we have $(T_1 > t)$. Using density function of Poisson distribution, we have,

$$P(T_1 > t) = P(X = 0) = \exp(-\nu t)$$

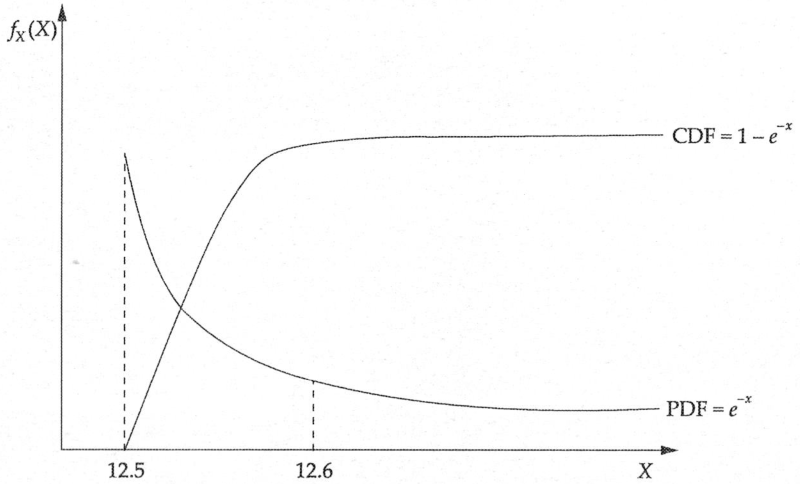

Fig. 3.1 Exponential PDF and CDF function

The CDF is given by the condition

$$F_{T_1}(t) = P(T_1 \le t) = 1 - \exp(-\nu t) \qquad ...(3.8)$$

The PDF can be evaluated by differentiating the CDF, hence

$$f_{T_1}(t) = \frac{dF}{dt} = \nu \exp(-\nu t) \qquad ...(3.9)$$

The mean and variance of exponential distribution are

$$\text{Mean } \mu_{T_1} = \frac{1}{\nu} \quad ...(3.10)$$

$$\text{Var } \sigma^2_{T_1} = \frac{1}{\nu^2} \quad ...(3.11)$$

Example 3.6 The building safety in an earthquake-prone area is under study. The records indicate that there were four strong earthquakes in the area over the last 100 years. If the occurrence of such earthquakes is assumed to follow Poisson process, what is the probability that (i) there will be such strong earthquakes in the area within next 3 years; or (ii) there will be *no* such strong earthquakes in the area within next 6 years.

Solution: The mean number of earthquakes per unit time (year) = $\nu = \frac{\lambda}{t} = \frac{4}{100}$.

(i) The probability that there will be such strong earthquakes in the area in the next 3 years is $P(T_1 \le t) = 1 - \exp(-\nu t)$

$$P(T_1 \le 3) = 1 - \exp\left(-\frac{4}{100}(3)\right) = 0.113$$

The probability that there will be *no* such strong earthquakes in the area in the next 6 years is

$$P(T_1 > t) = P(X = 0) = \exp(-\nu t)$$

or

$$P(T_1 > 10) = \exp\left(-\frac{4}{100}(10)\right)$$

$$= 0.67$$

3.5 UNIFORM DISTRIBUTION

The PDF function has a constant value for all possible values of random variable within the range [*a*, *b*]. This means that all numbers are equally likely to appear.
The PDF of the variable following uniform distribution is

$$f(x) = \begin{cases} \dfrac{1}{(b-a)} & a \le x \le b \\ 0 & \text{otherwise} \end{cases} \quad ...(3.12)$$

where *a* and *b* define the lower and upper bound values of the random variable.
The CDF is given by

$$F_X(x) = \begin{cases} 0 & x \le a \\ \dfrac{(x-a)}{(b-a)} & a \le x \le b \\ 1 & \text{otherwise} \end{cases} \quad ...(3.13)$$

The CDF for a uniform random variable is shown in Fig. (3.2)
The mean and variance of a uniform distribution are as follows:

$$\text{Mean } \mu_X = \left(\frac{a+b}{2}\right) \quad \text{...(3.14)}$$

$$\text{Standard deviation } \sigma_X = \frac{(b-a)^2}{12} \quad \text{...(3.15)}$$

Knowing the mean and variance the parameters *a* and *b* are given by

$$a = \mu_X - \sqrt{3}\sigma_X \quad \text{...(3.16)}$$

$$b = \mu_X + \sqrt{3}\sigma_X \quad \text{...(3.17)}$$

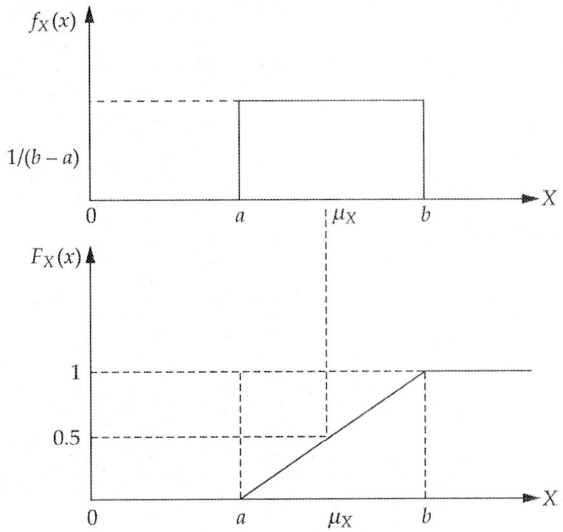

Fig. 3.2 PDF and CDF of uniform random variable

Example 3.7 Find the parameters of the uniform distribution, with a mean of 32.5 N/mm^2, and a standard deviation of 4.6398 N/mm^2.

Solution: The parameters of distribution is given by Eqs. 3.16 and 3.17

$$a = \mu - \sqrt{3}\sigma = 32.5 - \sqrt{3} \times 4.6398 = 24.46$$

$$b = \mu + \sqrt{3}\sigma = 32.5 + \sqrt{3} \times 4.6598 = 40.54$$

3.6 NORMAL DISTRIBUTION

Normal distribution is also known as Gaussian distribution (Gaussian law of errors) after Karl Friedrich (1777–1855). This distribution is most widely and commonly used in engineering. This distribution is important because:

(i) It represents the distribution of random errors in many kinds of measurements.
(ii) Even if individual observed errors in the data do not follow normal distribution, the mean of these errors follows normal distribution.

(iii) Many sets of experimental results in practice turn out to follow the normal distribution.

(iv) The probability of positive and negative errors is 0.5. This fact is of fundamental importance and will be used in deriving the equation for the normal distribution from binomial distribution.

(v) Small errors occur more frequently than large errors.

(vi) Positive errors are as frequent as negative errors.

The probability density function (PDF) of normal distribution is

$$f_X(x) = \frac{1}{\sigma \times \sqrt{2\pi}} e^{-\frac{1}{2}\left(\frac{x-\mu}{\sigma}\right)^2} \quad \text{for } -\infty < x < \infty \qquad ...(3.18)$$

where $\pi = 3.1415926$ and e = 2.71828 (natural logarithm)

The CDF is given by

$$P(X \le x) = F_X(x) = \int_{-\infty}^{x} f_X(x)\,dx \qquad ...(3.19)$$

The transformation from x to z which is linear

$$\text{If } z = \frac{x-\mu}{\sigma} \text{ then } dz = \frac{dx}{\sigma} \qquad ...(3.20)$$

where μ is called the position factor and σ is called the scaling factor. Substituting Eq. 3.20 in Eq. 3.18, we have,

$$f_Z(z) = \frac{1}{\sqrt{2\pi}} e^{-\frac{z^2}{2}} \qquad ...(3.21)$$

The CDF after transformation becomes

$$F_Z(z) = \int_{-\infty}^{\frac{x-\mu}{\sigma}} f_Z(z)dz = \Phi(z) = \Phi\left(\frac{x-\mu}{\sigma}\right) \qquad ...(3.22)$$

The value of $\Phi(z)$ is found using standard normal distribution tables for different standard normal variables, z.

Now,
$$E(z) = E\left(\frac{x-\mu}{\sigma}\right) = \frac{1}{\sigma}[E(x) - E(\mu)]$$

$$= \frac{1}{\sigma}[\mu - \mu] = 0$$

$$\text{Var}(z) = E[z - E(z)]^2 = E(z^2)$$

$$= E\left[\left(\frac{x-\mu}{\sigma}\right)^2\right] = \frac{1}{\sigma^2}\left[E(x-\mu)^2\right]$$

$$= \frac{\sigma^2}{\sigma^2} = 1.0$$

The standard equation of the transformed density function (Eq. 3.21) has variance of value equal to one. Hence, the transformed Eq. (3.21) has a mean equal to zero and standard deviation equal to one and is called *standard normal distribution.*
The mean and variance of a normal distribution are found as follows:

$$E(x) = \int_{-\infty}^{\infty} x f_X(x)dx = \int_{-\infty}^{\infty} (x)\left(\frac{1}{\sqrt{2\pi}} e^{-\frac{1}{2}\left(\frac{x-\mu}{\sigma}\right)^2}\right)dx$$

Transforming $z = \dfrac{x-\mu}{\sigma}$, $dz = \dfrac{dx}{\sigma}$ then $x = \mu + z\sigma$

$$E(x) = \int_{-\infty}^{\infty} (\mu + z\sigma)\frac{1}{\sqrt{2\pi}} e^{-\frac{z^2}{2}} dz$$

$$= \int_{-\infty}^{\infty} \mu \frac{1}{\sqrt{2\pi}} e^{-\frac{z^2}{2}} dz + \int_{-\infty}^{\infty} \sigma \frac{1}{\sqrt{2\pi}} z e^{-\frac{z^2}{2}} dz$$

$$= \mu \int_{-\infty}^{\infty} \frac{1}{\sqrt{2\pi}} e^{-\frac{z^2}{2}} dz + \sigma \int_{-\infty}^{\infty} \frac{1}{\sqrt{2\pi}} z e^{-\frac{z^2}{2}} dz$$

$$= (\mu)(1) + (\sigma)(0)$$

$$= \mu$$

Hence, the expectation of the PDF corresponding to normal distribution is equal to the mean, μ of the random variable.
Similarly, the variance of the PDF corresponding to normal distribution is given by

$$E(X-\mu)^2 = \int_{-\infty}^{\infty} (x-\mu)^2 f_X(x)dx$$

Since $z = \dfrac{x-\mu}{\sigma}$, $(x-\mu) = z\sigma$

$$E(X-\mu)^2 = \int_{-\infty}^{\infty} z^2\sigma^2 \frac{1}{\sqrt{2\pi}} e^{-\frac{z^2}{2}} dz$$

$$= \sigma^2 \frac{1}{\sqrt{2\pi}} \int_{-\infty}^{\infty} z^2 e^{-\frac{z^2}{2}} dz$$

$$= \sigma^2 \left\{ \left| \frac{-ze^{-\frac{z^2}{2}}}{\sqrt{2\pi}} \right|_{-\infty}^{\infty} + \int_{\infty}^{\infty} \frac{1}{\sqrt{2\pi}} e^{-\frac{z^2}{2}} dz \right\} = \sigma^2\{0 + 1\} = \sigma^2$$

$\text{Var}(X) = \sigma^2$

Summarising the mean and standard deviation of a random variable following normal distribution are μ and σ respectively. Normal distribution which has mean, μ and standard deviation, σ is also denoted by $N[\mu, \sigma]$.

3.6.1 Curvature of Normal Distribution

The curvature of PDF of a normal distribution is given by the following equation,

$$\frac{d^2 f_X(x)}{dX^2} = -\frac{1}{\sigma^3\sqrt{2\pi}}\left(e^{-\frac{x^2}{2\sigma^2}} - \frac{x^2}{\sigma^2}e^{-\frac{x^2}{2\sigma^2}}\right) = 0$$

or

$$\left(1 - \frac{x^2}{\sigma^2}\right) = 0 \quad \text{and} \quad x = \sigma$$

The point of inflexion lies on either side of the mean at a distance equal to the standard deviation, i.e., $z = 1.0$.

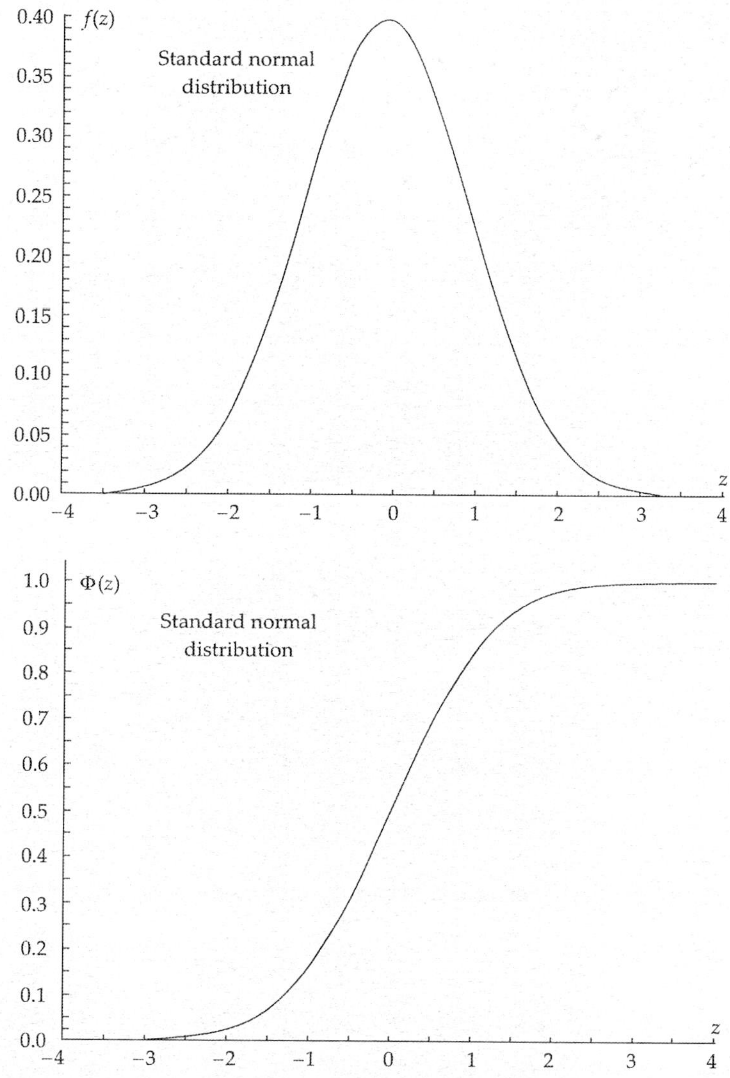

Fig. 3.3 Density function and distribution function of the normal distribution

The total area under the normal distribution curve is one. However, the area under the normal distribution curve on either side of the mean is as follows:

(i) Between $x = (\mu - \sigma)$ and $x = (\mu + \sigma)$, area = 68.26%

(ii) Between $x = (\mu - 2\sigma)$ and $x = (\mu + 2\sigma)$, area = 95.44%

(iii) Between $x = (\mu - 3\sigma)$ and $x = (\mu + 3\sigma)$, area = 99.73%

Since the area under PDF defines the probability, and suppose a normal variable X with distribution $N[\mu, \sigma]$ the probability

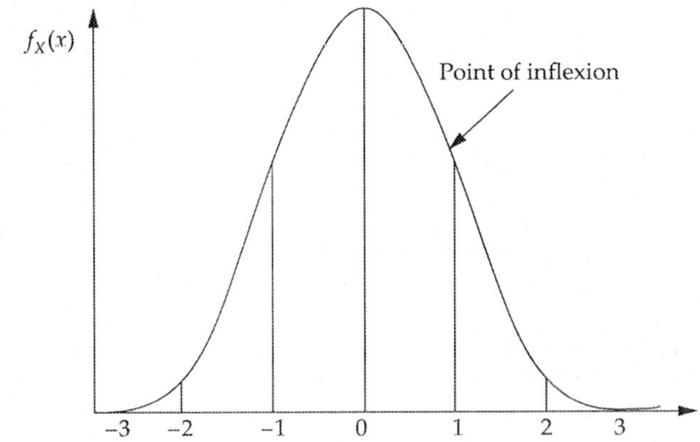

$z = -3$	$z = -2$	$z = -1$	$z = 0$	$z = 1$	$z = 2$	$z = 3$
$x = -3\sigma$	$x = -2\sigma$	$x = -\sigma$	$x = \mu$	$x = \sigma$	$x = 2\sigma$	$x = 3\sigma$
			68.26 %			
			95.44%			
			99.73%			

Fig. 3.4 Density function and distribution function of the normal distribution

$$P(a \le x \le b) = \int_a^b f_X(x)dx$$

$$= \int_a^b \frac{1}{\sigma\sqrt{2\pi}} e^{-\frac{1}{2}\left(\frac{x-\mu}{\sigma}\right)^2} dx \quad \text{...(3.23)}$$

If $z = \dfrac{x-\mu}{\sigma}, \; dz = \dfrac{dx}{\sigma}$

When $x = a, \; z_1 = \dfrac{a-\mu}{\sigma}$

When $x = b, \; z_2 = \dfrac{b-\mu}{\sigma}$

Equation 3.23 can then be written as

$$P(a \le x \le b) = \int_{(a-\mu)/\sigma}^{(b-\mu)/\sigma} \frac{1}{\sigma\sqrt{2\pi}} e^{-\frac{z^2}{2}} \sigma dz$$

$$= \frac{1}{\sqrt{2\pi}} \int_{\frac{(a-\mu)}{\sigma}}^{\frac{(b-\mu)}{\sigma}} e^{-\frac{z^2}{2}} dz \quad ...(3.24)$$

Which is the area under standard normal density function $(a - \mu)/\sigma$ and $(b - \mu)/\sigma$ and Eq. 3.24 can be written as

$$P(a \le x \le b) = \Phi\left(\frac{b-\mu}{\sigma}\right) - \Phi\left(\frac{a-\mu}{\sigma}\right) \quad ...(3.25)$$

The value of $\Phi(z)$ is found using standard normal distribution given in Table A1.

Example 3.8 The finished inside diameter of a ring is normally distributed with a mean of 4.5 cm and a standard deviation of 0.005 cm. What is the probability of obtaining a diameter exceeding 4.51 cm?

Solution: Mean, μ = 4.5 cm

Standard deviation, σ = 0.005 cm

We have, $z = \dfrac{x-\mu}{\sigma}$

$$\frac{(4.51 - 4.5)}{0.005} = 2.0$$

$$\Phi(z) = \Phi\left(\frac{x-\mu}{\sigma}\right) = \Phi(2.0)$$

Referring to standard normal charts the value of $\Phi(z)$ = 0.9772.

The probability of obtaining a diameter exceeding 4.51 cm is given by $P(z > 2)$ = area under the standard normal curve between z = (2.00 and $+\infty$)

= 1 – 0.9772

= 0.0228 (a chance 1 in 44)

= 2.28%

Example 3.9 The income of officers in a certain company follows normal distribution. The average income of an officer is ₹ 15000. The standard deviation of the income of officers is ₹ 5000, there were 242 officers drawing salary above ₹ 18500. How many officers were there in the company?

Solution: Given: $P(X > 18500) = 242$

We have $z = \dfrac{x-\mu}{\sigma}$

$$= \frac{(18500 - 15000)}{5000}$$
$= 0.7$

The value of $\Phi(z) = 1 - \Phi(0.7) = 1 - 0.7580 = 0.242$

Hence, $P(X > 18500) = P(Z > 0.7)$

= area under the standard normal curve between z = (0.7 and $+\infty$)

= 0.242 or 24.2%

Number of officers = 1000

Now, 0.242 × Number of officers = 0.242 × 1000 = 242

Example 3.10 A teacher of a class assigns grades using a system commonly known as grading on the curve in the following manner, if

(a) $\text{marks} > (\overline{x} + 1.6\sigma)$ — Grade A
(b) $(\overline{x} + 0.4\sigma) \le \text{marks} \le (\overline{x} + 1.6\sigma)$ — Grade B
(c) $(\overline{x} - 0.4\sigma) \le \text{marks} \le (\overline{x} + 0.4\sigma)$ — Grade C
(d) $(\overline{x} - 1.6\sigma) \le \text{marks} \le (\overline{x} - 0.4\sigma)$ — Grade D
(e) $\text{marks} < (\overline{x} - 1.6\sigma)$ — Grade E

What is the percentage of marks considered for grading?

Solution: Given condition for grade *A*, $\text{marks} > (\overline{x} + 1.6\sigma)$

i.e., $P(z > 1.6)$, which defines the area between z = (1.6 and $+\infty$)

The value of $\Phi(1.6)$ = 0.4452, hence area under standard normal curve = 0.0548

Therefore, Grade *A* = (0.0548 × 100) = 5.48%

Similarly, for grade *B*, it corresponds to the area in the interval between z = 1.6 and 0.4

$$= \Phi(1.6) - \Phi(0.4) = (0.9452 - 0.6554) = 0.2898$$
$$= 0.2898 \text{ or } 28.98\%$$

For Grade *C*, it corresponds to the area in the interval between z = 0.4 and –0.4

$$= \Phi(0.4) - \Phi(-0.4) = 0.6554 - 0.3446 = 0.3108$$
$$= 31.08\%$$

For Grade *D*, it corresponds to the area in the interval between z = –0.4 and –1.6

$$= \Phi(-0.4) - \Phi(-1.6) = (0.3446 - 0.0548)$$
$$= 0.2898 \text{ or } 28.98\%$$

For Grade *F*, it corresponds to area in the interval between $z = -\infty$ and –1.6 i.e., $P(z < -1.6)$

$$= \Phi(-1.6) = 0.0548 \quad \text{or } 5.48\%$$

Check: 0.0548 + 0.2898 + 0.3108 + 0.2898 + 0.0548 = 1.00 which is the area under standard normal curve.

Example 3.11 The total annual rainfall at a place is found to be normal, N (600 mm, 150 mm). What is the probability of the following events?

(a) That in the next year the annual rainfall will be between 400 mm and 700 mm.

(b) That the annual rainfall will be at least 300 mm.

Solution: Given parameters of normal distribution, mean, μ = 600 mm
Standard deviation = 150 mm

(a) Probability that in the next year the annual rainfall will be between 400 mm and 700 mm.

$$P(400 < X \leq 700) = \Phi\left(\frac{700-600}{150}\right) - \Phi\left(\frac{400-600}{150}\right)$$

$$= \Phi(0.67) - \Phi(-1.33)$$
$$= \Phi(0.67) - [1 - \Phi(1.33)]$$
$$= 0.7486 - [1 - 0.9082]$$
$$= 0.6568$$

(b) Probability that the annual rainfall will be at least 300 mm.

$$P(X \geq 300) = \Phi(+\infty) - \Phi\left(\frac{300-600}{150}\right)$$

$$= 1 - \Phi(-2.00)$$
$$= 1 - [1 - \Phi(2.00)]$$
$$= \Phi(2.00)$$
$$= 0.9772$$

3.7 LOG-NORMAL DISTRIBUTION

A random variable, X is log-normal if natural logarithm of X is normal. A log-normal random variable is defined for position values only ($x \geq 0$). The PDF of a log-normal random variable is given by

$$f_X(x) = \frac{1}{\xi x\sqrt{2\pi}} e^{-\frac{1}{2}\left(\frac{\ln x - \lambda}{\xi}\right)^2} \quad \text{for } x \geq 0 \qquad \text{...(3.26)}$$

where $\lambda = E(\ln X)$ and $\xi = \sqrt{\text{Var}(\ln X)}$

The probability that the log-normal variable can assume values in the interval (a, b) is

$$P(a < X \le b) = \int_a^b \frac{1}{\xi x \sqrt{2\pi}} e^{-\frac{1}{2}\left(\frac{\ln x - \lambda}{\xi}\right)^2} dx \qquad ...(3.27)$$

Using transformations let

$z = \dfrac{\ln x - \lambda}{\xi}$, $dz = \dfrac{dx}{x \cdot \xi}$ substituting in Eq. 3.26

$$P(a < X \le b) = \int_{\frac{\ln a - \lambda}{\xi}}^{\frac{\ln b - \lambda}{\xi}} \frac{1}{\sqrt{2\pi}} e^{-\frac{1}{2}(z)^2} dz = \Phi\left(\frac{\ln b - \lambda}{\xi}\right) - \Phi\left(\frac{\ln a - \lambda}{\xi}\right) \qquad ...(3.28)$$

As the natural logarithm random variable is normally distributed, the standard normal tables can be used to find the probabilities associated with log normal variable.

The mean of variable following log normal distribution is given by

$$\mu_X = E(X) = \frac{1}{\xi x \sqrt{2\pi}} \int_0^\infty x e^{-\frac{1}{2}\left(\frac{\ln x - \lambda}{\xi}\right)^2} dx \qquad ...(3.29)$$

If $y = \ln x$, then $x = e^y$, and $dx = xdy$, substituting in Eq. 3.29

$$\mu_X = \frac{1}{\xi \sqrt{2\pi}} \int_{-\infty}^{\infty} e^y e^{-\frac{1}{2}\left(\frac{y-\lambda}{\xi}\right)^2} \cdot dy$$

$$= \frac{1}{\xi \sqrt{2\pi}} \int_{-\infty}^{\infty} e^{\left[y - \frac{1}{2}\left(\frac{y-\lambda}{\xi}\right)^2\right]} \cdot dy$$

Expanding the square of the exponent, we have,

$$= \left[\frac{1}{\xi \sqrt{2\pi}} \int_{-\infty}^{\infty} e^{\left\{-\frac{1}{2}\left(\frac{y - (\lambda + \xi^2)}{\xi}\right)^2\right\}} \times dy\right] \times e^{\left(\lambda + \frac{\xi^2}{2}\right)}$$

The expression within the brackets is the total unit area of normal distribution

$N[\lambda + \xi^2, \xi]$. Therefore $\mu_X = e^{\left(\lambda + \frac{\xi^2}{2}\right)}$ ...(3.30)

or $\ln \mu_x = \lambda + \dfrac{\xi^2}{2}$ and further simplifying $\lambda = \ln \mu_x - \dfrac{\xi^2}{2}$ ...(3.31)

Similarly, the variance if X can be found as follows,

$$E(X^2) = \frac{1}{\xi\sqrt{2\pi}} \int_{-\infty}^{\infty} e^{2y} e^{\left[-\frac{1}{2}\left(\frac{y-\lambda}{\xi}\right)^2\right]} \times dy$$

$$= \frac{1}{\xi\sqrt{2\pi}} \int_{-\infty}^{\infty} e^{\left[-\frac{1}{2\xi^2}(y^2 - 2(\lambda + 2\xi^2)y) + \lambda^2\right]} \times dy = \left[\frac{1}{\xi\sqrt{2\pi}} \int_{-\infty}^{\infty} e^{-\frac{1}{2}\left(\frac{y-(\lambda+2\xi^2)}{\xi}\right)^2} dy\right] e^{\{2(\lambda+\xi^2)\}}$$

$$= e^{\{2(\lambda+\xi^2)\}} \qquad \text{...(3.32)}$$

As $\text{Var}(X) = E(X^2) - [E(X)]^2$, using Eqs 3.30 and 3.32 we have,

$\text{Var}(X) = e^{\{2(\lambda+\xi^2)\}} + e^{2\left(\lambda + \frac{\xi^2}{2}\right)} = \mu_X^2 (e\xi - 1)$ on further simplified we obtain

$$\xi^2 = \ln\left(1 + \frac{\sigma_X^2}{\mu_X^2}\right) \qquad \text{...(3.33)}$$

If $\left(\sigma_X / \mu_X\right)$ is less than 0.30, then $\ln\left(1 + \frac{\sigma_X^2}{\mu_X^2}\right) \approx \left(\frac{\sigma_X^2}{\mu_X^2}\right)$. In such circumstances, we have

$$\xi \approx \frac{\sigma_X}{\mu_X} = \text{CoV}_X \qquad \text{...(3.34)}$$

which is the coefficient of variation of variable, X.

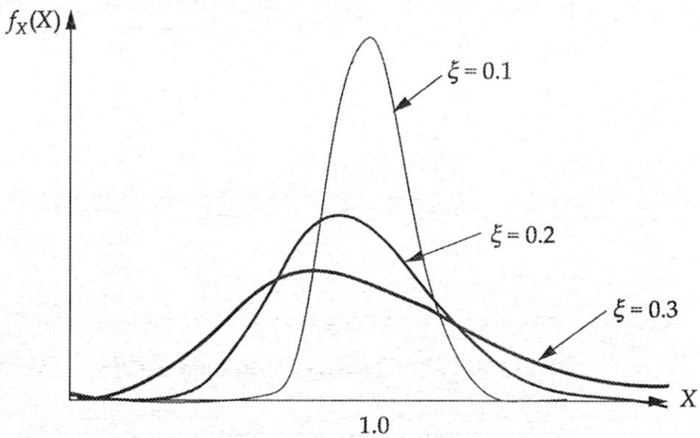

Fig. 3.5 Log-normal density function

Example 3.12 The mean strength of weld is 1.2 kN/mm^2 and its standard deviation is 0.4 kN/mm^2. If the weld is designed for strength of 1.5 kN/mm^2, what is the probability of survival of the welds, if the weld strength follows (a) normal and (b) log-normal distribution?

Solution:

(a) Assuming normal distribution, we have

$$P(X > 1.5) = 1 - P(X \leq 1.5)$$

$$= 1 - \Phi\left(\frac{1.5 - 1.2}{0.4}\right)$$

$$= (1 - 0.7734) = 0.2266 \text{ or } 22.66\%$$

(b) Assuming log-normal distribution, the parameters, ξ and λ are evaluated.

$$\xi^2 = \ln\left(1 + \frac{\sigma_X^2}{\mu_X^2}\right) = \ln\left(1 + \frac{0.4^2}{1.2^2}\right)$$

$$= 0.105$$

or $$\xi = 0.324$$

$$\lambda = \ln(\mu_x) - \left(\frac{\xi^2}{2}\right) = \ln(1.2) - \left(\frac{0.324^2}{2}\right)$$

$$= 0.130$$

$$P(X > 1.5) = 1 - P(X \leq 1.5)$$

$$= 1 - \Phi\left(\frac{\ln X - \lambda}{\xi}\right) = 1 - \Phi\left(\frac{\ln 1.5 - 0.130}{0.324}\right)$$

$$= (1 - 0.8023)$$
$$= 0.1977$$

3.8 GAMMA DISTRIBUTION

While the occurrences of an event constitute a Poisson process, then the *time* till k^{th} event is described by the *gamma probability distribution*. If T_k denotes the time till the kth event, then probability density function for a gamma distribution is given by

$$f_{Tk}(t) = \frac{v(vt)^{k-1}}{Fact(k-1)} e^{-vt} \quad \text{for } t \geq 0 \qquad ...(3.35)$$

where v is the average number of occurrences of the event per unit time and *Fact* $(k - 1)$ denotes factorial of $(k - 1)$, equals $(k - 1)!$

Terms v and k define the parameters of gamma distribution. The mean time, μ_{Tk} till the occurrence of kth event and the variance, σ_{Tk} are

$$\mu_{Tk} = E(T_k) = \frac{k}{v} \qquad ...(3.36)$$

$$\sigma_{Tk} = \text{Var}\,(T_k) = \frac{k}{v^2} \qquad ...(3.37)$$

The parameter k need not be an integer. When k is a non-integer the term $Fact(k - 1)$ in Eq. (3.35) is replaced the gamma function (k) given as

$$\Gamma(k) = \int_0^\infty x^{(k-1)} e^{-x} dx \quad \text{for } (k > 0) \qquad ...(3.38)$$

Hence, the general gamma density function, $f_{Tk}(t)$ is

$$f_{Tk}(t) = \frac{v(vt)^{k-1}}{\Gamma(k)} e^{-vt} \quad \text{for } (k \geq 0) \qquad ...(3.39)$$

The mean and variance remain same k/v and k/v^2 respectively. The probability values involving gamma distribution can be found using tables of incomplete gamma function. The CDF of the Eq. (3.39) is given by

$$F_{Tk}(t) = \int_0^t f_{Tk}(t)\,dt$$

$$= \int_0^t \frac{v(vt)^{k-1}}{\Gamma(k)} e^{-vt} dt$$

Substituting $y = vt$ and $dy = vdt$

$$F_{Tk}(t) = \int_0^{vt} \frac{y^{k-1}}{\Gamma(k)} e^{-y} dy = \frac{\Gamma(vt, k)}{\Gamma(k)} \qquad ... (3.40)$$

For a case, if T has a gamma function with $a \geq 0$, $b \geq 0$

$$P(a < T \leq b) = \int_a^b \frac{v(vt)^{k-1}}{\Gamma(k)} e^{-vt} dt$$

Substituting $y = vt$ and $dy = vdt$

$$P(a < T \leq b) = \int_0^{v \times b} \frac{y^{k-1}}{\Gamma(k)} e^{-y} dy - \int_0^{v \times a} \frac{y^{k-1}}{\Gamma(k)} e^{-y} dy$$

$$= \frac{\Gamma(vb, k)}{\Gamma(k)} - \frac{\Gamma(va, k)}{\Gamma(k)}$$

Hence, the incomplete gamma function ratio is the CDF of gamma probability distribution with the parameter v of unit value.

The following simplifications may be used for evaluation the value of gamma functions.

(i) $\Gamma(k) = (k-1)\Gamma(k-1)$

(ii) $\Gamma(k+1) = (k)\Gamma(k)$

Likewise

$$\begin{aligned}\Gamma(k+2) &= (k+1)\Gamma(k+1)\\ &= (k+1)\times k\times\Gamma(k)\\ \Gamma(k+3) &= (k+2)\times\Gamma(k+2)\\ &= (k+2)(k+1)\times\Gamma(k+1)\\ &= (k+2)(k+1)k\times\Gamma(k)\end{aligned}$$

The value of $\Gamma(k)$ can also be evaluated by the following expression

$$\Gamma(k)=\frac{\Gamma(k+1)}{k}=\frac{\Gamma(k+2)}{(k+1)k}=\frac{\Gamma(k+3)}{(k+2)(k+1)k}=\ldots\ldots=\frac{\Gamma(k+n)}{(k+n-1)(k+n-2)(k+n-3)\ldots..(k+1)k}$$

For example if the value of $\Gamma(3.5)$ is to be evaluated, then

$$\Gamma(3.5)=(2.5)(1.5)\Gamma(1.5)$$

Value of $\Gamma(1.5)$ = 0.8862

Therefore, $\Gamma(3.5)$ = (2.5) (1.5) (0.8862) = 3.3232

Gamma function is described by one parameter, two parameters and three parameters. They are given as under:

Gamma (single parameter *a*) PDF called standard gamma distribution

$$f_X(x)=\frac{1}{\Gamma(b)}y^{b-1}e^{-y} \qquad \ldots(3.41)$$

$$y={}^{x}\!/_{a} \qquad \ldots(3.42)$$

If b is a positive integer the distribution reduces to an exponential distribution.

Gamma (two parameter *a*, *c*) function

$$f_X(x)=\frac{1}{a\times\Gamma(b)}\left(\frac{x}{a}\right)^{b-1}\exp\left(-\frac{x}{a}\right) \quad \text{for } (x>0),\ (a,b>0) \qquad \ldots(3.43)$$

$\Gamma(.)$ is the gamma function and $\Gamma(b)=\int_0^\infty y^{b-1}\exp(-y)dy$ ...(3.44)

Used for hydrologic frequency analysis

$$\text{Mean } \mu_X = ab \qquad \ldots(3.45)$$

$$\text{Var}(X)\ \sigma_X^2 = a^2b \qquad \ldots(3.46)$$

$$\gamma_X=\frac{2}{\sqrt{b}} \qquad \ldots(3.47)$$

Gamma (three parameter *a*, *b*, *c*) function

$$f_X(x)=\frac{1}{a\times\Gamma(b)}\left(\frac{x-c}{a}\right)^{b-1}\exp\left(-\frac{(x-c)}{a}\right) \quad \text{for } (x>c),\ (a,b>0) \qquad \ldots(3.48)$$

Mean $\mu_X = c + ab$ ...(3.49)

Var (X) $\sigma_X^2 = a^2 b$ ...(3.50)

$$\gamma_X = \frac{2}{\sqrt{b}} \quad \text{...(3.51)}$$

The parameters of the gamma distribution are

$$a = \frac{\sigma_x^2}{\mu_x}; b = \left(\frac{\mu_x}{\sigma_x}\right)^2 \quad \text{... (3.52)}$$

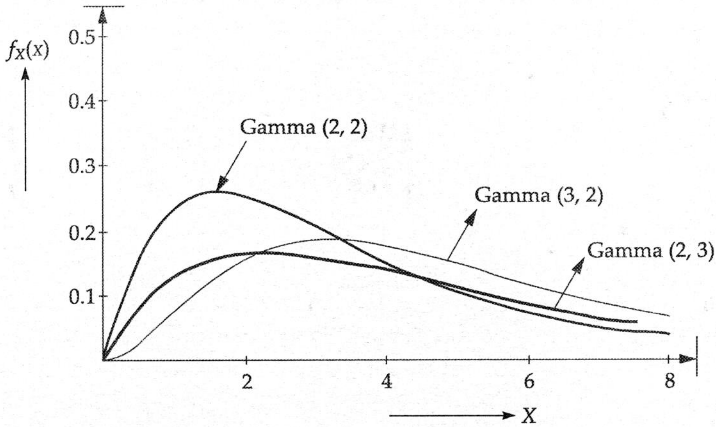

Fig. 3.6 Gamma density function

Example 3.13 The compressive strength of concrete is found to follow a gamma distribution with an expected value concrete 28.284 MPa and a standard deviation of 20 MPa. The minimum strength was found to be 7 MPa. What is the probability that a sample chosen at random will possess a compressive strength less than 17 MPa?

Solution: Expected value $\mu_X = ab$

Variance $\sigma_X^2 = a^2 b$

Solve for a and b for the Gamma PDF

$$a = \frac{\sigma_x^2}{\mu_x}; \ b = \left(\frac{\mu_x}{\sigma_x}\right)^2$$

$$a = \frac{20^2}{28.284} = 14.142; b = 2$$

$$f_X(x) = \frac{1}{a \cdot \Gamma(b)} \cdot \left(\frac{x - c}{a}\right)^{b-1} \exp\left(-\frac{(x - c)}{a}\right)$$

$$F(x)=\int_{7}^{r}\int\frac{1}{14.142^2}(x-7)\exp\left(-\frac{(x-7)}{14.142}\right)dx$$

$F_X(17) = 1 - 0.837 = 0.163$

3.9 BETA DISTRIBUTION

This distribution is used when bounds of the random values lie between the finite limits, say a and b. Many random values of engineering significance like stress in concrete or steel, vehicles count in traffic studies, etc., may be bounded by two limits and beta distribution could be appropriate. The beta distribution is used to obtain approximations for data sets, such as material parameters, whose measures must always be positive quantities and whose ranges are of rather limited extent. The beta distribution requires only numerical values of their parameters to make quantitative assessments. Frequently, probability distributions cannot be determined on a purely theoretical basis. What is known is only that certain experiments have produced certain outcomes. Data are available and conclusions must be drawn. Often graphical representations, such as a histogram or frequency polygon, refined to remove roughness, suffice. However, such presentations lack the manipulative and analytical advantages of a mathematical statement. Among the many advantages of an explicit formulation are the following: (i) Distribution parameters are distinctly determined thus providing valuable index characteristics. (ii) The inherent roughness of histograms is removed and missing data are supplied. (iii) The objective can be achieved by probabilistic statements.

The PDF of beta distribution is given by

$$f_X(x)=\begin{cases}\dfrac{1}{B(q,r)}\times\dfrac{(x-a)^{q-1}(b-x)^{r-1}}{(b-a)^{q+r-1}} & \text{for } (a\le x\le b)\\ 0 & \text{elsewhere}\end{cases} \qquad \text{... (3.53)}$$

where $B(q, r)$ is the beta function in which q and r are the parameters of the distribution. The beta function $B(q, r)$ is given by

$$B(q,r)=\int_0^1 x^{(q-1)}\times(1-x)^{(r-1)}\times dx \qquad \text{... (3.54)}$$

Its relation to the gamma function is given by

$$B(q,r)=\frac{\Gamma(q)\times\Gamma(r)}{\Gamma(q+r)} \qquad \text{... (3.55)}$$

The mean and the variance of a beta distribution are

$$\text{Mean}=E(X)=\mu_X=a+\frac{q}{(q+r)}\times(b-a) \qquad \text{... (3.56)}$$

$$\text{Var (X)}=\sigma_X^2=\frac{q\times r}{(q+r)^2\times(q+r+1)}\times(b-a)^2 \qquad \text{... (3.57)}$$

If the limits of the random variable following beta distribution are between 0 and 1 ($a = 0$, $b = 1$) then the Eq. (3.53) reduces to

$$f_X(x) = \begin{cases} \frac{1}{B(q,r)} \times x^{q-1} \times (1-x)^{r-1} & \text{for } (a \le x \le b) \\ 0 & \text{elsewhere} \end{cases} \quad \text{... (3.58)}$$

Equation (3.58) is called the standard beta distribution. When q and r are both equal to one, the resulting equation reduces to uniform distribution.

The associated probability of random variable following beta distribution can be evaluated in terms of incomplete beta function given by

$$B_y(q,r) = \int_0^y x^{(q-1)} \times (1-x)^{(r-1)} \times dx \qquad \text{for } (0 < x < a) \quad \text{...(3.59)}$$

As an illustration to evaluate the probability between lower limit $x = y_1$ and upper limit $x = y_2$, then

$$P(y_1 < X \le y_2) = \int_{y_1}^{y_2} \frac{1}{B(q,r)} \frac{(x-a)^{q-1} \times (b-x)^{r-1}}{(b-a)^{q+r-1}} \times dx$$

On further simplification if $y = \frac{x-a}{b-a}$ so that $dy = \frac{dx}{b-a}$ and $(1-y) = \frac{b-x}{b-a}$ the above equation with change of variable becomes

$$P(y_1 < X \le y_2) = \int_0^{\frac{(y_2-a)}{(b-a)}} \frac{1}{B(q,r)} \times y^{q-1} \times (1-y)^{(r-1)} dy - \int_0^{\frac{(y_1-a)}{(b-a)}} \frac{1}{B(q,r)} \times y^{q-1} \times (1-y)^{(r-1)} \times dy$$

It may be noted that the last two integrals in the above expression are incomplete beta functions.

If $u = \frac{y_2 - a}{b-a}$ any $v = \frac{y_1 - a}{b-a}$, then the incomplete beta functions in the integrals of the above expression are $B_u(q,r)$ and $B_v(q,r)$. Hence, we have

$$P(y_1 < X \le y_2) = \frac{1}{B(q,r)} (B_u(q,r) - B_v(q,r)) \quad \text{...(3.60)}$$

The value of incomplete beta functions can be found using table for beta distribution similar to standard normal Table A1.

3.9.1 Alternative Beta Distribution

The probability density function has been defined with respect to mean, standard deviation, minimum, and maximum. $\alpha > -1$, and $\beta > -1$ then normalising the constant, a is minimum and b is maximum.

$$f_X(x) = C(x-a)^{\alpha} (b-x)^{\beta}$$

$$C = \frac{1}{B(\alpha+1, \beta+1)(b-a)^{(\alpha+\beta+1)}}$$

where $\alpha > -1$; $\beta > -1$ and $a \leq x \leq b$ ...(3.61)

The beta function $(B(\alpha + 1, \beta + 1)$ is related to the gamma function as

$$B(\alpha+1, \beta+1) = \frac{\Gamma(\alpha+1)\Gamma(\beta+1)}{\Gamma(\alpha+\beta+2)} \quad ...(3.62)$$

Expected value as

$$E(x) = \alpha + \frac{(\alpha+1)}{(\alpha+\beta+2)}(b-a) \quad ...(3.63)$$

Variance as

$$V(x) = \frac{(b-a)^2 (\alpha+1)(\beta+1)}{(\alpha+\beta+2)^2 (\alpha+\beta+3)} \quad ...(3.64)$$

The shape coefficient α, β

$$\alpha = \frac{X^2}{Y^2}(1-X) - (1+X) \quad ...(3.65)$$

$$\beta = \frac{(\alpha+1)}{X} - (\alpha+2) \quad ...(3.66)$$

The variable X and Y are expressed in terms of mean, standard deviation, and minimum and maximum. A procedure presented to approximate a beta distribution by simply determining the mean, coefficient of variation or standard deviation. The third and fourth moments are not necessary but they are desirable to understand the characteristic of the distribution.

$$X = \frac{(\bar{x}-a)}{(b-a)} \quad ...(3.67)$$

$$Y = \frac{\sigma}{(b-a)} \quad ...(3.68)$$

The skewness coefficient for the beta distribution is given as

$$\sqrt{\beta(1)} = \frac{2(\beta-\alpha)}{(\alpha+\beta+4)}\sqrt{\frac{(\alpha+\beta+3)}{(\alpha+1)(\beta+1)}} \quad ...(3.69)$$

The coefficient of kurtosis as

$$\beta(2) = \frac{3(3+\beta+\alpha)[2(\alpha-\beta)^2 + (\alpha+1)(\beta+1)(\alpha+\beta+4)]}{(\alpha+1)(\beta+1)(\alpha+\beta+4)(\alpha+\beta+5)} \quad ...(3.70)$$

A listing of the effects of the parameters α, β on the shapes of the distribution and it follows as given below.

Example 3.14 Obtain the parameters of the beta distribution for the data, having a mean of 5, variance of 0.25, having a minimum of 4 and maximum of 6.5 units.

Solution: To avoid undue complication we shall first assume α, β are integers.

$$X = \frac{(\bar{x} - a)}{(b - a)} = \frac{(5 - 4)}{(6.5 - 4)} = 0.4$$

$$Y = \frac{\sigma}{(b - a)} = \frac{0.5}{(6.5 - 4)} = 0.2$$

$$\alpha = \frac{X^2}{Y^2}(1 - X) - (1 + X) = \frac{0.4^2}{0.2^2}(1 - 0.4) - (1 + 0.4) = 1.0$$

$$\beta = \frac{(\alpha + 1)}{X} - (\alpha + 2) = \frac{(1 + 1)}{0.4} - (1 + 2) = 2.0$$

$$C = \frac{1}{B(\alpha + 1, \beta + 1)(b - a)^{(\alpha + \beta + 1)}} = \frac{1}{B(1 + 1, 2 + 1)(6.5 - 4)^{(1+2+1)}} = \frac{1}{B(2, 3)(2.5)^4}$$

$$B(2, 3) = \frac{\Gamma(2)\Gamma(3)}{\Gamma(5)} = \frac{1 \times \Gamma(1) \times 2 \times 1 \times \Gamma(1)}{4 \times 3 \times 2 \times \Gamma(1)} = \frac{2}{24}$$

$$C = \frac{1}{B(2,3)(2.5)^4} = \frac{24}{2 \times 39.0625} = 0.3072$$

The probability density function is given as

$$f_X(x) = 0.3072(x - 4)^1 (6.5 - x)^2$$

Example 3.15 The strength of concrete data has a mean of 32.6 N/mm^2, standard deviation of 4.7 N/mm^2, having a minimum of 21 N/mm^2 and maximum of 45 N/mm^2. Obtain the parameters of the beta distribution.

Solution:

$$X = \frac{(\bar{x} - a)}{(b - a)} = \frac{(32.6 - 21)}{(43 - 21)} = 0.5272$$

$$Y = \frac{\sigma}{(b - a)} = \frac{4.7}{(43 - 21)} = 0.2136$$

$$\alpha = \frac{X^2}{Y^2}(1 + X) - (1 - X) = \left[\frac{0.5272^2}{0.2136^2}(1 - 0.5272)\right] - (1 + 0.5272) = 1.3530$$

$$\beta = \frac{(1.3530 + 1)}{0.5272} - (1.3530 + 2) = \frac{(2.3530)}{0.5272} - (3.3530) = 1.1102$$

$$C = \frac{1}{B(\alpha + 1, \beta + 1)(b - a)^{(\alpha + \beta + 1)}} = \frac{1}{B(2.3530, 2.1102)(22)^{(3.4632)}}$$

From Table A5 of gamma function the values are chosen and computed

$$B(2.3530, 2.1102) = \frac{\Gamma(2.3530)\Gamma(2.1102)}{\Gamma(4.4632)} = \frac{1.3530 \times \Gamma(1.0353) \times 1.1102 \times \Gamma(1.1102)}{3.4632 \times 2.4632 \times 1.4632 \times \Gamma(1.4632)}$$

$$= \Gamma(1.3530) = 0.89115;\ \Gamma(1.1102) = 0.9474;\ \Gamma(1.4632) = 0.8856$$

$$C = \frac{1.2682}{11.0539 \times (22)^{(3.4632)}} = \frac{1.2682}{492712.7034} = 2.5739 \times 10^{-6}$$

The probability density function is given as

$$f_X(x) = 2.5739 \times 10^{-6}(x - 21)^{1.3530}(43 - x)^{1.1102}$$

Note: 1 Gamma function for a value greater than 2 can be evaluated using equation

$$\Gamma(a) = \frac{\Gamma(a+1)}{a} = \frac{\Gamma(a+2)}{a(a+1)} = \frac{\Gamma(a+2)}{a(a+1)(a+2)} = \cdots = \frac{\Gamma(a+k+1)}{a(a+1)(a+2)\ldots(a+k)}$$

Example 1 Evaluate $\Gamma(3.5)$

Solution: Here $a + k + 1 = 3.5$ whereas we wish to reduce argument to $1 \le a \le 2$, k must be 1 (an integer) and $a = 1.5$. From Table A5, $\Gamma(1.5) = 0.8862$ and hence

$$\Gamma(3.5) = (a)(a + 1)\Gamma(a) = (1.5)(2.5)(0.8862) = 3.323$$

Example 2 Evaluate $\Gamma(-2.5)$

Solution: Here we require $1 \le (a + k + 1) \le 2$ or $1 \le (-2.5 + k + 1) \le 2$ or $k = 3$ Therefore,

$$\Gamma(a) = \frac{\Gamma(a+k+1)}{a(a+1)(a+2)\ldots(a+k)} = \frac{\Gamma(1.5)}{(-2.5)(-1.5)(-0.5)(0.5)} = \frac{0.8862}{0.9375} = -0.9450$$

Note: 2 The beta function $B(a, b)$ is related to gamma function as

$$B(a, b) = \frac{\Gamma(a)\Gamma(b)}{\Gamma(a+b)}$$

Example 3 Evaluate $\Gamma(3.5, -2.5)$
Using results of examples 1 and 2, we have

$$B(3.5, -2.5) = \frac{\Gamma(3.5)\Gamma(-2.5)}{\Gamma(1)} = \frac{3.323 \times -0.945}{1} = -3.140$$

3.10 TRIANGULAR DISTRIBUTION

The triangular distribution has the following characteristics visually; it is a simple distribution as shown in Fig. 3.7. It can be symmetric or skewed distribution. The X values must lie between a and c, i.e., $[a, c]$. It has a single peak or mode at b. The triangular distribution is designated at T (min, mode, max).

$$f_X(x) = \frac{2(x-a)}{(b-a)(c-a)} \quad \text{for } a \le x \le b \qquad \ldots (3.71)$$

$$f_X(x) = \frac{2(c-x)}{(c-a)(c-b)} \quad \text{for } b \le x \le c \qquad \ldots (3.72)$$

$$P(X \le x) = F_X(x) = \frac{(x-a)^2}{(b-a)(c-a)} \qquad \text{for } a \le x \le b \qquad \dots (3.73)$$

$$P(X \le x) = F_X(x) = 1 - \frac{(c-x)^2}{(c-a)(c-b)} \qquad \text{for } b \le x \le c \qquad \dots (3.74)$$

Domain $a \le b \le c$

$$\text{Mean} = \frac{(a+b+c)}{3} \qquad \dots (3.75)$$

$$\text{Standard deviation} = \sqrt{\frac{a^2 + b^2 + c^2 - ab - ac - bc}{18}} \qquad \dots (3.76)$$

Shape positively skewed $b < \dfrac{(a+c)}{2}$ ... (3.77)

Negatively skewed $b > \dfrac{(a+c)}{2}$ ... (3.78)

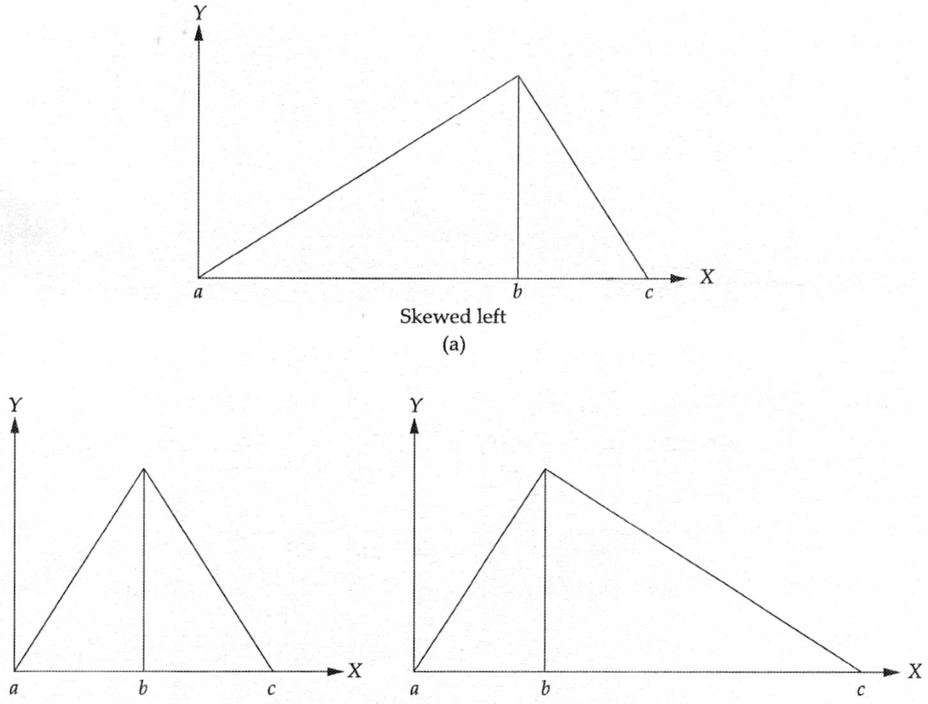

Skewed left
(a)

Symmetric
(b)

Skewed right
(c)

Fig. 3.7 Triangular distribution skewed left, symmetric, and right

Example 3.16 A triangular PDF, distribution with a lower bound of 8.5, peak 12.0 and upper bound of 15.0 is given. Find k, mean and standard deviation of the triangular distribution.

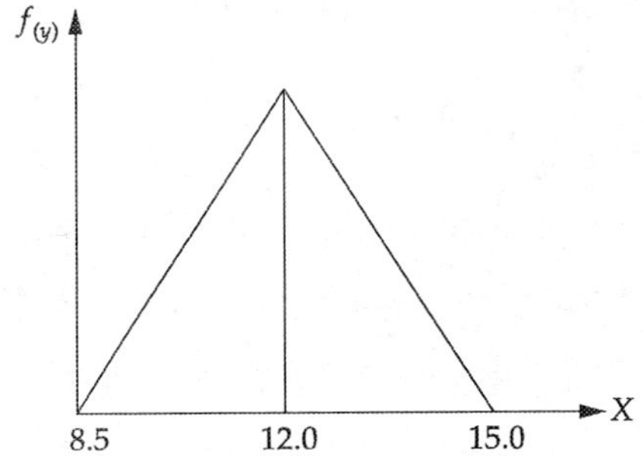

Fig. 3.8 Triangular distribution

The solution to triangular distribution is as follows.

$$f(y) = \begin{cases} k(y - 8.5)/3.5 & 8.5 \le y \le 12.0 \\ k(15.0 - y)/3.0 & 12.0 \le y \le 15.0 \\ 0 & \text{elsewhere } (y < 8.5, y > 15.0) \end{cases}$$

Solution: Choose k by choosing area under the density curve is unity.

Area below 12.0 is 0.5((12.0 – 8.5)k) = 1.75k

Area above 12.0 is 0.5((15.0 – 12.0)k) = 1.5k

Hence, total area is 3.25k, therefore k = 1/3.25

$$f(y) = \begin{cases} (y - 8.5)/11.375 & 8.5 \le y \le 12.0 \\ (15.0 - y)/9.75 & 12.0 \le y \le 15.0 \\ 0 & \text{elsewhere} \end{cases}$$

On integration PDF we get CDF

$$F(y) = \begin{cases} 0 & y < 8.5 \\ 3.175824 - 0.747253y + 0.043956y^2 & 8.5 \le y \le 12 \\ -10.538462 + 1.538462y - 0.051282y^2 & 12 \le y \le 15 \\ 1 & y > 15 \end{cases}$$

$$E(Y) = \int_{-\infty}^{\infty} yf(y)dy = \int_{8.5}^{12} y\left[\frac{y-8.5}{11.375}\right]dy + \int_{12}^{15} y\left[\frac{15-y}{9.75}\right]dy = \left[\frac{y^3}{34.125} - \frac{8.5y^2}{22.5}\right]_{8.5}^{12} + \left[\frac{15y^2}{19.5} - \frac{y^3}{29.25}\right]_{12}^{15}$$

$$= \left[\left(\frac{12^3}{34.125} - \frac{8.5(12)^2}{22.5}\right) - \left(\frac{8.5^3}{34.125} - \frac{8.5^3}{22.5}\right)\right] + \left[\left(\frac{15^3}{19.5} - \frac{15^3}{29.25}\right) - \left(\frac{15(12)^2}{19.5} - \frac{12^3}{29.25}\right)\right]$$

$$= 11.84$$

$$E(Y^2)=\int_{-\infty}^{\infty} y^2 f(y)dy=\int_{8.5}^{12} y^2\left[\frac{y-8.5}{11.37}\right]dy+\int_{12}^{15} y^2\left[\frac{15-y}{9.75}\right]dy=\left[\frac{y^4}{45.5}-\frac{8.5y^3}{34.125}\right]_{8.5}^{12}+\left[\frac{15y^3}{29.25}-\frac{y^4}{39}\right]_{12}^{15}$$

$$=\left[\left(\frac{12^4}{45.5}-\frac{8.5(12)^3}{34.125}\right)-\left(\frac{8.5^4}{45.5}-\frac{8.5^4}{34.125}\right)\right]+\left[\left(\frac{15^4}{29.25}-\frac{15^4}{39}\right)-\left(\frac{15(12)^3}{29.25}-\frac{12^4}{39}\right)\right]$$

$$=141.21$$

$$\Rightarrow V(Y)=E(Y^2)-[E(Y)]^2=141.21-11.84^2=141.21-140.19=1.02$$

$$\Rightarrow \sigma=\sqrt{1.02}=1.01$$

3.11 EXTREME VALUE DISTRIBUTIONS

Reliability analysis often focuses on statistical characteristics of extreme events. The design of flood control structures may be concerned with distribution of largest events over the recorded period. The establishment of a drought management plan or water quality management scheme might be interested in the statistical properties of minimum flow over a specified period.

The asymptotic distributions of extremes turn out to be independent of sample size n and largely depend on the tail behaviour of parent distributions in either direction towards the extreme.

Mathematically, the maxima and minima can be expressed as

$$\lim_{n\to\infty} F_{\max,n}(x)=F_{\max}(x)$$

$$\lim_{n\to\infty} F_{\min,n}(x)=F_{\min}(x)$$

Three types of asymptotic distributions of extremes are:

Type I: Parent distributions are unbounded in the direction of extremes and all statistical moments exist.

Examples: (i) Normal distribution—largest and smallest extremes.
(ii) Log-normal distribution, gamma distribution—largest extremes.

Type II: Parent distributions are unbounded in the direction of extremes but all statistical moments do not exist.

Example: Cauchy distribution.

Type III: Parent distributions are bounded in the direction of the desired extremes.

Example: (i) Beta distribution (largest and smallest extreme)
(ii) Log-normal distribution, gamma distribution—smallest extremes.

$$F_{\min}(x)=1-F_{\max}(-x)$$

3.11.1 Extreme Value Type I (Gumbel) Distribution

Extreme value distributions are useful to characterise the probabilistic nature of extreme values (largest and smallest) of some phenomenon overtime. For example, consider a time interval of duration n. Each time the interval may be one year. During each year, there may be a *maximum value* of some phenomenon (such as wind speed). This is referred to as Gumbel distribution, Fisher-Tippet distribution and double exponential distribution.

Suppose we want to determine the probability distribution for largest annual wind speeds during the year one to n, then X = max (W_1, W_2, W_n) might be characterized as Extreme Type I random variable. Type I asymptotic extreme-value distribution has been used by a number of investigators to describe the strength of brittle materials. The model is also commonly adopted to describe hydrological phenomena such as the maximum daily flow in a year or the annual peak hourly discharge during flood.

The PDF and CDF for this random variable are as follows:

For largest extreme value or maxima

$$f_X(x) = a\exp\{-a(x-b)-\exp(-a(x-b)\}$$

$$F_X(x) = \exp\{-\exp[-x(x-b)\times a]\} \text{ for } (-\infty \le x \le \infty), (a>0) \text{ and } (b<\infty)$$

Mean, $\mu_X = b + \dfrac{\gamma}{a}$

where γ is Euler constant and an infinite sequence of numbers with the first three numbers whose value equals $\gamma = 0.5772$.

$$\mu_X = b + \frac{0.5772}{a} \quad \text{...(3.79)}$$

$$\text{Var}\ \sigma_X^2 = \frac{1.645}{a^2} \quad \text{...(3.80)}$$

Solving the parameters of Gumbel largest, we have

$$a = \frac{1.282}{\sigma_X}$$

$$b = \mu_x - 0.451\sigma_X \quad \text{...(3.81)}$$

For smallest value or minima, the PDF and CDF for this random variable are

$$f_X(x) = a\exp\{a(x-b)-\exp(a(x-b))\}$$

$$F_X(x) = 1-\exp\{-\exp[(x-b)\times a]\}$$

$$\text{Mean}\ \mu_X = b - \frac{0.5772}{a} \quad \text{...(3.82)}$$

$$\text{Var}\ (X)\sigma_X^2 = \frac{\pi^2}{6\times a^2} \quad \text{...(3.83)}$$

Solving for a and b, we have,

$$a = \frac{1.282}{\sigma_X}, \quad b = \mu_x + 0.451\sigma_X \quad \text{...(3.84)}$$

Example 3.17 The records of meteorological department between 1922–1970, comprises maximum snow fall depth has annual maxima with a mean value of 475 mm and standard deviation of 292.1 mm and coefficient of variation of 0.6149. Determine the parameters a and b of the Gumbel distribution for annual maxima.

Solution: The Gumbel largest extreme distribution, namely, cumulative and the probability density function, measures of location, dispersion, are estimated by the parameters a and b from Eq. 3.84.

$$a = \frac{1.282}{\sigma_X} = \frac{1.282}{292.1} = 4.3889 \times 10^{-3}$$

$$b = \mu_x - 0.451 \times \sigma_X = 475 - (0.451 \times 292.1) = 343.4866$$

For largest extreme value or maxima–PDF

$$f_X(x) = a\exp\{-a(x-b) - \exp(-a(x-b)\}$$

$$f_X(x) = 0.0043889\exp\{-0.0043889(x - 343.4866) - \exp(-0.0043889 \times (x - 343.4866))\}$$

For largest extreme value or maxima–CDF

$$F_X(x) = \exp\{-\exp[-x(x-b)\times a]\} \text{ for } (-\infty \le x \le \infty), (a > 0) \text{ and } (b < \infty)$$

$$F_X(x) = \exp\{-\exp[-x(x - 343.4866)\times 0.0043889]\}, \text{ for } (-\infty \le x \le \infty), (a > 0) \text{ and } (b < \infty)$$

An engineer has good estimates of the mean and variance of the peak annual flow in a small stream and that he models the distribution is of the extreme-value type I because the annual peak flow is the smallest, with a mean value of 100 and standard deviation of 50. Determine the parameters a and b of the Gumbel distribution for annual minimum.

3.11.2 Extreme Value Type II (Fisher-Tippet II, Fréchet Distribution)

The Extreme Type II distribution sometimes gives the best approximation of the distribution for maximum seismic load applied to the structure. This distribution is limited on the left at zero, but unlimited to the right in the tail of interest. The relation between this distribution and type I distribution is identical to that between the log-normal and normal distributions. This type II distribution of largest values has been used to represent annual maximum winds, owing to the physically meaningful limitation of the random variable, to positive values of annual maximum wind velocity as being the largest of daily values, or largest of many storms maxima and gust velocities. This is used in the application of seismic loads acting on structure.

The PDF and CDF are called Fisher-Tippet II, Fréchet and are given by

$$f_X(x) = \frac{a}{b} \times \left(\frac{b}{x}\right)^{a+1} \times \exp\left[-\left(\frac{b}{x}\right)^{a}\right] \quad \text{for } (x \ge 0) \text{ and } (a, b > 0) \qquad ...(3.85)$$

$$F_X(x) = \exp\left[-\left(\frac{b}{x}\right)^{a}\right] \quad \text{for } (0 \le x \le \infty) \qquad ...(3.86)$$

$$\text{Mean X, } \mu_X = b\Gamma\left(1 - \frac{1}{a}\right) \qquad ...(3.87)$$

$$\text{Var}(X),\ \sigma_X^2 = b^2\left\{\Gamma\left(1-\frac{2}{a}\right)-\Gamma^2\left(1-\frac{1}{a}\right)\right\} \quad ...(3.88)$$

$$1+\text{CoV}^2 = \frac{\Gamma(1-2/a)}{\Gamma^2(1-1/a)} \quad ...(3.89)$$

Example 3.18 The records of meteorological department, comprises maximum wind velocity has annual maxima with a mean value of 13.74 m/s and standard deviation of 8.1 m/s and coefficient of variation of 0.5895. Determine the parameters a and b of the Fréchet distribution for annual maxima wind speed.

Computing mean, standard deviation and coefficient of variation, a and b are obtained by solving a set of equation: where $b > 0$, scale parameter and $a > 0$, shape parameter

$$\mu_X = b\Gamma\left(1-\frac{1}{a}\right)$$

$$1+\text{CoV}^2 = \frac{\Gamma(1-2/a)}{\Gamma^2(1-1/a)}$$

$$\text{CoV} = \sqrt{\left[\frac{\Gamma\left(1-2/a\right)}{\Gamma^2\left(1-1/a\right)}\right]-1}$$

To find the parameters, we first find the coefficient of variation;

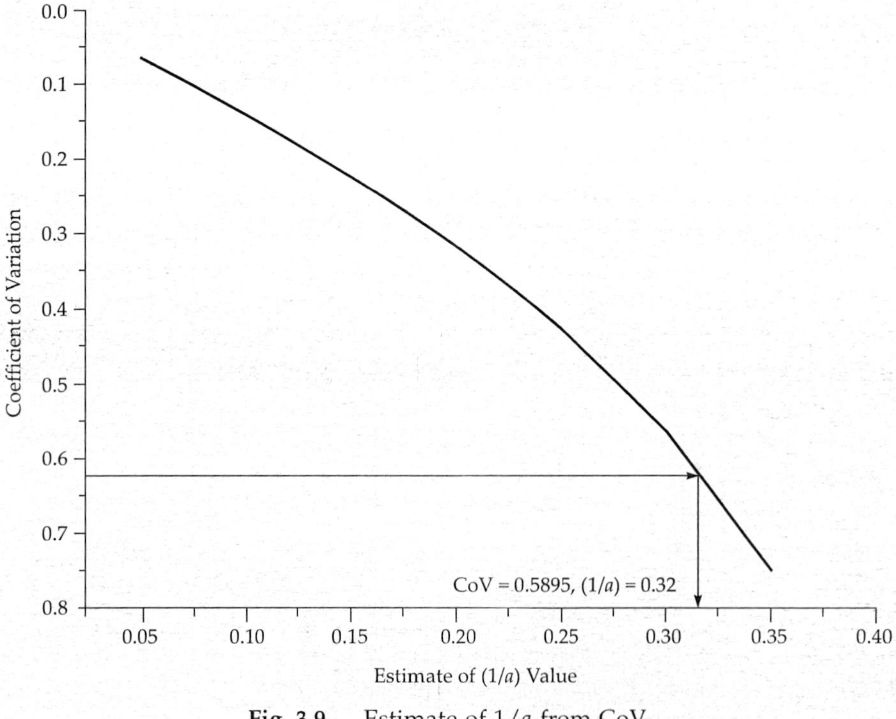

Fig. 3.9 Estimate of $1/a$ from CoV

CoV = 0.5895, from the graph it is found that, $1/a$ = 0.32, hence, from tables of Γ function

$$\mu_X = b\Gamma\left(1-\frac{1}{a}\right)$$

$$\mu_X/b = \Gamma\left(1-\frac{1}{a}\right) = \Gamma(1-0.32) = \Gamma(0.68) = 1.32;$$

$$a = 3.125,$$

$$b = \mu_X/1.32 = 13.74/1.32 = 10.40$$

$$F_X(x) = \exp\left[-\left(\frac{10.40}{x}\right)^{3.125}\right]$$

3.11.3 Extreme Value Type III Distribution Defined by Three Parameters

The purpose of this distribution is used to determine the material strength and fatigue, where c represents the lower bounds on strength and on number of cycles before which failures occur, in case of experimental investigation. This is often referred to as Weibull distribution and it is commonly employed in statistical reliability studies of lifetimes of components and systems.

The PDF and CDF for the minima or smallest value are

$$f_X(x) = \frac{a}{b}\times\left(\frac{x-c}{b}\right)^{a-1}\times\exp\left[-\left(\frac{x-c}{b}\right)^{a}\right] \quad \text{for } (c < x < \infty),$$

$$(a,\ b > 0) \quad \text{and } (c > -\infty) \qquad \text{...(3.90)}$$

$$F_X(x) = 1-\exp\left[-\left(\frac{x-c}{b}\right)^{a}\right] \qquad \text{...(3.91)}$$

$$\text{Mean } \mu_X = c + b\times\Gamma\left(1+\frac{1}{a}\right) \qquad \text{...(3.92)}$$

$$\text{Var }(X),\ \sigma_X^2 = b^2\left[\Gamma\left(1+\frac{2}{a}\right)-\Gamma^2\left(1+\frac{1}{a}\right)\right] \qquad \text{...(3.93)}$$

If c = 0 in Eq. 3.92, the coefficient of variation is given by the following relationship

$$\text{CoV}^2 = \frac{\Gamma\left(1+\frac{2}{a}\right)}{\Gamma^2\left(1+\frac{1}{a}\right)} - 1 \qquad \text{...(3.94)}$$

3.11.4 Two-Parameter Weibull Distribution—(Type III)

$$f_X(x) = \left(\frac{a}{b}\right) \times \left(\frac{b}{x}\right)^{a+1} \times \exp\left[-\left(\frac{b}{x}\right)^a\right] \quad \text{for } (x \geq 0), (a, b > 0) \qquad ...(3.95)$$

$$F_X(x) = \exp\left[-\left(\frac{b}{x}\right)^a\right] \quad \text{for } (0 < x \leq \infty) \qquad ...(3.96)$$

$$\text{Mean } \mu_X = b \times \Gamma\left(1 - \frac{1}{a}\right) \qquad ...(3.97)$$

$$\text{Var } (X),\ \sigma_X^2 = b^2\left[\Gamma\left(1 - \frac{2}{a}\right) - \Gamma^2\left(1 - \frac{1}{a}\right)\right] \qquad ...(3.98)$$

The three distributions provide a valuable link between observed data of extreme events and mathematical models which aid an engineer in evaluating past results and predicting future outcomes. It is difficult to determine which of the three distributions applies.

Example 3.19 The load capacity of a steel section is to be modelled with type III distribution of smallest value, with a mean of 300 kN and a CoV of 0.15. Determine the parameters of the distribution. The minimum strength of steel section has a lower bound of 100 kN.

Solution: The lower bound value of c = 100 kN in Eq. 3.92, and compute the coefficient of variation from Eq. 3.94

$$\text{CoV}^2 = \frac{\Gamma\left(1 + \frac{2}{a}\right)}{\Gamma^2\left(1 + \frac{1}{a}\right)} - 1 = \frac{\text{standard deviation}}{\text{mean} - c} = \frac{0.15 \times 300}{300 - 100} = 0.225$$

CoV = 0.225; from Fig. 3.9, $1/a$ = 0.14 and a = 7.14

Using equation, mean, $\mu_X = c + b \times \Gamma\left(1 + \frac{1}{a}\right)$

$$300 = 100 + b\Gamma\left(1 + \frac{1}{5}\right),\ b = \frac{200}{\Gamma(1.2)} = 217.82$$

The parameter b = 217.82, c = 100 kN, and a = 5

$$F_X(x) = 1 - \exp\left[-\left(\frac{x - c}{b}\right)^a\right] = 1 - \exp\left[-\left(\frac{200 - 100}{217.82}\right)^5\right] = 0.02018$$

3.12 GENERAL EXPRESSIONS OF RELIABILITY

3.12.1 Fundamental Case

Now examine how to determine the probability of failure for a relatively simple performance function given by

$$G(R, S) = R - S \qquad \text{...(3.99)}$$

The following cases exist:

If $g(x_1, x_2, x_3, \ldots x_n) > 0$, indicates safety

If $g(x_1, x_2, x_3, \ldots x_n) = 0$, indicates limit state (boundary between safe and unsafe state)

If $g(x_1, x_2, x_3 \ldots x_n) < 0$, indicates failure.

The structure fails when the load effect exceeds the resistance. If R is equal to a specific value say r_i, then the probability of failure is equal to the probability that the load is greater than the resistance or $P(S > r_i)$.

The probability of failure is composed of all possible combination of $(R = r_i)$ and $(S > r_i)$ which is written as

$$P_f = \sum P(R = r_i \cap S > r_i)$$

$$= \sum P((S > R) \mid (R = r_i))\, P(R = r_i) \qquad \text{... (3.100)}$$

Conditional probability has been used in Eq. (3.100). For the continuous case, the summation then becomes integral.

$P((S > R) \mid (R = r_i))$ is simply $[1 - P((S > R) \mid (R = r_i))]$ or equal to $[1 - F_S(r_i)]$.

In the limit, the probability, $P(R = r_i) \approx f_R(r_i)dr_i$. Combining all these modifications into Eq. (3.100), we have

$$P_f = \int_{-\infty}^{\infty} [1 - F_S(r_i)] f_R(r_i)\, dr_i = 1 - \int_{-\infty}^{\infty} F_S(r_i) f_R(r_i)\, dr_i \qquad \text{...(3.101)}$$

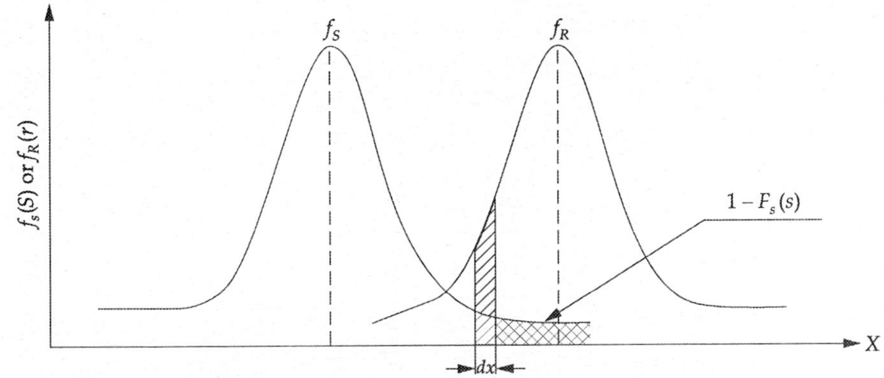

Fig. 3.10 Load and resistance factor interference

3.12.2 Alternative Formulation

If the load effect, S, is equal to a specific value, s_i, then the probability is equal to the probability that the resistance is less than the load or $P(R < s_i)$. However, since S is a random variable, there is a probability associated with each value of s_i. Therefore, the probability of failure is composed of all possible combinations of $S = s_i$ and $R < s_i$ which can be written as

$$P_f = \sum P((S = s_i) \cap (R < s_i)) = \sum P((R < S) \mid (S = s_i)) \times P(S = s_i) \qquad \ldots(3.102)$$

Following the same logic as earlier, this can be written in an integral form as

$$P_f = \int_{-\infty}^{\infty} F_R(s_i) f_S(s_i)\, ds_i \qquad \ldots(3.103)$$

Although Eqs. 3.101 and 3.103 appear rather straightforward, but it is difficult to evaluate these integrals in general. This integration requires special numerical techniques, and accuracy of these techniques may not be adequate. In practice the probability of failure is calculated indirectly using other procedures.

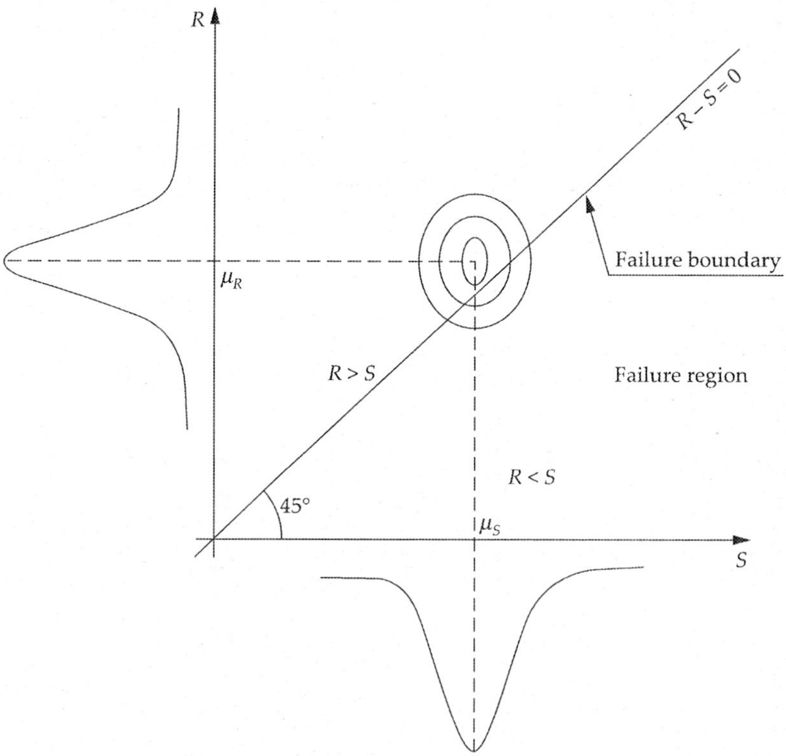

Fig. 3.11 Safe and failure domain in 2D state space

3.12.3 Computation of Probability of Failure

A simple problem of two independent design variables, R and S, each described by a common quantity, X, for example, stress. The probability of failure can be computed by first computing the conditional probability of failure, that is, the probability that one quantity is an interval, dx and that failure has occurred, then summing this conditional probability of failure over all dx values using integral form. The probability that S is in the interval x to $(x + dx)$ is the PDF of S at x times dx, thus

$$P(x < S < (x + dx)) = f_S(x)\, dx$$

The probability that R is less than S is the CDF of R at x, which is $F_R(x)$. The probability that both R and S are in the dx interval of failure and that failure occurs when $P((R < S)$ and $(x \le S \le (x + dx))$ is $F_R(x) f_s(x)$.

Figure 3.10 illustrates probability of failure calculation for load effects, S and resistance model.

The total probability of failure is given by

$$P_f = \int_{-\infty}^{\infty} F_R(x) f_S(x) dx \qquad \text{...(3.104)}$$

An equivalent expression results by considering the probability that R is in the interval; $f_R(x)$ and S is above x, $[1 - F_S(x)]$.

$$P_f = \int_{-\infty}^{\infty} f_R(x)[1 - F_S(x)] dx \qquad \text{...(3.105)}$$

3.12.3.1 First Expression of P_f

Let A be the event representing load effect such that, $S \in [x, x + dx]$ and event B representing resistance, R such that $[R < x]$. The probability of failure for load effect ranging between x and $(x + dx)$ is shown in Fig. 3.9.

The probability of the event $(A \cap B)$ is as follows

$$P(A) = f_S(x)dx$$

$$P(B) = \int_{-\infty}^{\infty} f_R(\tau)d\tau = F_R(x)$$

The events are independent, therefore

$$dP_f = P(A \cap B) = P(A)P(B)$$

$$= f_s(dx) \cdot F_R(x) \quad \text{for all possible values}$$

$$P_f = \int_{-\infty}^{\infty} f_S(dx) F_R(x)\, dx \qquad \ldots(3.106)$$

3.12.3.2 Second Expression of P_f

Let event, $A = \{\text{Resistance}, R \in [x, x + dx]\}$

$$P(A) = f_R(x)\, dx$$

Let event, $B = \{\text{stress}, S < x\}$

$$P(B) = \int_{-\infty}^{\infty} f_S(\tau)\, d\tau = F_S(x)$$

Failure is the intersection of events A and $\overline{B}$ [$\overline{B}$ is complementary to B]

$$P_f = \int_{-\infty}^{\infty} [1 - F_S(x)] f_R(x)\, dx \quad (R, S \text{ independent}) \qquad \ldots(3.107)$$

If probability of failure is given by $P_f = \int_{-\infty}^{\infty} f_S(x) F_R(x)\, dx$

Then the probability of survival, $P_s = 1 - P_f$

$$P_S = 1 - \int_{-\infty}^{\infty} f_S(x) F_R(x)\, dx$$

Alternatively,

$$P_S = 1 - \int_{-\infty}^{\infty} [1 - F_S(x)] f_R(x)\, dx$$

$$= 1 - \int_{-\infty}^{\infty} f_R(x)\, dx + \int_{-\infty}^{\infty} F_S(x) f_R(x)\, dx$$

$$P_f = \int_{-\infty}^{\infty} F_S(x) f_R(x)\, dx$$

The expressions for probability of failure, P_f and probability of survival, P_S are given in Table 3.1.

Table 3.1 Probability of failure and safety of probability distribution

Probability of failure, P_f	Probability of survival, $P_S = 1 - P_f$
$P_f = \int_{-\infty}^{\infty} f_S(x)F_R(x)dx$	$P_S = 1 - \int_{-\infty}^{\infty} f_S(x)F_R(x)dx$
$P_f = \int_{-\infty}^{\infty} [1 - F_S(x)] f_R(x)dx$	$P_S = 1 - \int_{-\infty}^{\infty} [1 - F_S(x)] f_R(x)dx$ $= 1 - \int_{-\infty}^{\infty} f_R(x)dx + \int_{-\infty}^{\infty} F_S(x)f_R(x)dx$ $= \int_{-\infty}^{\infty} F_S(x)f_R(x)dx$
$P_f = 1 - \int_{-\infty}^{\infty} F_S(x)f_R(x)dx$	$P_S = \int_{-\infty}^{\infty} F_S(x)f_R(x)dx$

3.12.4 Computation of Probability of Failure by Direct Integration

If the resistance, R is uniform in the interval $[a, b]$ and load S is uniform in the interval $[c, d]$, with $(c < a < d < b)$ given

$$F_R(x) = \begin{cases} 0 & \text{if } (x \le a) \\ \dfrac{x-a}{b-a} & \text{if } x \in [a, b] \\ 1 & \text{if } x > b \end{cases}$$

and the density function of S

$$f_S(x) = \begin{cases} 0 & \text{if } (x \le c) \\ \dfrac{1}{d-c} & \text{if } x \in [c, d] \\ 0 & \text{if } x > d \end{cases}$$

The intersection of the two defined domains of the density function is $[a, d]$ and $f_S(x) = \dfrac{1}{d-c}$ for $x \in [a, d]$.

$$P_f = \int_{-\infty}^{\infty} f_S(x)F_R(x)dx$$

$$= \int_a^d \frac{1}{(d-c)} \times \frac{(x-a)}{(b-a)} dx$$

$$= \frac{(d-a)^2}{2 \times (d-c)(b-a)}$$

Example 3.20 A bar is subjected to an axial force by the application of a force P. The cross-sectional area is A and the yield limit state is f_y. Assume all the variables as uniformly distributed and independent.

What is the probability of failure?

Solution

(i) Given action = load P

(ii) Geometrical data = cross section, A

(iii) Stress $\sigma = \frac{P}{A}$

(iv) Resistance $= f_y$

(v) Design rule $= \frac{P}{A} < f_y$

The basic vector of variables is $[X] = \{P, A, f_y\}$

The limit state $G(X) = (f_y A - P)$

The performance function $g(x) = -\sigma + f_y$

$= (R - S)$

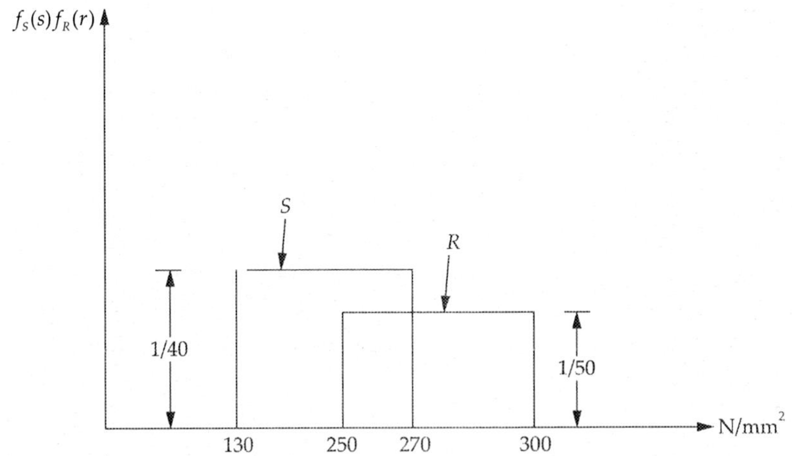

Fig. 3.12

$a = 250$, $b = 300$; $c = 130$, $d = 270$.

Probability of failure, P_f is given by

$$P_f = \frac{(d-a)^2}{2\times(d-c)(b-a)} = \frac{(270-250)^2}{2(270-130)\times(300-250)}$$

$$= 0.0285$$

Alternatively, the probability of failure can be found using the expression

$$P_f = \int_{-\infty}^{\infty} [1 - F_S(x)] f_R(x) dx$$

Since the density function of f_s given by

$$f_S(x) = \begin{cases} 0 & \text{if } (x \le c) \\ \dfrac{1}{d-c} & \text{if } x \in [c, d] \\ 0 & \text{if } x > d \end{cases} = \begin{cases} 0 & \text{if } (x \le 130) \\ \dfrac{1}{(270-130)} & \text{if } (130 \le x \le 270) \\ 0 & \text{if } (x > 270) \end{cases}$$

The CDF of $f_s(x)$ is given by, $F_S(x)$

$$F_S(x) = \int_{-\infty}^{x} f_s(x) dx = \begin{cases} 0 & \text{if } (x \le 130) \\ \dfrac{(x-130)}{140} & \text{if } (130 \le x \le 270) \\ 0 & \text{if } x > 270 \end{cases}$$

The PDF of resistance is given by

$$f_R(x) = \frac{d(F_R(x))}{dx} = \begin{cases} 0 & \text{if } (x \le a) \\ \dfrac{1}{b-a} & \text{if } (250 \le x \le 300) \\ 0 & \text{if } (x > 300) \end{cases}$$

$$= \begin{cases} 0 & \text{if } (x \le a) \\ \dfrac{1}{(300-250)} & \text{if } (250 \le x \le 300) \\ 0 & \text{if } (x > 300) \end{cases}$$

The integral of probability of failure is separated into five regions, i.e.

$$P_f = \int_{-\infty}^{\infty} [1 - F_S(x)] f_R(x) dx$$

$$= \left\{ \int_{-\infty}^{130} [1-F_S(x)] \times f_R(x)dx + \int_{130}^{250} [1-F_S(x)] \times f_R(x)dx + \int_{250}^{270} [1-F_S(x)] \times f_R(x)dx \right.$$

$$\left. + \int_{270}^{300} [1-F_S(x)] \times f_R(x)dx + \int_{300}^{\infty} [1-F_S(x)] \times f_R(x)dx \right\}$$

Substituting the expressions for $F_S(x)$ and $F_R(x)$, the probability of failure is 0.029 which is same as obtained earlier.

Example 3.21 (Probability of failure and normal distribution) The mean and standard deviation of yield stress of a steel member are 275 MPa and 14.434 MPa respectively. The member is loaded such that the induced stress has a mean of 200 MPa and standard deviation of 46.188 MPa. Compute the probability of failure assuming that the material yield stress and load induced stress are independent and both are normally distributed.

Solution:

$$\mu_R = f_y = 275 \text{ MPa}, \qquad \sigma_R = 14.43 \text{ MPa}$$

$$\mu_S = f_{\text{actual}} = 200 \text{ MPa}, \quad \sigma_S = 46.188 \text{ MPa}$$

Central safety factor $= \dfrac{\mu_R}{\mu_s} = 275/200 = 1.375$

The mean margin of safety $\mu_{MS} = \mu_R - \mu_s$

$$= (275 - 200) = 75$$

$$\sigma_{MS} = \sqrt{\sigma_R^2 + \sigma_S^2}$$

Standard deviation of margin of safety $= \sqrt{14.434^2 + 46.188^2}$

$$= 48.391$$

Reliability index $\beta = \dfrac{\mu_{MS}}{\sigma_{MS}} = \dfrac{75}{48.391} = 1.550$

$$P_f = 1 - \Phi(\beta) = 1 - \Phi(1.55) = 1 - 0.9394 = 0.0606$$

Probability of failure, P_f is 6.06% and probability of survival = 93.94%

Example 3.22 (Probability of failure and log-normal distribution) If the variables are log-normally distributed find the probability of failure.

Solution: The reliability index when variables are log-normally distributed is

$$\beta_{LN} = \frac{\ln \mu_R - \ln \mu_S}{\sqrt{(\text{CoV})_R^2 + (\text{CoV})_S^2}}$$

$$\ln(\mu_R) = \ln(275) = 5.617$$

$$\ln(\mu_S) = \ln(200) = 5.298$$

$$(\text{CoV})_R = \frac{\sigma_R}{\mu_R} = \frac{14.434}{2.75} = 0.052$$

$$(\text{CoV})_S = \frac{\sigma_S}{\mu_S} = \frac{46.188}{200} = 0.231$$

$$\beta_{LN} = \frac{5.617 - 5.298}{\sqrt{0.052^2 + 0.231^2}} = 1.34724$$

$$P_f = 1 - \Phi(\beta) = 1 - \Phi(1.34) = 1 - 0.9099 = 0.0901$$

Probability of failure, P_f is 9.01% and probability of survival = 90.99%

3.13 PROBABILITY OF FAILURE (BOTH *R* AND *S* ARE NORMAL)

The random variable R representing resistance and load S with known mean, standard deviation, μ_R, μ_S and σ_R, σ_S are considered to have normal PDF. The failure is defined when the load induced action on the structure exceeds the strength of the structure, i.e.,

$$S \geq R \qquad \ldots(3.108)$$

If margin of safety, Z is defined by

$$Z = R - S \qquad \ldots(3.109)$$

Then Z is a new random variable which is a function of R and S and failure occurs when Z is less than or equal to zero.

Case 1: The PDF for R and S assuming a normal distribution is given by

$$f_S(s) = \frac{1}{\sigma_s\sqrt{2\pi}} \exp\left[-\frac{1}{2}\left(\frac{s-\mu_s}{\sigma_s}\right)^2\right] \qquad \ldots(3.110)$$

$$f_R(r) = \frac{1}{\sigma_r\sqrt{2\pi}} \exp\left[-\frac{1}{2}\left(\frac{r-\mu_r}{\sigma_r}\right)^2\right] \qquad \ldots(3.111)$$

The random variable, Z has a normal density function because it is a linear combination of two normally distributed random variables. Hence,

$$\mu_Z = \mu_R - \mu_S \quad ...(3.112)$$

$$\sigma_Z^2 = \sigma_R^2 + \mu_S^2 \quad ...(3.113)$$

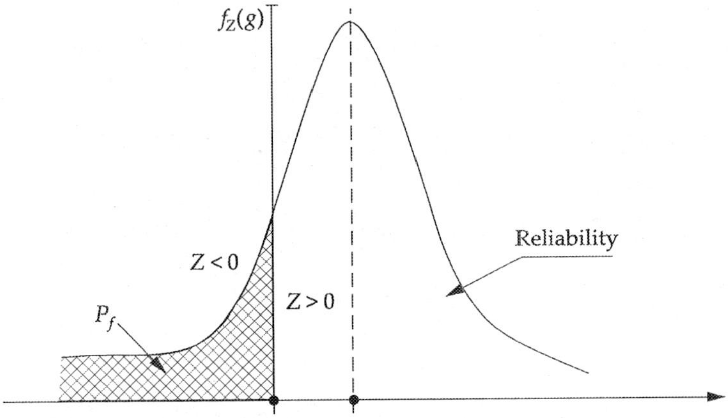

Fig. 3.13 Safe and failure probability in margin of safety

The PDF of variable Z is given by

$$f_z(z) = \frac{1}{\sigma_z\sqrt{2\pi}} \exp\left[-\frac{1}{2}\left(\frac{z-\mu_z}{\sigma_z}\right)^2\right] \quad ...(3.114)$$

The probability of failure, P_f is given when

$$P[Z < 0] = \int_{-\infty}^{0} f_z(z)dz \quad ...(3.115)$$

The central safety factor is given by

$$\text{Central safety factor (CSF)} = \frac{\mu_R}{\mu_S} \quad ...(3.116)$$

At limit state $Z = \mu_Z - \beta\sigma_Z = 0$

The β value is directly related to the probability of failure. Therefore, we have

$$\beta = \frac{\mu_R - \mu_S}{\sqrt{\sigma_R^2 + \sigma_S^2}} = \frac{\mu_Z}{\sigma_Z} \quad ...(3.117)$$

where β is called the reliability index or safety index.

$$\text{if } u = \frac{z-\mu_z}{\sigma_z} \quad \text{...(3.118)}$$

Then Eq. (3.114) can be transformed into

$$f(u) = \frac{1}{\sqrt{2\pi}} \exp\left[-\frac{1}{2}(u)^2\right] \quad \text{...(3.119)}$$

The probability of failure, P_f is given when

$$P[Z < 0] = F_Z(0) = \phi\left(\frac{0-\mu_Z}{\sigma_Z}\right)$$

The probability of failure is also given $P_f = \int_{-\infty}^{-\beta} f(u)du$

Normally, strength failure, P_f shall be of the order of 10^{-4} ($\beta = 3.72$) and serviceability failure shall be of order 10^{-2} ($\beta = 2.32$).

Case 2: A random variable follows a log-normal distribution when natural logarithm of the random variable is normally distributed. If R and S are independent and follow a log-normal distribution, the ratio of R divided by S, that is

$$\ln\left(\frac{R}{S}\right) = \ln(R) - \ln(S) \quad \text{...(3.120)}$$

If $X = \ln(R)$ and $Y = \ln(S)$, then

$$Z = X - Y \quad \text{...(3.121)}$$

where X, Y and Z are normally distributed. Therefore,

$$\mu_Z = \mu_X - \mu_Y \quad \text{...(3.122)}$$

and

$$\sigma_Z^2 = \sigma_X^2 + \mu_Y^2 \quad \text{...(3.123)}$$

Hence, the PDF of the random variable, Z can be written as

$$f_{\ln(Z)}(z) = \frac{1}{\sigma_Z . \sqrt{2\pi}} \exp\left\{-\frac{1}{2}\left(\frac{z-\mu_Z}{\sigma_Z}\right)^2\right\} \quad \text{...(3.124)}$$

The power series expansion of

$$\ln(1+x) = x - \frac{x^2}{2} + \frac{x^3}{3} - \frac{x^4}{4} + \ldots \text{ for } (-1 < x < 1)$$

$$\mu_X = \ln \mu_R$$

$$\mu_Y = \ln \mu_S$$

If the value of x is small, then the square of co-efficient of variation is small, and therefore this approximation is acceptable.

$$\bar{X} = \ln \bar{R}$$

Hence, $\sigma_X^2 = \mathrm{COV}_X^2$

$$\text{if } U = \frac{z - \mu_z}{\sigma_z}$$

$$\beta = \frac{\ln_R - \ln_S}{\sqrt{\mathrm{CoV}_R^2 + \mathrm{CoV}_S^2}} = \frac{\mu_Z}{\sigma_Z}$$

Example 3.23 (Probability of failure, normal and log-normal distributions) A cantilever beam of span 1.8 m is subjected to a load of 17.1248 kN at its free end. Assume variables normally distributed, central safety factor of 1.5, coefficient of variation for load effect and resistance effect as 15.58% and 15% respectively. Compute the following:

(a) the probability of failure if the induced moment at support exceeds the bending capacity of beam; and
(b) the probability of failure is required to be less than 0.5% and mean bending capacity remain at 15%.

Solution: Mean load capacity, $\mu_{MS} = (17.1248 \times 1.8) = 30.8246$ kN-m

Case (a)

Step 1: Central safety factor $= \dfrac{\mu_R}{\mu_S}$

$$= \frac{\text{Mean bending moment capacity}}{\text{Mean load capacity}} = \frac{\mu_{MR}}{\mu_{MS}}$$

$$= 1.5$$

$$\mu_{MR} = (1.5 \times 30.8246) = 46.2369 \text{ kN-m}$$

Coefficient of variation, $\mathrm{CoV} = \dfrac{\sigma}{\mu}$

$$\sigma_{MS} = (\mathrm{CoV})_{MS} \times \mu_{MS} = 0.1558 \times 30.8246 = 4.8025 \text{ kN-m}$$

Step 2: Standard deviation of mean bending capacity,

$$\sigma_{MR} = (\mathrm{CoV})_{MR} \times \mu_{MR} = 0.15 \times 46.2369 = 6.9355 \text{ kN-m}$$

Step 3: Failure is defined to occur when *MR* is less than *MS*. If the performance function is given by $Z = MR - MS$, then

$$\mu_Z = \mu_{MR} - \mu_{MS}$$

$$\sigma_Z^2 = \sigma_{MR}^2 + \sigma_{MS}^2$$

$$\beta = \frac{(46.2369 - 30.8246)}{\sqrt{4.8025^2 + 6.9355^2}} = 1.8270$$

Case (b)

$$P_f = 0.005 \quad \text{and} \quad \beta = 2.71 \text{ (Assume)}$$

$$\beta = \frac{\mu_Z}{\sigma_Z} = \frac{\mu_{MR} - \mu_{MS}}{\sqrt{\sigma_{MR}^2 + \sigma_{MS}^2}} = \frac{\mu_{MR} - 30.8246}{\sqrt{(0.15 \times \mu_{MR})^2 + 6.9355^2}}$$

$$2.71 = \frac{\mu_{MR} - 30.8246}{\sqrt{(0.15 \times \mu_{MR})^2 + 6.9355^2}}$$

Solving the equation, roots are 61.4189 and 12.7001 (not accepted)

$$\mu_{MR} < \mu_S$$

$$\mu_{MR} = 61.4189$$

$$\beta^* = \frac{61.4189 - 30.8246}{\sqrt{(0.15 \times 61.4189)^2 + 6.9355^2}}$$

$$= 2.6530$$

UNSOLVED PROBLEMS

1. An engineer has good estimates of the mean and variance of the peak annual flow in a small stream and he models the distribution as external value type 1, because the annual peak flow is the smallest, with a mean value of 100 and standard deviation of 50. Determine the parameters *a* and *b* of the Gumbel distribution for annual minimum.
2. The data on maximum wind velocity V_n at a site have been recorded; its mean and standard deviation are estimated to be 100 km/hr and 12 km/hr respectively. Assuming that V_n has type I extreme value distribution, what will be the probability that the maximum wind velocity will exceed 160 km/hr in any given year.
3. One hundred year records of earthquakes in a region show that there were 12 earthquakes of intensity 6 or more. It is known that the occurrence of such high intensity earthquakes in this region follows a Poisson process. Calculate: (i) the

probability that such earthquakes will occur in the next 3 years; (ii) the probability that no such earthquake will occur in the next 10 years; and (iii) the return period of an intensity-6 earthquake.

4. The probability of a weld being defective is 0.1, what is the probability of obtained exactly 0, 1, 2 and 3 defective welds in a sample of 5 welds? What is the probability of obtaining less than 2 defective welds? If a structure has 100 welds, how many of them would you expect to be defective? What is the probability of there being at least 90 welds free from defect in a structure containing 100 welds?
5. The probability of a structure collapsing after 30 years of service is 0.01 find the probability that out of 10 such structures the following will collapse after 30 years of service. (a) none; (b) exactly one; (c) not more than one; (d) more than one; and (e) at least one.
6. The probability of a concrete beam failing in compression is 0.05; use the Poisson approximation to obtain the probability from a sample of 50 beams: (i) at least 3 will fail in compression; and (ii) no beam will fail in compression.
7. The maximum annual river flow in Kaveri river in Mysore for the years 1930 to 1960, the distribution model was considered a gamma function. The parameters are to be estimated given the mean flow as 7.2689 m^3/sec and standard deviation as 5.3802 m^3/sec. What is the probability of flow less than 11.3267 m^3/sec in any year?

4

CHAPTER

First Order Second Moment Method

The First Order Second Method (FOSM), also referred to as the mean value FOSM (MVFOSM), simplifies the functional relationship and alleviates the complexities of the probability-of-failure calculation. The name "first-order" comes from the first-order expansion of the function. As implied, inputs and outputs are expressed as the mean and standard deviation. Higher moments, which might describe skewness and kurtosis of the distribution, are ignored.

In the MVFOSM method, the limit-state function is represented as the first-order Taylor series expansion at the mean value point. Assuming that the variables X are statistically independent, the approximate limit-state function at the mean is written as

$$g(X) \approx g(\mu_X) + \nabla g(\mu_X)^T (X_i - \mu_{X_i}) \qquad \ldots(4.1)$$

where, $\mu_X = \{\mu_{x_1}, \mu_{x_2}, \ldots, \mu_{x_n}\}^T$, and $\nabla g(\mu_X)$ is the gradient of g evaluated at μ_X,

$$\nabla g(\mu_X) = \left\{ \frac{\partial g(\mu_X)}{\partial x_1}, \frac{\partial g(\mu_X)}{\partial x_2}, \ldots, \frac{\partial g(\mu_X)}{\partial x_n} \right\}^T \qquad \ldots(4.1a)$$

The mean value of the approximate limit-state function $g(X)$ is

$$\mu_X \approx E[g(X)] = g(\mu_X) \qquad \ldots(4.2)$$

Because

$$\text{Var}\,[g(\mu_X)] = 0, \text{Var}\,[\nabla g(\mu_X)] = 0 \qquad \ldots(4.3)$$

$$\begin{aligned} \text{Var}\,[\nabla g(\mu_X)^T (X - \mu_X)] &= \text{Var}\,[\nabla g(\mu_X)^T X] - \mu_X \text{Var}\,[\nabla g(\mu_X)] \\ &= \text{Var}\,[\{\nabla g(\mu_X)^T\} X] \\ &= [\nabla g(\mu_X)^T]^2\, \text{Var}\,(X) \end{aligned} \qquad \ldots(4.4)$$

The variance of the approximate limit-state function $g(X)$ is

$$\text{Var}[g(X)] \approx \text{Var}[g(\mu_X)] + \text{Var}[\nabla g(\mu_X)^T (X - \mu_X)] \qquad \ldots(4.5)$$

Therefore, the standard deviation of the approximate limit-state function is

$$\sigma_{g(X)} = \sqrt{\text{Var}[g(X)]} = \sqrt{[\nabla g(\mu_X)^T]^2\, \text{Var}(X)}$$

$$= \left[\sum_{i=1}^{n} \left(\frac{\partial g(\mu_X)}{\partial x_i} \right)^2 \sigma_{x_i}^2 \right]^{\frac{1}{2}} \quad \text{...(4.6)}$$

The reliability index β is computed as:

$$\beta = \frac{\mu_{g(X)}}{\sigma_{g(X)}} \quad \text{...(4.7)}$$

If the limit-state function is nonlinear, the approximate limit-state surface is obtained by linearising the original limit-state function at the mean value. Therefore, this method is called the mean-value method, and β given in Eq. 4.7 is called MVFOSM reliability index.

In a general case with independent variables of n-dimensional space, the failure surface is a hyper plane and can be defined as a linear-failure function:

$$g(X) = c_0 + \sum_{i=1}^{n} c_i x_i \quad \text{...(4.8)}$$

The reliability index given in Eq. 4.7 can still be used for this (n-dimensional) case, in which

$$\mu_{g(X)} = c_0 + c_1 \mu_{x_1} + c_2 \mu_{x_2} + \cdots + c_n \mu_{x_n} \quad \text{...(4.9)}$$

$$\sigma_{g(X)} = \sqrt{\sum c_i^2 \sigma_{x_i}^2} \quad \text{...(4.10)}$$

The MVFOSM method changes the original complex probability problem into a simple problem. This method directly establishes the relationship between the reliability index and the basic parameters (mean and standard deviation) of the random variables via Eq. 4.7. However, there are two serious drawbacks in the MVFOSM method.

(1) Evaluation of reliability by linearising the limit-state function about the mean values leads to erroneous estimates for performance functions with high nonlinearity, or for large coefficients of variation. This can be seen from the following mean value calculation of $g(X)$, which assumes that truncation of the Taylor series expansion for a case of only one random variable at the first three terms is

$$g(X) \approx g(\mu_X) + (X - \mu_X) \nabla g(\mu_X) + \frac{(X - \mu_X)^2}{2} \nabla g^2 (\mu_X) \quad \text{...(4.11)}$$

The mean value of the approximate limit-state function $g(X)$ can be calculated as

$$\mu_{g(X)} \approx E[g(\mu_X)] + E[(X - \mu_X) \nabla g(\mu_X)] + E\left[\frac{(X - \mu_X)^2}{2} \nabla g^2 (\mu_X) \right] \quad \text{...(4.12)}$$

Because

$$E[g(X)] = g(\mu_X) \quad \text{...(4.13a)}$$

$$E[(X - \mu_X) \nabla g(\mu_X)] = E[(X \nabla g(\mu_X)] - E[(\mu_X) \nabla g(\mu_X)]$$

$$= \nabla g(\mu_X) E(X) - (\mu_X) \nabla g(\mu_X) = 0 \quad \text{...(4.13b)}$$

$$E\left[\frac{(X-\mu_X)^2}{2}\nabla^2 g(\mu_X)\right]=\frac{1}{2}\nabla g^2(\mu_X)\,E\,[(X-\mu_X)^2]$$

$$=\frac{1}{2}\nabla g^2(\mu_X)\,\mathrm{Var}(X) \qquad \ldots(4.13c)$$

From Eq. 4.13c, it is obvious that the third term on the right side of Eq. 4.11 depends on the variance of X and the second-order gradients of the limit-state function. If the variance of X is small or the limit-state function is close to linear, the third term of Eq. 4.11 can be ignored and the mean value of $g(X)$ is the same as Eq. 4.2. Otherwise, large errors in the mean value estimation will result.

(2) The MVFOSM method fails to be invariant with different mathematically equivalent formulations of the same problem. This is a problem not only for nonlinear forms of $g(X)$, but also for certain linear forms.

Example 4.1 (Beam Element) A steel beam, whose cross section is classified as a compact section, this implies that its moment carrying capacity is the plastic moment computed as $M_p = f_y\, Z$, where Z = plastic section modulus and f_y = yield strength. The total load effect (demand) is denoted by M, which is the maximum moment demand on the beam due to the applied loading. The strength limit state function is $g(x) = f_y Z - M$. Given the following data.

μ_{f_y} = 275.8 N/mm^2, μ_Z = 819353.2 mm^3, μ_M = 1.1298 × 10^8 N-mm, $(\mathrm{CoV})_{f_y}$ = 0.125, $(\mathrm{CoV})_Z$ = 0.05 and $(\mathrm{CoV})_M$ = 0.20, respectively.

Solve the problem by FOSM method and compute the following:

(a) Reliability index (β)
(b) The resistance factor (ϕ)
(c) The load factor (γ)
(d) The risk or probability of failure (P_f)

Solution: The limit-state function $g(x) = f_y\, Z - M$

where f_y = Yield strength
Z = Plastic section modulus
M = Total load effect on beam

Here, $R = f_y\, Z$, $S = M$ $\Rightarrow g(x) = R - S$

where, R = Resistance/Strength, S = Load

Here $f_y\, Z$ and M are random variables.

Now, standard deviations for random variables are,

Standard deviation (σ) = Mean value (μ) × CoV

$\sigma_{f_y} = \mu_{f_y} = (\mathrm{CoV})_{f_y} = 275.8 \times 0.125 = 34.475$ N/m^2

$\sigma_Z = \mu_Z = (\mathrm{CoV})_Z = 819353.2 \times 0.05 = 40967.66$ mm^3

$\sigma_M = \mu_M = (\mathrm{CoV})_M = 1.1298 \times 10^8 \times 0.20 = 22.596 \times 10^6$ N-mm

Partially differentiating $g(x)$ w.r.t random variables, we get,

$$\frac{\partial g(x)}{\partial f_y} = Z = \mu_Z = 819353.2$$

$$\frac{\partial g(x)}{\partial Z} = f_y = \mu_{f_y} = 275.8$$

$$\frac{\partial g(x)}{\partial M} = -1.0$$

Now,

Mean value,

$$\mu_R = \mu_{f_y} = \mu_Z = 275.8 \times 819353.2 = 225.977 \times 10^6 \text{ N-mm}$$

$$\mu_S = \mu_M = 112.98 \times 10^6 \text{ N-mm}$$

Standard deviation,

$$\sigma_R^2 = \left(\frac{\partial g(x)}{\partial f_y} \times \sigma_{f_y}\right)^2 + \left(\frac{\partial g(x)}{\partial Z} \times \sigma_Z\right)^2$$

$$\Rightarrow \quad \sigma_R^2 = (819353.2 \times 34.475)^2 + (275.8 \times 40967.66)^2$$

$$\Rightarrow \quad \sigma_R^2 = 9.255 \times 10^{14}$$

$$\therefore \quad \sigma_R = \sqrt{9.255 \times 10^{14}} = 30.423 \times 10^6 \text{ N-mm}$$

Similarly,

$$\sigma_S^2 = \left(\frac{\partial g(x)}{\partial M} \times \sigma_M\right)^2 = \left((-1.0) \times 22.596 \times 10^6\right)^2$$

$$\Rightarrow \quad \sigma_S^2 = 5.105 \times 10^{14}$$

$$\therefore \quad \sigma_S = \sqrt{5.105 \times 10^{14}} = 22.596 \times 10^6 \text{ N-mm}$$

Now,

(a) *Reliability index (β),*

$$\beta = \frac{\mu_R - \mu_S}{\sqrt{\sigma_R^2 + \sigma_S^2}} = \frac{225.977 \times 10^6 - 112.98 \times 10^6}{\sqrt{(9.255 \times 10^{14} + 5.105 \times 10^{14})}} = 2.9817$$

(b) *The resistance factor (ϕ),*

$$\phi = [1 - 0.75 \times \beta \times (\text{CoV})_R]$$

$$\therefore \quad (\text{CoV})_R = \frac{\sigma_R}{\mu_R} = \frac{30.423 \times 10^6}{225.977 \times 10^6} = 0.1346$$

$$\Rightarrow \quad \phi = [1 - 0.75 \times 2.9817 \times 0.1346] = 0.698$$

(c) *The load factor (γ),*

$$\gamma = [1 + 0.75 \times \beta \times (\text{CoV})_S]$$

$$\therefore \quad (\text{CoV})_S = \frac{\sigma_S}{\mu_S} = \frac{22.596 \times 10^6}{112.98 \times 10^6} = 0.20$$

$$\Rightarrow \quad \gamma = [1 + 0.75 \times 2.9817 \times 0.20] = 1.447$$

(d) *The risk or probability of failure (P_f),*
From the standard normal distribution Table A1 (refer Appendix 1)

$$\Phi(-\beta) = 1 - \Phi(\beta)$$

Probability of failure, $P_f = 1 - \Phi(\beta) = 1 - \Phi(2.9817) = 1 - 0.9986$

$\therefore$ $P_f = 0.00140$ or [1 in 714.28]

Probability of success, $P_s = 1 - P_f = 1 - 0.00140 = 0.9986$

Example 4.2 (Tension Element) Analyze a tension element of circular cross section, subjected to external load.

Given the following data:

Sl. No.	Variables	Mean value (μ)	CoV	Standard deviation (σ)
1	Load, P (N)	2.6689×10^4	0.0666	1.7774×10^3
2	Yield, f_y (N/m^2)	1.0755×10^9	0.0275	2.9576×10^7
3	Diameter, d (m)	0.5842×10^{-2}	0.0130	0.0759×10^{-3}

Compute the reliability index, probability of success and probability of failure.

Solution: The limit state function, $g(x) = f_y A - P$

Here, $R = f_y A$, $S = P$ $\qquad \Rightarrow g(x) = R - S$

Now, area of member for circular cross section

$$A = \frac{\pi \times d^2}{4} \text{ m}^2$$

Now,

Mean value,

$$\mu_A = \frac{\pi \times \mu_{d^2}}{4} = \frac{\pi \times (0.5842 \times 10^{-2})^2}{4} = 2.6804 \times 10^{-5} \text{ m}^2$$

Now, partially differentiating A w.r.t d, we get,

$$\frac{\partial A}{\partial d} = \frac{\pi \times 2 \times d}{4} = \frac{\pi \times d}{2} = \frac{\pi \times \mu_d}{2} = \frac{\pi \times 0.5842 \times 10^{-2}}{2} = 9.176 \times 10^{-3}$$

Standard deviation,

$$\sigma_A^2 = \left(\frac{\partial A}{\partial d} \times \sigma_d\right)^2 = \left(9.176 \times 10^{-3} \times 0.0759 \times 10^{-3}\right)^2 = 4.851 \times 10^{-13}$$

$\therefore$

$$\sigma_A = \sqrt{4.851 \times 10^{-13}} = 6.965 \times 10^{-7} \text{ m}^2$$

Now, partially differentiating $g(x)$ w.r.t f_y, A and P, we get,

$$\frac{\partial g(x)}{\partial f_y} = A = \mu_A = 2.6804 \times 10^{-5}$$

$$\frac{\partial g(x)}{\partial A} = f_y = \mu_{f_y} = 1.0755 \times 10^9$$

$$\frac{\partial g(x)}{\partial P} = -1.0$$

Mean value,

$$\mu_R = \mu_{f_y} \times \mu_A = 1.0755 \times 10^9 \times 2.6804 \times 10^{-5} = 28.827 \times 10^3 \text{ N}$$

$$\mu_S = \mu_P = 26.689 \times 10^3 \text{ N}$$

Standard deviation,

$$\sigma_R^2 = \left(\frac{\partial g(x)}{\partial f_y} \times \sigma_{f_y}\right)^2 + \left(\frac{\partial g(x)}{\partial A} \times \sigma_A\right)^2$$

$$= \left(2.6804 \times 10^{-5} \times 2.9576 \times 10^7\right)^2 + \left(1.0755 \times 10^9 \times 6.965 \times 10^{-7}\right)^2$$

$$\sigma_R^2 = 11.895 \times 10^5$$

Similarly,

$$\sigma_R^2 = \left(\frac{\partial g(x)}{\partial P} \times \sigma_P\right)^2 = \left((-1.0) \times 1.774 \times 10^3\right)^2 = 31.47 \times 10^5$$

Now,

Reliability index (β),

$$\beta = \frac{\mu_R - \mu_S}{\sqrt{\sigma_R^2 + \sigma_S^2}} = \frac{28.827 \times 10^3 - 26.689 \times 10^3}{\sqrt{(11.895 \times 10^5 + 31.47 \times 10^5)}} = 1.0266$$

Probability of success, $P_s = \Phi(\beta) = \Phi(1.0266)$

From the standard normal distribution Table A1 (refer Appendix 1), $\Phi(1.0266)$, we get,

$\therefore$ $$P_s = \Phi(1.0266) = 0.8476$$

The risk or probability of failure (P_f), $P_f = 1 - P_s = 1 - 0.8476 = 0.1524$

$\therefore$ $$P_f = 0.1524 \text{ or } [1 \text{ in } 6.56]$$

Example 4.3 (Settlement in Consolidated Clay) The settlement of a point A (Fig. 4.1) is caused by the construction of a structure can be shown to be primarily by the consolidation of the clay layer assuming the contribution of settlement due to secondary consolidation is negligible, for normally loaded clay, the settlement $g(x)$ is given by

$$g(x) = 63.5 - N\frac{C_c}{1+e_o}H \log\left(\frac{P_o + \Delta P}{P_o}\right)$$

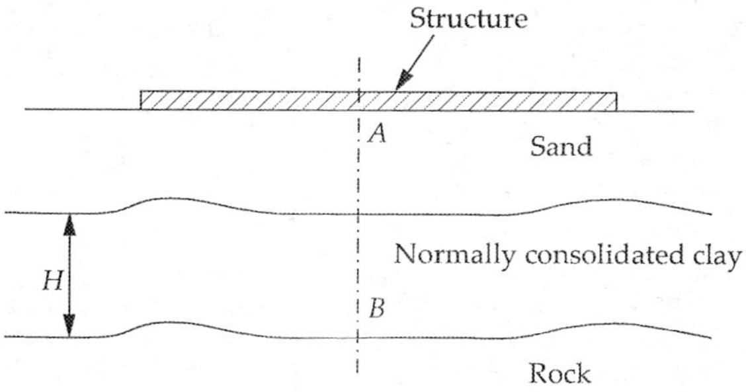

Fig. 4.1 Soil layer

The performance require that the settlement be less than 63.5 mm and the variables have the following statistics:

Sl. No.	Parameters	Mean value (μ)	CoV	Standard deviation (σ)
1	N	1.00	0.10	0.10
2	C_c	0.3960	0.25	0.099
3	e_0	1.19	0.15	0.1785
4	H, (mm)	4267.20	0.05	213.36
5	P_o, (N/mm^2)	0.1711	0.052	0.0089
6	ΔP, (N/mm^2)	0.0239	0.20	0.0047

Solution: The settlement in consolidated clay $g(x)$ is given by,

$$g(x) = 63.5 - N\frac{C_c}{1+e_o}H\log\left(\frac{P_o+\Delta P}{P_o}\right)$$

Since the settlement should be restricted to within 63.5 mm. Here, N, C_c, e_0, H, P_o and ΔP are random variables.

Now, the mean and standard deviations are

Mean value,

$$\mu_{g(x)} = 63.5 - \mu_N\frac{\mu_{C_c}}{1+\mu_{e_0}}\mu_H\log\left(\frac{\mu_{P_0}+\mu_{\Delta P}}{\mu_{P_0}}\right)$$

$$\Rightarrow \quad \mu_{g(x)} = 63.5 - (1.0)\times\frac{0.3960}{(1+1.1900)}\times(4267.2)\times\log\left(\frac{0.1711+0.0239}{0.1711}\right)$$

$$\therefore \quad \mu_{g(x)} = 63.5 - 771.603\times\log(1.1396) = 19.709$$

Now, partially differentiating $g(x)$, w.r.t variables, we get,

$$\frac{\partial g(x)}{\partial N} = \frac{-C_c}{1+e_o}H\log\left(\frac{P_0+\Delta P}{P_o}\right) = -\frac{\mu_{C_c}}{1+\mu_{e_0}}\mu_H\log\left(\frac{\mu_{P_0}+\mu_{\Delta P}}{\mu_{P_0}}\right)$$

$$= -\frac{0.3960}{(1+1.1900)} \times (4267.2) \times \log\left(\frac{0.1711+0.0239}{0.1711}\right) = -43.78$$

$$\frac{\partial g(x)}{\partial C_c} = \frac{-N}{1+e_o} H \log\left(\frac{P_o+\Delta P}{P_o}\right) = -\frac{\mu_N}{1+\mu_{e_o}} \mu_H \log\left(\frac{\mu_{P_o}+\mu_{\Delta P}}{\mu_{P_o}}\right)$$

$$= -\frac{1.0}{(1+1.1900)} \times (4267.2) \times \log\left(\frac{0.1711+0.0239}{0.1711}\right) = -110.57$$

$$\frac{\partial g(x)}{\partial e_o} = \frac{NC_c}{(1+e_o)^2} H \log\left(\frac{P_o+\Delta P}{P_o}\right) = \frac{\mu_N \mu_{C_c}}{(1+\mu_{e_o})^2} \mu_H \log\left(\frac{\mu_{P_o}+\mu_{\Delta P}}{\mu_{P_o}}\right)$$

$$= \frac{1.0 \times 0.3960}{(1+1.1900)^2} \times (4267.2) \times \log\left(\frac{0.1711+0.0239}{0.1711}\right) = 19.99 \approx 20.00$$

$$\frac{\partial g(x)}{\partial H} = \frac{-NC_c}{1+e_o} \log\left(\frac{P_o+\Delta P}{P_o}\right) = -\frac{\mu_N \mu_{C_c}}{1+\mu_{e_o}} \log\left(\frac{\mu_{P_o}+\mu_{\Delta P}}{\mu_{P_o}}\right)$$

$$= -\frac{1.0 \times 0.3960}{(1+1.1900)} \times \log\left(\frac{0.1711+0.0239}{0.1711}\right) = -0.01025$$

$$\frac{\partial g(x)}{\partial P_o} = \frac{-NC_c}{1+e_o} H\left[\frac{-1}{\ln(10)}\left(\frac{\Delta P}{(P_o+\Delta P)P_o}\right)\right] = \frac{\mu_N \mu_{C_c}}{1+\mu_{e_o}} \mu_H \left[\frac{-1}{\ln(10)}\left(\frac{\mu_{\Delta P}}{(\mu_{P_o}+\mu_{\Delta P})\mu_{P_o}}\right)\right]$$

$$= \frac{1.0 \times 0.3960}{(1+1.1900)} \times (4267.2) \times 0.4343 \times \left(\frac{0.0239}{(0.1711+0.0239)0.1711}\right) = 240.17$$

$$\frac{\partial g(x)}{\partial \Delta P} = \frac{-NC_c}{1+e_o} H \frac{1}{\ln(10)}\left(\frac{1}{P_o+\Delta P}\right) = -\frac{\mu_N \mu_{C_c}}{1+\mu_{e_o}} \mu_H \frac{1}{\ln(10)}\left(\frac{1}{\mu_{P_o}+\mu_{\Delta P}}\right)$$

$$= -\frac{1.0 \times 0.3960}{(1+1.1900)} \times (4267.2) \times 0.4343 \times \left(\frac{1}{(0.1711+0.0239)}\right) = -1717.401$$

Standard deviation,

$$\sigma_{g(x)}^2 = \left(\frac{\partial g(x)}{\partial N} \times \sigma_N\right)^2 + \left(\frac{\partial g(x)}{\partial C_c} \times \sigma_{C_c}\right)^2 + \left(\frac{\partial g(x)}{\partial e_o} \times \sigma_{e_o}\right)^2 + \left(\frac{\partial g(x)}{\partial H} \times \sigma_H\right)^2 + \left(\frac{\partial g(x)}{\partial P_o} \times \sigma_{P_o}\right)^2 + \left(\frac{\partial g(x)}{\partial \Delta P} \times \sigma_{\Delta P}\right)^2$$

$$\sigma^2_{g(x)} = (-43.78 \times 0.10)^2 + (-110.57 \times 0.099)^2 + (20.00 \times 0.1785)^2 + (-0.01025 \times 213.36)^2 + (240.17 \times 0.0089)^2 + (-1717.401 \times 0.0047)^2$$

$$\Rightarrow \sigma^2_{g(x)} = 19.16 + 119.82 + 12.74 + 4.78 + 4.56 + 65.15 = 226.21$$

$$\therefore \quad \sigma_{g(x)} = \sqrt{226.21} = 15.04$$

Now,

Reliability index (β),

$$\beta = \frac{\mu_{g(x)}}{\sigma_{g(x)}} = \frac{19.709}{15.04} = 1.310$$

Probability of success, $P_s = \Phi(\beta) = \Phi(1.310)$

From the standard normal distribution Table (A_1), $\Phi(1.310)$, we get,

$$\therefore \quad P_s = \Phi(1.310) = 0.9049$$

The risk or probability of failure (P_f), $P_f = 1 - P_s = 1 - 0.9049 = 0.0951$

$$\therefore \quad P_f = 0.0951 \text{ or } [1 \text{ in } 10.51]$$

Note: $y = \log_b u$ where u is a differentiable function of x, $b > 0$ and $b \neq 1$

$$\therefore \quad \frac{dy}{dx} = \frac{1}{u} \times \frac{1}{\ln(10)} \times \frac{du}{dx}$$

Example 4.4 (Spillway Hydrology) Determine the probability of an unsatisfactory spillway to carry inflow water during an extreme flood. The discharge over a spillway with an uncontrolled overflow ogee crest is given by $Q_c = NCLH^{3/2}$. The inflow rate at the spillway can be modelled by $Q_L = RQ_I$, where Q_I is the peak flow upstream to the reservoir crest and R is the attenuation factor to account for the volume effect of the reservoir.

Sl. No.	Variables	Mean value (μ)	CoV	Standard Deviation (σ)	Distribution
1	N	1.00	0.20	0.20	Normal
2	C	3.85	0.07	0.2695	Normal
3	L, (m)	28.46	0.06	1.707	Normal
4	H, (m/sec)	3.65	0.14	0.511	Normal
5	R	0.70	0.14	0.098	Normal
6	Q_I (m^3/sec)	258.98	0.35	90.643	Type 1

Solution: The discharge over a spillway with an uncontrolled overflow ogee crest is given by

$$Q_C = NCLH^{3/2}$$

The given function, $g(x) = Q_C - Q_L$

$$\Rightarrow \quad g(x) = NCLH^{3/2} - RQ_I$$

Now, the mean and standard deviations are

Mean value,

$$\mu_{g(x)} = \mu_N \mu_C \mu_L \mu_{H^{3/2}} - \mu_R \mu_{Q_I}$$

$$\Rightarrow \quad \mu_{g(x)} = 1.00 \times 3.85 \times 28.46 \times (3.65)^{3/2} - 0.70 \times 258.98 = 582.78$$

Now, partially differentiating $g(x)$, w.r.t variables, we get,

$$\frac{\partial g(x)}{\partial N} = CLH^{3/2} = \mu_C \mu_L \mu_{H^{3/2}} = 3.85 \times 28.46 \times (3.65)^{3/2} = 764.07$$

$$\frac{\partial g(x)}{\partial C} = NLH^{3/2} = \mu_N \mu_L \mu_{H^{3/2}} = 1.00 \times 28.46 \times (3.65)^{3/2} = 198.46$$

$$\frac{\partial g(x)}{\partial L} = NCH^{3/2} = \mu_N \mu_C \mu_{H^{3/2}} = 1.00 \times 3.85 \times (3.65)^{3/2} = 26.84$$

$$\frac{\partial g(x)}{\partial H} = (3/2) \times NCLH^{1/2} = (3/2) \times \mu_N \mu_C \mu_L \mu_{H^{1/2}}$$

$$= 1.5 \times 1.00 \times 3.85 \times 28.46 \times (3.65)^{1/2} = 314.00$$

$$\frac{\partial g(x)}{\partial R} = -Q_I = -\mu_{Q_I} = -258.98$$

$$\frac{\partial g(x)}{\partial Q_I} = -R = -\mu_R = -0.70$$

Standard deviation,

$$\sigma^2_{g(x)} = \left(\frac{\partial g(x)}{\partial N} \times \sigma_N\right)^2 + \left(\frac{\partial g(x)}{\partial C} \times \sigma_C\right)^2 + \left(\frac{\partial g(x)}{\partial L} \times \sigma_L\right)^2 + \left(\frac{\partial g(x)}{\partial H} \times \sigma_H\right)^2 + \left(\frac{\partial g(x)}{\partial R} \times \sigma_R\right)^2 + \left(\frac{\partial g(x)}{\partial Q_I} \times \sigma_{Q_I}\right)^2$$

$$\Rightarrow \quad \sigma^2_{g(x)} = (764.07 \times 0.20)^2 + (198.46 \times 0.2695)^2 + (26.84 \times 1.707)^2 + (314 \times 0.511)^2 + (-258.98 \times 0.098)^2 + (-0.70 \times 90.643)^2$$

$$\Rightarrow \quad \sigma^2_{g(x)} = 23352.11 + 2860.64 + 2099.09 + 25745.48 + 644.14 + 4025.91 = 58727.37$$

$$\therefore \quad \sigma_{g(x)} = \sqrt{58727.37} = 242.33$$

Now,

Reliability index (β),

$$\beta = \frac{\mu_{g(x)}}{\sigma_{g(x)}} = \frac{582.78}{242.33} = 2.404$$

Probability of success, $P_s = \Phi(\beta) = \Phi(2.404)$

From the standard normal distribution Table A1 (refer Appendix 1), $\Phi(2.404)$, we get,

$\therefore$ $$P_s = \Phi(2.404) = 0.9918$$

The risk or Probability of failure (P_f), $P_f = 1 - P_s = 1 - 0.9918 = 0.0082$

$\therefore$ $$P_f = 0.0082 \text{ or } [1 \text{ in } 121.95]$$

Example 4.5 (Highway Engineering) Determine the probability of collision of a vehicle B following another vehicle A in traffic flow at a distance S apart (Fig. 4.2). The driver in vehicle A suddenly applies the brakes because of an unusual traffic jam ahead. Suppose the driver in vehicle B drives T seconds to react to driver A's action and to his surprise he discovers that his brakes had failed, so he has to rely completely on the friction of the road pavement to slow down his vehicle.

Collision between vehicle A and B will occur if the stopping distance of vehicle B exceeds the gap S and the stopping distance of vehicle A. Given the following data:

Sl. No.	Parameters	Mean value (μ)	CoV	Standard deviation (σ)	Type of distribution
1	r	$0.2g = 0.2 \times 9.81$ 1.962	0.10	0.1962	Normal
2	S(m)	60.960	0.80	48.768	Gamma
3	T(sec)	0.90	0.33	0.297	Normal
4	f	$0.05g = 0.05 \times 9.81$ 0.4905	0.05	0.00245	Normal

Velocity of vehicles A and B, $V = V_A = V_B = \mu_A = \mu_B = 11.176$ m/sec

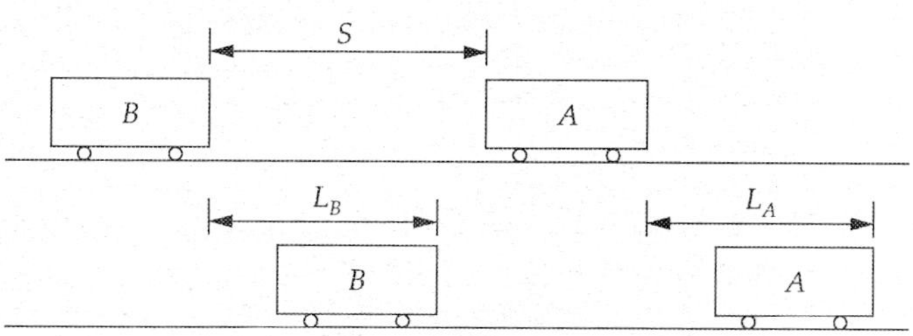

Fig. 4.2 Two vehicles in a traffic flow

From Fig. 4.2 the stopping distance is given by,

$$g(x) = (L_A + S - L_B)$$

or

$$g(x) = \frac{V^2}{2r} + S - \left(VT + \frac{V^2}{2f} \right)$$

Solution: The stopping distance is given by,

$$g(x) = \frac{V^2}{2r} + S - \left(VT + \frac{V^2}{2f} \right)$$

Mean value,

$$\mu_{g(x)} = \frac{\mu_{V^2}}{2\mu_r} + \mu_S - \left(\mu_V \mu_T + \frac{\mu_{V^2}}{2\mu_f} \right)$$

$$\Rightarrow \quad \mu_{g(x)} = \frac{(11.176)^2}{2\times 1.962} + 60.960 - \left(11.176 \times 0.90 + \frac{(11.176)^2}{2\times 0.4905} \right)$$

$$\therefore \quad \mu_{g(x)} = 31.83 + 60.960 - 137.38 = -44.59$$

Now, partially differentiating $g(x)$, w.r.t variables, we get,

$$\frac{\partial g(x)}{\partial r} = -\frac{V^2}{2r^2} = -\frac{\mu_{V^2}}{2\mu_{r^2}} = -\frac{(11.176)^2}{2\times(1.962)^2} = -16.22$$

$$\frac{\partial g(x)}{\partial S} = 1.0$$

$$\frac{\partial g(x)}{\partial T} = -V = -\mu_V = -11.176$$

$$\frac{\partial g(x)}{\partial f} = \frac{V^2}{2f^2} = \frac{\mu_{V^2}}{2\mu_{f^2}} = \frac{(11.176)^2}{2\times(0.4905)^2} = 259.57$$

Standard deviation,

$$\sigma^2_{g(x)} = \left(\frac{\partial g(x)}{\partial r} \times \sigma_r \right)^2 + \left(\frac{\partial g(x)}{\partial S} \times \sigma_S \right)^2 + \left(\frac{\partial g(x)}{\partial T} \times \sigma_T \right)^2 + \left(\frac{\partial g(x)}{\partial f} \times \sigma_f \right)^2$$

$$\Rightarrow \quad \sigma^2_{g(x)} = (-16.22\times 0.1962)^2 + (1.0\times 48.768)^2 + (-11.176\times 0.297)^2 + (259.57\times 0.0245)^2$$

$$\Rightarrow \quad \sigma^2_{g(x)} = 10.127 + 2378.31 + 11.017 + 40.44 = 2439.894$$

$$\therefore \quad \sigma_{g(x)} = \sqrt{2439.894} = 49.395$$

Now,

Reliability index (β),

$$\beta = \frac{\mu_{g(x)}}{\sigma_{g(x)}} = \frac{-44.59}{49.395} = -0.902$$

Probability of success (P_s), $P_s = \Phi(-\beta) = 1 - \Phi(\beta)$

From the standard normal distribution Table A1 (refer Appendix 1), $\Phi(-0.902)$, we get,

$$\therefore \quad P_s = \Phi(-0.902) = 0.1841$$

The risk or probability of failure (P_f), $P_f = 1 - P_s = 1 - 0.1841 = 0.8159$

$$\therefore \quad P_f = 0.8159 \text{ or } [1 \text{ in } 1.22]$$

Comment: It is deceleration of the vehicles which turns out to be negative quantity. Hence, reliability index (β) can be negative. If absolute value is chosen then it can be positive.

Example 4.6 Consider a steel beam whose cross section is classified as a compact section. This implies that its moment carrying capacity is the plastic moment calculated as $M_p = f_y Z$, where Z = plastic section modulus and f_y is the yield stress. Let the total load effect (demand) be denoted by M_{ext}, which is the maximum moment demand on the beam due to the applied loading. The limit state function is $g(x) = \frac{M \times y}{I} - f_y$. Compute the probability of failure. Given the following data.

Sl. No.	Variables	Mean value (μ)	CoV	Standard deviation (σ)
1	f_y, (kN/cm^2)	4.00	30%	1.20
2	y, (cm)	20.00	0.5%	1.00
3	M, (kN-cm)	20.00	25%	5.00
4	I, (cm^4)	90.00	10%	9.00

Solution: The limit state function is given by

$$g(x) = \frac{M \times y}{I} - f_y$$

Now, the mean and standard deviations are

Mean value,

$$\mu_{g(x)} = \frac{\mu_M \times \mu_y}{\mu_I} - \mu_{f_y} = \frac{20.0 \times 20.0}{90.0} - 4.00 = 0.44$$

Now, partially differentiating $g(x)$, w.r.t variables M, y, I and f_y we get,

$$\frac{\partial g(x)}{\partial M} = \frac{y}{I} = \frac{\mu_y}{\mu_I} = \frac{20.0}{90.0} = 0.22$$

$$\frac{\partial g(x)}{\partial y} = \frac{M}{I} = \frac{\mu_M}{\mu_I} = \frac{20.0}{90.0} = 0.22$$

$$\frac{\partial g(x)}{\partial I} = -\frac{M \times y}{I^2} = -\frac{\mu_M \times \mu_y}{\mu_{I^2}} = -\frac{20.0 \times 20.0}{(90.0)^2} = 0.049 \approx 0.05$$

$$\frac{\partial g(x)}{\partial f_y} = -1.0$$

Standard deviation,

$$\sigma^2_{g(x)} = \left(\frac{\partial g(x)}{\partial M} \times \sigma_M\right)^2 + \left(\frac{\partial g(x)}{\partial y} \times \sigma_y\right)^2 + \left(\frac{\partial g(x)}{\partial I} \times \sigma_I\right)^2 + \left(\frac{\partial g(x)}{\partial f_y} \times \sigma_{f_y}\right)^2$$

$$\Rightarrow \quad \sigma^2_{g(x)} = (0.22 \times 5.00)^2 + (0.22 \times 1.00)^2 + (0.05 \times 9.00)^2 + (-1.00 \times 1.20)^2$$

$$\Rightarrow \quad \sigma^2_{g(x)} = 1.21 + 0.0484 + 0.2025 + 1.44 = 2.9009$$

$$\therefore \quad \sigma_{g(x)} = \sqrt{2.9009} = 1.703$$

Now,

Reliability index (β),

$$\beta = \frac{\mu_{g(x)}}{\sigma_{g(x)}} = \frac{0.44}{1.703} = 0.258$$

Probability of success (P_s), $P_s = \Phi(\beta) = \Phi(0.258)$

From the standard normal distribution Table A1 (refer Appendix 1), $\Phi(0.258)$, we get,

$$\therefore \quad P_s = \Phi(0.258) = 0.6018$$

The risk or probability of failure (P_f), $P_f = 1 - P_s = 1 - 0.6018 = 0.3982$

$$\therefore \quad P_f = 0.3982 \text{ or } [1 \text{ in } 2.51]$$

Example 4.7 Compute the reliability index against the limit state of collapse in flexure, for the beam shown in Fig. 4.3 subjected to an uniform distributed load Q_1 and a point load of Q_2 at the midspan. The resistance of the beam is R. Given are the following data.

$\mu_{Q1} = 400$ N; $\sigma_{Q1} = 10$ N; $\mu_{Q2} = 5000$ N; $\sigma_{Q2} = 2000$ N;

$\mu_R = 10000$ N; $\sigma_R = 1000$ N.

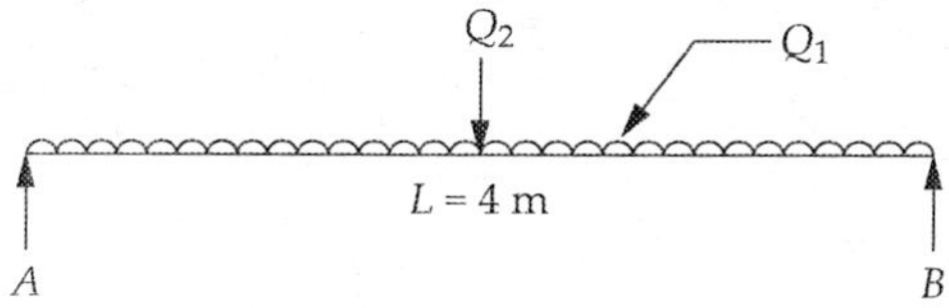

Fig. 4.3 Simply supported beam carrying UDL and central concentrated load

Compute the reliability index for the following cases.

(a) Only Q_1 is acting on the structure.

(b) Only Q_2 is acting on the structure.

(c) Both Q_1 and Q_2 is acting on the structure.

Comment on your results and suggest if the principle of superposition is valid for this problem.

Solution:

Case (a): Only Q_1 is acting on the structure.

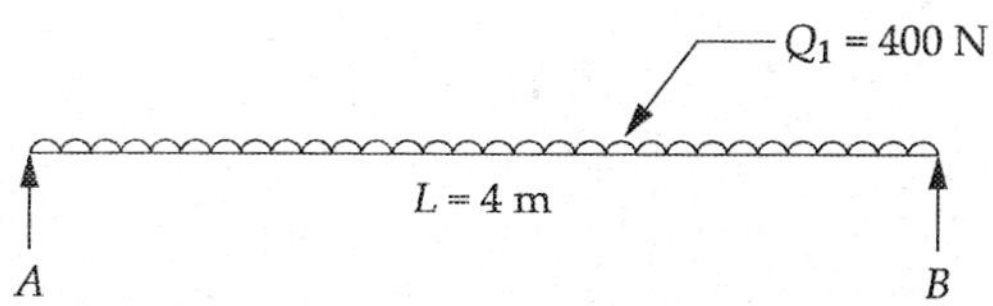

Maximum bending moment

$$M = \frac{WL^2}{8} \Rightarrow S = \frac{Q_1L^2}{8}$$ for simply supported beam carrying UDL per metre

Here, S = Resistance

Now,

Mean value,

$$\mu_s = \frac{Q_1L}{8} = \frac{\mu_{Q_1}\mu_L}{8} = \frac{400 \times 4}{8} = 200 \text{ N}$$

Standard deviation,

$$\sigma_s = \frac{\sigma_{Q_1}}{2} = \frac{10}{2} = 5.0 \text{ N}$$

Reliability index, β is given by

$$\beta = \frac{\mu_R - \mu_S}{\sqrt{\sigma_R^2 + \sigma_S^2}} = \frac{10000 - 200}{\sqrt{1000^2 + 5^2}} = 9.8$$

Case (b): Only Q_2 is acting on the structure.

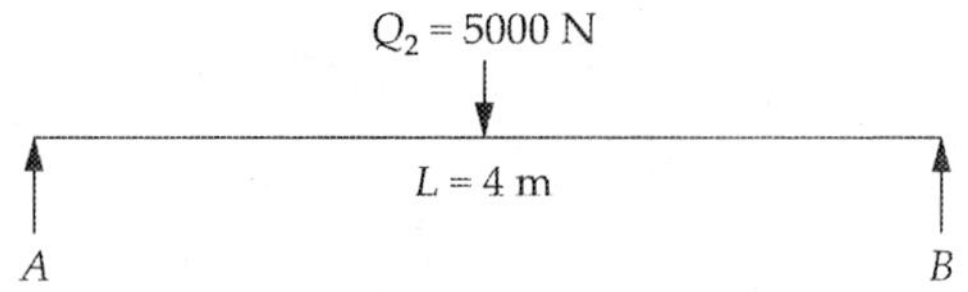

Max bending moment

$M = \frac{PL}{4} \Rightarrow S = \frac{Q_2 L}{4}$ for simply supported beam carrying central concentrated load per metre

Now,

Mean value,

$$\mu_S = \frac{Q_2 \cdot L}{4} = \frac{\mu_{Q_2} \mu_L}{4} = \frac{5000 \times 4}{4} = 5000 \text{ N}$$

Standard deviation,

$$\sigma_S = \sigma_{Q_2} = 2000 \text{ N}$$

Reliability index, β is given by

$$\beta = \frac{\mu_R - \mu_S}{\sqrt{\sigma_R^2 + \sigma_S^2}} = \frac{10000 - 5000}{\sqrt{1000^2 + 2000^2}} = 2.23$$

Case (c): Both Q_1 and Q_2 are acting on the structure.

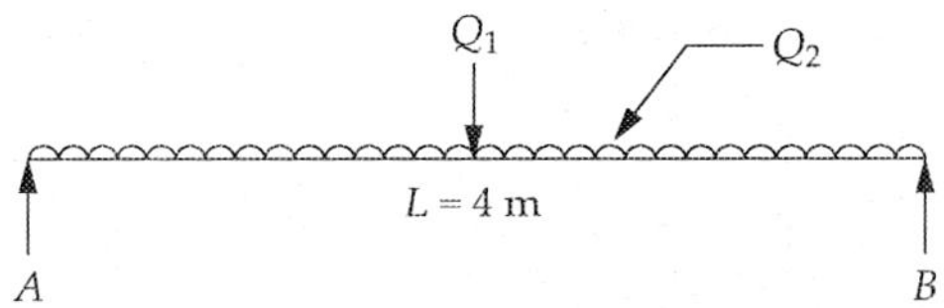

Maximum bending moment

$$M = \frac{WL^2}{8} + \frac{PL}{4}$$

$\Rightarrow S = \frac{Q_1 L^2}{8} + \frac{Q_2 L}{4}$ for simply supported beam carrying UDL and concentrated load per metre

Mean value,

$$\mu_s = \frac{Q_1 L}{8} + \frac{Q_2 L}{4} = \frac{\mu_{Q1} \mu_L}{8} + \frac{\mu_{Q_2} \mu_L}{4}$$

$$= \frac{400 \times 4}{8} + \frac{5000 \times 4}{4} = 5200 \text{ N}$$

Standard deviation,

$$\sigma_S = \frac{\sigma_{Q_1}}{2} + \sigma_{Q_2} = \frac{10}{2} + 2000 = 2005 \text{ N}$$

Reliability index, β is given by

$$\beta = \frac{\mu_R - \mu_S}{\sqrt{\sigma_R^2 + \sigma_S^2}} = \frac{10000 - 5200}{\sqrt{1000^2 + 2005^2}} = 2.142$$

Comment: It has so far been assumed that the failure function is a linear combination of the basic variables. However, this may not be true most of the times in practical cases. Also the principle of superposition is not valid for this problem.

Example 4.8 Determine the reliability index for a simply supported beam for the limit state of shear $g(x) = \tau_s t_w d - 0.5P$. The beam is subjected to point load P at the mid-span; where d is the depth of the beam, t_w is the thickness of the web, τ_s is the shear strength of the material. $\mu_P = 4000$ N, $\sigma_P = 1000$ N, $\mu_{\tau s} = 95$ N/mm^2, $\sigma_{\tau s} = 10$ N/mm^2, $\mu_d = 50$ mm, $\sigma_d = 2.5$ mm and $d/t_w = 40$. Solve the problem by FOSM method.

$$g(x) = \tau_s t_w d - 0.5P \qquad \frac{d}{t_w} = 40 \qquad \therefore t_w = \frac{d}{40}$$

$$\Rightarrow \qquad g(x) = \tau_s \frac{d^2}{40} - 0.5P$$

Solution: The limit state of shear $g(x) = \tau_s t_w d - 0.5P$

Given $\quad \dfrac{d}{t_w} = 40$

$\therefore \quad t_w = \dfrac{d}{40}$

$$g(x) = \tau_s \frac{d^2}{40} - 0.5P$$

$$\Rightarrow \qquad g(x) = R - S$$

Here, $\quad R = \tau_s \dfrac{d^2}{40} \quad S = 0.5P$

Now, the mean and standard deviations are

Mean value,

$$\mu_R = \frac{\mu_{\tau_s} \times \mu_{d^2}}{40} = \frac{95 \times (50)^2}{40} = 5937.5 \text{ N}$$

$$\mu_S = 0.5 \times \mu_P = 0.5 \times 4000 = 2000 \text{ N}$$

Now, partially differentiating R and S w.r.t variables t_s, d and P, we get,

$$\frac{\partial R}{\partial \tau_s} = \frac{d^2}{40} = \frac{\mu_{d^2}}{40} = \frac{(50)^2}{40} = 62.5$$

$$\frac{\partial R}{\partial d} = \tau_s \times \frac{1}{40} \times 2 \times d = \mu_{\tau_s} \times \frac{\mu_d}{20} = 95 \times \frac{50}{20} = 237.5$$

$$\frac{\partial S}{\partial P} = 0.5$$

Standard deviation,

$$\sigma_R^2 = \left(\frac{\partial R}{\partial \tau_s} \times \sigma_{\tau_s}\right)^2 + \left(\frac{\partial R}{\partial d} \times \sigma_d\right)^2 = (62.5 \times 10)^2 + (237.5 \times 2.5)^2 = 74.316 \times 10^4 \text{ N}$$

$$\sigma_S^2 = \left(\frac{\partial S}{\partial P} \times \sigma_P\right)^2 = (0.5 \times 1000)^2 = 25 \times 10^4 \text{ N}$$

Now,

Reliability index (β),

$$\beta = \frac{\mu_R - \mu_S}{\sqrt{\sigma_R^2 + \sigma_S^2}} = \frac{5937.5 - 2000}{\sqrt{(74.316 \times 10^4 + 25 \times 10^4)}} = 3.951$$

Example 4.9 The limit state of failure of $g(x) = Z f_y - M_{ext}$ of the steel beam for strength formulation is given. Determine the section modulus of the section for a reliability index of 3.09. Solve by FOSM method given the following data.

Sl. No.	Variables	Characteristic values	CoV	Distribution
1	f_y	25 kN/cm^2	0.05	Normal
2	M_{ext}	200 kN/cm^2	0.20	Normal
3	Z	?	0.05	Normal

Solution: The limit state of failure of $g(x) = f_y Z - M_{ext}$

Standard deviation = Mean value × CoV

i.e.,

$$\sigma_Z = \mu_Z \times (\text{CoV})_Z$$

$$\mu_Z = \frac{\sigma_Z}{(\text{CoV})_Z} = \frac{\sigma_Z}{0.05} = 20\sigma_Z$$

Here, strengths are underestimated and loads are overestimated.

Strengths are f_y and Z, load is M_{ext}

$$\mu_{f_y} = \frac{x_{f_y}}{[1-1.645\times(\text{CoV})_{f_y}]} = \frac{25}{[1-1.645\times0.05]} = 27.240 \text{ kN/cm}^2$$

$$\mu_Z = \frac{x_Z}{[1-1.645\times(\text{CoV})_Z} = \frac{x_Z}{[1-1.645\times0.05]} = 1.0896\, x_Z \text{ cm}^3$$

$$\mu_{M_{ext}} = \frac{x_{M_{ext}}}{[1+1.645\times(\text{CoV})_{M_{ext}}} = \frac{200}{[1+1.645\times0.20]} = 150.489 \text{ kN-cm}$$

Sl. No.	Variables	Characteristic values	CoV	Mean value (μ)	Standard deviation (σ)
1	f_y	25 kN/cm^2	0.05	27.240	1.362
2	M_{ext}	200 kN/cm^2	0.20	150.489	30.10
3	Z	?	0.05	$20\sigma_z$	–

Now,

$$g(x) = R - S$$

Here, $R = f_y Z \quad S = M_{ext}$

Now,

Mean value,

$$\mu_R = \mu_{f_y} \times \mu_Z = 27.240 \times \mu_Z = 27.240\times 20\sigma_Z = 544.80\sigma_Z \text{ kN-cm}$$

$$\mu_S = \mu_{M_{ext}} = 150.489 \text{ kN-cm}$$

Now, partially differentiating R and S w.r.t variables f_y, Z and M_{ext}, we get,

$$\frac{\partial R}{\partial f_y} = Z = \mu_Z = 20\sigma_Z$$

$$\frac{\partial R}{\partial Z} = f_y = \mu_{f_y} = 27.240$$

$$\frac{\partial g(x)}{\partial M_{ext}} = 1.0$$

Standard deviation,

$$\sigma_R^2 = \left(\frac{\partial R}{\partial f_y}\times\sigma_{f_y}\right)^2 + \left(\frac{\partial R}{\partial Z}\times\sigma_Z\right)^2 = (\mu_Z\times1.362)^2 + (27.240\times\sigma_Z)^2$$

Here, $\mu_Z = 20\sigma_Z$

$$\Rightarrow \quad \sigma_R^2 = (20\sigma_Z \times 1.362)^2 + (27.240\sigma_Z)^2 = (742.01\sigma_Z^2 + 742.01\sigma_Z^2)$$

$$\Rightarrow \quad \sigma_R^2 = 1484.03\sigma_Z^2 \text{ kN-cm}$$

Similarly,

$$\sigma_S^2 = \left(\frac{\partial R}{\partial M_{ext}} \times \sigma_{M_{ext}}\right)^2 = (1.0 \times 30.09)^2 = 905.40 \text{ kN-cm}$$

Now,

Reliability index (β) is given by

$$\beta = \frac{\mu_R - \mu_S}{\sqrt{\sigma_R^2 + \sigma_S^2}}$$

Given $\beta = 3.09$

$$\beta = \frac{(544.80\sigma_Z - 150.489)}{\sqrt{(1484.03\sigma_Z{}^2 + 905.40)}}$$

Now, squaring on both sides, we get,

$$\left[\beta \times \left(\sqrt{(1484.03\sigma_Z^2 + 905.40)}\right)\right]^2 = (544.80\sigma_Z - 150.489)^2$$

$$\Rightarrow \quad (3.09)^2 \times (1484.03\sigma_Z^2 + 905.40) = (544.80\sigma_Z - 150.489)^2$$

$$\Rightarrow \quad 14170.144\sigma_Z^2 + 8644.84 = (544.80\sigma_Z)^2 + (150.489)^2 - 2 \times 544.80\sigma_Z \times 150.489$$

$$\Rightarrow \quad 14170.144\sigma_Z^2 + 8644.84 = 296807.04\sigma_Z^2 + 22646.93 - 163972.81\sigma_Z$$

$$\Rightarrow \quad 14170.144\sigma_Z^2 - 296807.04\sigma_Z^2 + 163972.81\sigma_Z + 8644.84 - 22646.93 = 0$$

$$\Rightarrow \quad 282636.89\sigma_Z^2 - 163972.81\sigma_Z + 14002.09 = 0$$

By solving quadratic equation in σ_Z, we get,

$$\therefore \quad \sigma_{Z_{max.}} = 0.4760$$

$$\therefore \quad \sigma_{Z_{min.}} = 0.1040$$

We know that,

$$\mu_Z = 20\sigma_Z$$

$$\Rightarrow \quad \mu_Z = 20\sigma_{Z_{max.}} = 20 \times 0.4760 = 9.520 \text{ cm}^3$$

$$\Rightarrow \quad \mu_Z = 20\sigma_{Z_{min.}} = 20 \times 0.1040 = 2.080 \text{ cm}^3$$

Check:

$$\beta = \frac{\mu_R - \mu_S}{\sqrt{\sigma_R^2 + \sigma_S^2}} = \frac{544.80\sigma_Z - 150.489}{\sqrt{(1484.03\sigma_Z)^2 + 905.40}}$$

For, $\sigma_{Z_{max.}} = 0.4760$

$$\beta = \frac{544.80 \times 0.4760 - 150.489}{\sqrt{1484.03 \times (0.4760)^2 + 905.40}} = \frac{108.835}{35.23} = 3.08864$$

For, $\sigma_{Z_{min.}} = 0.1040$

$$\beta = \frac{544.80 \times 0.1040 - 150.489}{\sqrt{1484.03 \times (0.1040)^2 + 905.40}} = \frac{-93.829}{30.35} = -3.0910$$

Comment: If the chosen value σ is maximum, this results in positive reliability index. Otherwise for minimum value σ the reliability index is negative, i.e., the limit state is violated.

Example 4.10 (Limit states of strength, shear, serviceability and stress formulation) A simply supported beam, ISMB350 @ 52.4 kg/m at the limit state of flexure shown in Fig. 4.4. Characteristic values are given in the table below.

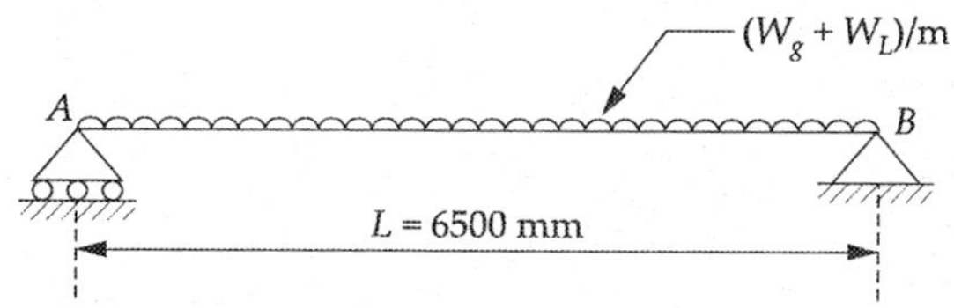

Fig. 4.4 Simply supported beam carrying UDL

Sl. No.	Variables	Characteristic values	Coefficient of variation (CoV)	Distribution
1	f_y (N/mm^2)	250	0.05	Normal
2	Z (mm^3)	7.7890×10^5	0.05	Normal
3	W_g (N/mm)	0.524	0.05	Normal
4	W_L (N/mm)	20.00	0.15	Normal
5	L, (mm)	6500	0.05	Normal
6	I, (mm^4)	1.3630×10^8	0.05	Normal
7	y, (mm)	175.00	0.05	Normal
8	E, (N/mm^2)	2.10×10^5	0.06	Normal
9	τ (N/mm^2)	100.00	0.05	Normal
10	t_w, (mm)	8.10	0.03	Normal
11	D, (mm)	350.00	0.05	Normal
12	A, (mm^2)	6.6710×10^3	0.05	Normal

Compute the limit state of the following performance function by FOSM method.

(i) Strength formulation

$$g(x) = f_y Z - \frac{(W_g + W_L)L^2}{8}$$

(ii) Stress formulation

$$g(x) = f_y - \frac{M}{I} y,$$

(iii) Serviceability formulation

$$g(x) = \Delta_{\text{lim}} - \Delta_{\text{act}} \qquad \Delta_{\text{lim}} = \frac{L}{360} \qquad \Delta_{\text{lim}} = \frac{5}{384}\frac{WL^4}{EI}$$

(iv) Shear formulation

$$g(x) = R - S \qquad R = \tau_s t_W d \qquad S = 0.5\,(W_g + W_L)$$

Find the following:

(a) Reliability index (β)
(b) Resistance factor (ϕ)
(c) Load factor (γ)
(d) Central safety factor (*CSF*)

Solution: The strengths are underestimated and loads are overestimated, the mean values are computed from the characteristic values and are tabulated as given below:

Sl. No.	Variables	Characteristic values	Coefficient of variation (CoV)	Mean value (μ)	Standard deviation (σ)
1	f_y, (N/mm^2)	250	0.05	272.40	13.62
2	Z, (mm^3)	7.7890×10^5	0.05	8.487×10^5	4.243×10^4
3	W_g (N/mm)	0.524	0.05	0.484	0.0242
4	W_L, (N/mm)	20.00	0.15	16.041	2.406
5	$W = (W_g + W_L)$	20.524	–	16.525	–
6	L, (mm)	6500	0.05	7082.53	354.12
7	I, (mm^4)	1.3630×10^8	0.05	1.4851×10^8	7.425×10^6
8	y, (mm)	175.00	0.05	190.68	9.534
9	E, (N/mm^2)	2.10×10^5	0.06	2.33×10^5	13.98×10^3
10	τ_s, (N/mm^2)	100.00	0.05	108.96	5.448
11	t_w, (mm)	8.10	0.03	8.52	0.2556
12	D, (mm)	350.00	0.05	381.36	19.068
13	A, (mm^2)	6.6710×10^3	0.05	7.268×10^3	363.40

(1) Strength formulation by FOSM method

$$g(x) = f_y Z - \frac{(W_g + W_L)L^2}{8}$$

$$\Rightarrow \quad g(x) = R - S, \quad R = f_y Z, \quad S = (W_g + W_L)\frac{L^2}{8}$$

Now, the mean and standard deviations are

Mean value,

$$\mu_R = \mu_{f_y} \times \mu_Z = 272.40 \times 8.487 \times 10^5 = 231.18 \times 10^6 \text{ N-mm}$$

$$\mu_S = \frac{(\mu_{W_g} \times \mu_{W_L}) \times \mu_{L^2}}{8} = \frac{(0.484 + 16.041) \times (7082.53)^2}{8} = 103.61 \times 10^6 \text{ N-mm}$$

Now, partially differentiating R and S w.r.t variables, we get,

$$\frac{\partial R}{\partial f_y} = Z = \mu_Z = 8.487 \times 10^5$$

$$\frac{\partial R}{\partial Z} = f_y = \mu_{f_y} = 272.40$$

Here, $W = W_g + W_L$

$$\Rightarrow \quad S = \frac{WL^2}{8}$$

$$\frac{\partial S}{\partial W} = \frac{L^2}{8} = \frac{\mu_{L^2}}{8} = \frac{(7082.53)^2}{8} = 6.27 \times 10^6$$

$$\frac{\partial S}{\partial L} = \frac{2WL}{8} = \frac{2\mu_W \mu_L}{8} = \frac{2 \times 16.525 \times 7082.53}{8} = 29.259 \times 10^3$$

Standard deviation,

$$\sigma_R^2 = \left(\frac{\partial R}{\partial f_y} \times \sigma_{f_y}\right)^2 + \left(\frac{\partial R}{\partial Z} \times \sigma_Z\right)^2 = (8.487 \times 10^5 \times 13.62)^2 + (272.40 \times 4.243 \times 10^4)^2$$

$$= 2.672 \times 10^{14}$$

$$\therefore \quad \sigma_R = \sqrt{2.672 \times 10^{14}} = 1.634 \times 10^7 \text{ N-mm}$$

Similarly,

$$\sigma_S^2 = \left(\frac{\partial S}{\partial W} \times \sigma_W\right)^2 + \left(\frac{\partial S}{\partial L} \times \sigma_L\right)^2$$

Here, $\sigma_W^2 = \sigma_{Wg}^2 + \sigma_{WL}^2 = (0.0242)^2 + (2.406)^2 = 5.789$

$$\Rightarrow \qquad \sigma_W = \sqrt{5.789} = 2.406$$

Now, $\sigma_S^2 = \left(6.27\times10^6 \times 2.406\right)^2 + \left(29.259\times10^3 \times 354.12\right)^2 = 3.349\times10^{14}$

$$\therefore \qquad \sigma_S = \sqrt{3.349\times10^{14}} = 1.830\times10^7 \text{ N-mm}$$

Now,

(a) *Reliability index (β),*

$$\beta = \frac{\mu_R - \mu_S}{\sqrt{\sigma_R^2 + \sigma_S^2}} = \frac{(231.18\times10^6 - 103.61\times10^6)}{\sqrt{(2.672\times10^{14} + 3.349\times10^{14})}} = 5.198$$

(b) *Resistance factor (ϕ),*

$$\phi = [1 - 0.75\times\beta\times(\text{CoV})_R]$$

$$\therefore \qquad (\text{CoV})_R = \frac{\sigma_R}{\mu_R} = \frac{1.634\times10^7}{231.18\times10^6} = 0.07$$

$$\Rightarrow \qquad \phi = [1 - 0.75\times5.198\times0.07] = 0.727$$

(c) *Load factor (γ),*

$$\gamma = [1 + 0.75\times\beta\times(\text{CoV})_S]$$

$$\therefore \qquad (\text{CoV})_S = \frac{\sigma_S}{\mu_S} = \frac{1.830\times10^7}{103.61\times10^6} = 0.176$$

$$\Rightarrow \qquad \gamma = [1 + 0.75\times5.198\times0.176] = 1.686$$

(d) *Central safety factor (CSF),*

$$CSF = \frac{\gamma}{\phi} = \frac{1.686}{0.727} = 2.319$$

(e) $$\alpha = \frac{\sqrt{\sigma_R^2 + \sigma_S^2}}{\sigma_R + \sigma_S} = \frac{\sqrt{(2.672\times10^{14} + 3.349\times10^{14})}}{1.634\times10^7 + 1.83\times10^7} = \frac{24.53\times10^6}{3.464\times10^7} = 0.708$$

(2) Stress formulation by the FOSM method

$$g(x) = f_y - \frac{M}{I}y$$

Here, $M = \dfrac{WL^2}{8}$ for simply supported beam carrying UDL

$$\Rightarrow \qquad g(x) = R - S$$

$$R = f_y,\ S = \frac{M}{I}y \quad \Rightarrow S = \frac{WL^2}{8I}y$$

Now, the mean and standard deviation are

Mean value,

$$\mu_R = \mu_{f_y} = 272.40 \text{ N/mm}^2$$

$$\mu_S = \frac{\mu_W \times \mu_{L^2}}{8\mu_I} \times \mu_y = \frac{16.525 \times (7082.53)^2}{8 \times 1.4851 \times 10^8} \times 190.68 = 133.03 \text{ N/mm}^2$$

Now, partially differentiating R and S w.r.t variables, we get,

$$\frac{\partial R}{\partial f_y} = 1.0$$

$$\frac{\partial S}{\partial W} = \frac{L^2 y}{8I} = \frac{\mu_{L^2}\mu_y}{8\mu_I} = \frac{(7082.53)^2 \times 190.68}{8 \times 1.4851 \times 10^8} = 8.05$$

$$\frac{\partial S}{\partial L} = \frac{2WLy}{8I} = \frac{2\mu_W \mu_L \mu_y}{8\mu_I} = \frac{2 \times 16.525 \times 7082.53 \times 190.68}{8 \times 1.4851 \times 10^8} = 0.0375$$

$$\frac{\partial S}{\partial y} = \frac{WL^2}{8I} = \frac{\mu_W \mu_{L^2}}{8\mu_I} = \frac{16.525 \times (7082.53)^2}{8 \times 1.4851 \times 10^8} = 0.6977$$

$$\frac{\partial S}{\partial I} = -\frac{WL^2 y}{8I^2} = -\frac{\mu_W \mu_{L^2} \mu_y}{8\mu_{I^2}} = -\frac{16.525 \times (7082.53)^2 \times 190.68}{8 \times (1.4851 \times 10^8)^2} = -8.957 \times 10^{-7}$$

Standard deviation,

$$\sigma_R^2 = \left(\frac{\partial R}{\partial f_y} \times \sigma_{f_y}\right)^2 = (1.0 \times 13.62)^2 = 185.50$$

$$\therefore \sigma_R = \sqrt{185.50} = 13.62 \text{ N/mm}^2$$

$$\sigma_S^2 = \left(\frac{\partial S}{\partial W} \times \sigma_W\right)^2 + \left(\frac{\partial S}{\partial L} \times \sigma_L\right)^2 + \left(\frac{\partial S}{\partial y} \times \sigma_y\right)^2 + \left(\frac{\partial S}{\partial I} \times \sigma_I\right)^2$$

$$\sigma_S^2 = (8.05 \times 2.406)^2 + (0.0375 \times 354.12)^2 + (0.6977 \times 9.534)^2 + \left(-8.957 \times 10^{-7} \times 7.425 \times 10^6\right)^2$$

$$\sigma_S^2 = 639.95$$

$$\therefore \sigma_S = \sqrt{639.95} = 25.29 \text{ N/mm}^2$$

Now,

(a) *Reliability index (β),*

$$\beta = \frac{\mu_R - \mu_S}{\sqrt{\sigma_R^2 + \sigma_S^2}} = \frac{272.40 - 133.03}{\sqrt{(185.50 + 639.95)}} = 4.85$$

(b) *Resistance factor* (ϕ),

$$\phi = [1 - 0.75 \times \beta \times (\text{CoV})_R]$$

$$\therefore \qquad (\text{CoV})_R = \frac{\sigma_R}{\mu_R} = \frac{13.62}{272.40} = 0.05$$

$$\Rightarrow \qquad \phi = [1 - 0.75 \times 4.85 \times 0.05] = 0.818$$

(c) *Load factor* (γ),

$$\gamma = [1 + 0.75 \times \beta \times (\text{CoV})_S]$$

$$\therefore \qquad (\text{CoV})_S = \frac{\sigma_S}{\mu_S} = \frac{25.29}{133.03} = 0.190$$

$$\Rightarrow \qquad \gamma = [1 + 0.75 \times 4.85 \times 0.190] = 1.691$$

(d) *Central safety factor (CSF)*,

$$CSF = \frac{\gamma}{\phi} = \frac{1.691}{0.818} = 2.067$$

(e) $$\alpha = \frac{\sqrt{\sigma_R^2 + \sigma_S^2}}{\sigma_R + \sigma_S} = \frac{\sqrt{(185.50 + 639.95)}}{13.62 + 25.29} = \frac{28.73}{38.91} = 0.738$$

(3) Serviceability formulation by the FOSM method

$$g(x) = \frac{L}{360} - \frac{5WL^4}{384EI}$$

$$\Rightarrow \qquad g(x) = R - S$$

Here, $$R = \frac{L}{360} \qquad S = \frac{5WL^4}{384EI}$$

Now, the mean and standard deviation are

Mean value,

$$\mu_R = \frac{\mu_L}{360} = \frac{7082.53}{360} = 19.67 \text{ mm}$$

$$\mu_S = \frac{5\mu_W \mu_{L^4}}{384\mu_E \mu_I} = \frac{5 \times 16.525 \times (7082.53)^4}{384 \times 2.33 \times 10^5 \times 1.4851 \times 10^8} = 15.64 \text{ mm}$$

Now, partially differentiating R and S w.r.t variables, we get,

$$\frac{\partial R}{\partial L} = \frac{1}{360} = 2.77 \times 10^{-3}$$

$$\frac{\partial S}{\partial W} = \frac{5L^4}{384EI} = \frac{5\mu_{L^4}}{384\mu_E \mu_I} = \frac{5 \times (7082.53)^4}{384 \times 2.33 \times 10^5 \times 1.4851 \times 10^8} = 0.9468$$

$$\frac{\partial S}{\partial L}=\frac{5\times 4\times WL^3}{384EI}=\frac{20\mu_W\mu_{L^3}}{384\mu_E\mu_I}=\frac{20\times 16.525\times (7082.53)^3}{384\times 2.33\times 10^5\times 1.4851\times 10^8}=8.836\times 10^{-3}$$

$$\frac{\partial S}{\partial E}=-\frac{5WL^4}{384E^2I}=-\frac{5\mu_W\mu_{L^4}}{384\mu_{E^2}\mu_I}=-\frac{5\times 16.525\times (7082.53)^4}{384\times (2.33\times 10^5)^2\times 1.4851\times 10^8}=-6.712\times 10^{-5}$$

$$\frac{\partial S}{\partial I}=-\frac{5WL^4}{384EI^2}=-\frac{5\mu_W\mu_{L^4}}{384\mu_E\mu_{I^2}}=-\frac{5\times 16.525\times (7082.53)^4}{384\times 2.33\times 10^5\times (1.4851\times 10^8)^2}=-1.053\times 10^{-7}$$

Standard deviation,

$$\sigma_R^2=\left(\frac{\partial R}{\partial L}\times\sigma_L\right)^2=(2.77\times 10^{-3}\times 354.12)^2=0.9621$$

$$\therefore\quad \sigma_R=\sqrt{0.9621}=0.9809\ \text{mm}$$

Similarly,

$$\sigma_S^2=\left(\frac{\partial S}{\partial W}\times\sigma_W\right)^2+\left(\frac{\partial S}{\partial L}\times\sigma_L\right)^2+\left(\frac{\partial S}{\partial E}\times\sigma_E\right)^2+\left(\frac{\partial S}{\partial I}\times\sigma_I\right)^2$$

$$\Rightarrow\quad \sigma_S^2=(0.9468\times 2.406)^2+\left(8.836\times 10^{-3}\times 354.12\right)^2+\left(-6.712\times 10^{-5}\times 13.98\times 10^3\right)^2+\left(-1.053\times 10^{-7}\times 7.425\times 10^6\right)^2$$

$$\sigma_S^2=16.471$$

$$\therefore\quad \sigma_S=\sqrt{16.471}=4.058\ \text{mm}$$

Now,

(a) *Reliability index (β),*

$$\beta=\frac{\mu_R-\mu_S}{\sqrt{\sigma_R^2+\sigma_S^2}}=\frac{19.67-15.64}{\sqrt{(0.9621+16.471)}}=0.9652$$

(b) *Resistance factor (ϕ),*

$$\phi=[1-0.75\times\beta\times(\text{CoV})_R]$$

$$\therefore\quad (\text{CoV})_R=\frac{\sigma_R}{\mu_R}=\frac{0.9809}{19.67}=0.0498$$

$$\Rightarrow\quad \phi=[1-0.75\times 0.9652\times 0.0498]=0.9639$$

(c) *Load factor (γ),*

$$\gamma=[1+0.75\times\beta\times(\text{CoV})_S]$$

$$\therefore\quad (\text{CoV})_S=\frac{\sigma_S}{\mu_S}=\frac{4.058}{15.64}=0.259$$

$$\Rightarrow\quad \gamma=[1+0.75\times 0.9652\times 0.259]=1.187$$

(d) *Central safety factor(CSF),*

$$\text{CSF} = \frac{\gamma}{\phi} = \frac{1.187}{0.9639} = 1.231$$

(e) $\alpha = \dfrac{\sqrt{\sigma_R^2 + \sigma_S^2}}{\sigma_R + \sigma_S} = \dfrac{\sqrt{(0.9621 + 16.471)}}{0.9809 + 4.058} = \dfrac{4.175}{5.038} = 0.828$

$$\text{CSF} = \frac{\gamma}{\phi} = \frac{1.187}{0.9639} = 1.231$$

(4) Shear formulation by the FOSM method

$$g(x) = \tau_s t_w d - 0.5(W_g + W_L)L$$

Here, $W = W_g + W_L$

$$\Rightarrow \qquad g(x) = \tau_s t_w d - 0.5\,WL$$

$$\Rightarrow \qquad g(x) = R - S$$

Here, $R = \tau_s t_w d,\ S = 0.5\,WL$

Now, the mean and standard deviations are

Mean value,

$$\mu_R = \mu_{\tau_s} \times \mu_{t_w} \times \mu_d = 108.96 \times 8.52 \times 381.36 = 354.031 \times 10^3 \text{ N}$$

$$\mu_S = 0.5\mu_W \mu_L = 0.5 \times 16.525 \times 7082.53 = 58.519 \times 10^3 \text{ N}$$

Now, partially differentiating R and S w.r.t variables, we get,

$$\frac{\partial R}{\partial \tau_s} = t_w d = \mu_{t_w}\mu_d = 8.52 \times 381.36 = 3249.18$$

$$\frac{\partial R}{\partial t_w} = \tau_s d = \mu_{\tau_s}\mu_d = 108.96 \times 381.36 = 41552.98$$

$$\frac{\partial R}{\partial d} = \tau_s t_w = \mu_{\tau_s}\mu_{t_w} = 108.96 \times 8.52 = 928.34$$

$$\frac{\partial S}{\partial W} = 0.5L = 0.5\mu_L = 0.5 \times 7082.53 = 3541.26$$

$$\frac{\partial S}{\partial L} = 0.5W = 0.5\mu_W = 0.5 \times 16.525 = 8.262$$

Standard deviation,

$$\sigma_R^2 = \left(\frac{\partial R}{\partial \tau_s} \times \sigma_{\tau_s}\right)^2 + \left(\frac{\partial R}{\partial t_w} \times \sigma_{t_w}\right)^2 + \left(\frac{\partial R}{\partial d} \times \sigma_d\right)^2$$

$$= (3249.18 \times 5.448)^2 + (41552.98 \times 0.2556)^2 + (928.34 \times 19.068)^2$$

$$= 739.49 \times 10^6$$

$$\therefore \quad \sigma_R = \sqrt{739.49 \times 10^6} = 27.193 \times 10^3 \text{ N}$$

Similarly,

$$\sigma_S^2 = \left(\frac{\partial S}{\partial W} \times \sigma_W\right)^2 + \left(\frac{\partial S}{\partial L} \times \sigma_L\right)^2$$

$$\Rightarrow \quad \sigma_S^2 = (3541.26 \times 2.406)^2 + (8.262 \times 354.12)^2 = 81.15 \times 10^6$$

$$\therefore \quad \sigma_S = \sqrt{81.15 \times 10^6} = 9.008 \times 10^3 \text{ N}$$

Now,

(a) *Reliability index* (*β*),

$$\beta = \frac{\mu_R - \mu_S}{\sqrt{\sigma_R^2 + \sigma_S^2}} = \frac{354.031 \times 10^3 - 58.519 \times 10^3}{\sqrt{(739.49 \times 10^6 + 81.15 \times 10^6)}} = 10.31$$

(b) *Resistance factor* (*ϕ*),

$$\phi = [1 - 0.75 \times \beta \times (\text{CoV})_R]$$

$$\therefore \quad (\text{CoV})_R = \frac{\sigma_R}{\mu_R} = \frac{27.193 \times 10^3}{354.031 \times 10^3} = 0.0768$$

$$\Rightarrow \quad \phi = [1 - 0.75 \times 10.31 \times 0.0768] = 0.4061$$

(c) *Load factor* (*γ*),

$$\gamma = [1 + 0.75 \times \beta \times (\text{CoV})_S]$$

$$\therefore \quad (\text{CoV})_S = \frac{\sigma_S}{\mu_S} = \frac{9.008 \times 10^3}{58.519 \times 10^3} = 0.1539$$

$$\Rightarrow \quad \gamma = [1 + 0.75 \times 10.31 \times 0.1539] = 2.19$$

(d) *Central safety factor(CSF),*

$$\text{CSF} = \frac{\gamma}{\phi} = \frac{2.19}{0.4061} = 5.392$$

(e) $$\alpha = \frac{\sqrt{\sigma_R^2 + \sigma_S^2}}{\sigma_R + \sigma_S} = \frac{\sqrt{(739.49 \times 10^6 + 81.15 \times 10^6)}}{27.193 \times 10^3 + 9.008 \times 10^3} = \frac{28646.81}{36.201 \times 10^3} = 0.791$$

Sl. No.	Formulations	Reliability index (β)	Resistance factor (ϕ)	Load factor (γ)	Central safety factor (CSF)	$\alpha = \frac{\sqrt{\sigma_R^2 + \sigma_S^2}}{\sigma_R + \sigma_S}$
1	Strength	5.198	0.727	1.686	2.319	0.708
2	Stress	4.85	0.818	1.691	2.067	0.738
3	Serviceability	0.9652	0.9639	1.187	1.231	0.828
4	Shear	10.31	0.4061	2.19	5.392	0.791

Comment: Reliability index for strength, stress, serviceability and shear formulations show different values. Hence, it shows lack of invariance by first order second moment method (FOSM).

UNSOLVED PROBLEMS

1. A performance function $Y = (3X_1 - 2X_2)$ whose mean $X_1 = 16.6$, mean $X_2 = 18.8$, $\sigma_{X_1} = 2.45$, $\sigma_{X_2} = 2.83$. Determine the probability of failure if X_1 and X_2 are uncorrelated.
2. Evaluate the probability of survival of a tension element whose resistance is given by $R = f_y A$ under the action of an axial load, P given the following data.

Sl. No.	Variables	Mean value (μ)	Coefficient of variation (CoV)
1	f_y	300 N/mm^2	10%
2	A	109.29 mm^2	5%
3	P	15 N	12%

3. A performance function of a beam of span L is given by $g(x) = f_y Z - \frac{WL^2}{8}$ where f_y = yield stress, Z = section modulus, W load on the beam. The mean values and the coefficient of variation of the variates are as follows.

Sl. No.	Variables	Mean value (μ)	Coefficient of variation (CoV)
1	f_y	250 N/mm^2	0.05
2	Z	8.00×10^5 mm^3	0.05
3	W	25 N/mm	0.20
4	L	5000 mm	0.05

Find the reliability of the beam under flexure corresponding to the performance function.

4. The capacity of a column under buckling is $\frac{\pi^2 EI}{l^2}$ where E modulus of elasticity, I moment of inertia and l is the effective length of the column. Determine the reliability of the column under buckling if the axial load on the column is P if the mean value and standard deviation are as follows.

Sl. No.	Variables	Mean value (μ)	Coefficient of variation (CoV)
1	E	200000 N/mm^2	0.1
2	I	(12.5×10^6) mm^2	0.05
3	l	5000 mm	0.05
4	P	700 kN	0.20

5. A reinforced concrete beam whose flexural strength given by

$$R = \left\{ f_y \times A_{st} \times d \left[1 - \frac{0.77 \times f_y \times A_{st}}{b \times d \times f_{ck}} \right] \right\}$$ is under dead load W_g and live load W_L. The span of the beam, L is 3 m. Evaluate the reliability index if A_{st}, W_g, b and d are deterministic. Assume the following given data.
$\mu_{W_L} = 6$ kN/m, $\sigma_{W_L} = 3$ kN/m, $\mu_{f_{ck}} = 30.28$ N/mm^2, $\sigma_{f_{ck}} = 4.5$ N/mm^2, $\mu_{f_y} = 320$ N/mm^2, $\sigma_{f_y} = 32$ N/mm^2, $A_{st} = 1400$ mm^2, $W_g = 3$ kN/m, $b = 240$ mm and $d =$ 480 mm.

6. A segment of brick deck is under axial compression, P_a and prestressing force P_f. The condition for safety corresponding to the stress at the top is given by $f_{ck} \geq \frac{P_a}{A} + \frac{P_f}{A} + \frac{P_f \times ey}{I} - \frac{M \times y}{I}$. The prestressing force P_f can be expressed as product of yield stress, f_y and area of tendons, A_{st}. Evaluate the safety if the characteristic values and coefficient of variation (CoV) are as follows.

Sl. No.	Variables	Mean value (μ)	Coefficient of variation (CoV)
1	f_{ck}	40 N/mm^2	0.05
2	f_y	1570 N/mm^2	0.05
3	P_a	57096×10^6 N	0.20
4	A_{st}	18780.44 mm^2	0.05
5	e	42.9 mm	0.05
6	y	568.4 mm	0.05
7	M	24651×10^6 N-mm	0.125
8	A	12259×10^3 mm^2	0.05
9	I	6.2282×10^{12} mm^4	0.05

5
CHAPTER

Point Estimate Method

5.1 POINT ESTIMATE METHOD (PEM)

It is a computationally direct application technique for the uncertainty analysis of engineering problems. Essentially, the method is based on Taylor series expansion about the mean of the stochastic random variables in a model. It can be used to estimate statistical moments of any order. The model response involves stochastic random variables that are either correlated or uncorrelated. The method was originally developed for handling stochastic basic variables that are symmetric (Rosenblueth 1975).

The Taylor series (FORM) method requires the evaluation of derivatives. This effort ranges from being simple to cumbersome and impossible in some cases. When the function is given implicitly such difficulties are easily circumvented by the point estimate method (PEM). The point estimate procedure was originally proposed by Rosenblueth (1975) and later extended by him in 1981. The PDF's of the random variables are simulated by "point" masses located at plus or minus one standard deviation from the mean values. This method is direct, gives reasonably accurate results quickly and has been discussed in McGuffey *et al.* (1981), Harr (1987), Wolff (1996), and Corps of Engineers (1997). In this section, details for determining the expected values and its variance will be presented for two and three random variables. For more than three variables, the presentation of the equations gets more complex, but the approach is the same.

An analogy can be established between the probability distribution and distributed vertical load on a horizontal rigid beam. The expected value is the analogy of the point of application of the equilibrant (of unit magnitude) or the centre of loading and the standard deviation as its radius of gyration, shown schematically in Fig. 5.1(a).

Consider a model $Y = g(x)$ which involves single random basic variable X, whose first three moments or PDF/PMF are known. The Rosenblueth point estimate method approximates the original PDF/PMF of the random variable X by assuming the entire probability mass of X is concentrated at two points X_- and X_+. Using two-point approximations, there are four unknown to be determined, namely, the locations of X_- and X_+ and corresponding probability masses P_- and P_+.

Conceptually, the expected value and the standard deviation supply information concerning the central tendency and scatter of the variate. Rosenblueth suggested that this information could be extracted from the beam analogy but with the rigid beam supported on two reactions (Fig. 5.1b), P_+ acting at $x = X_+$ and P_- acting at $x = X$.

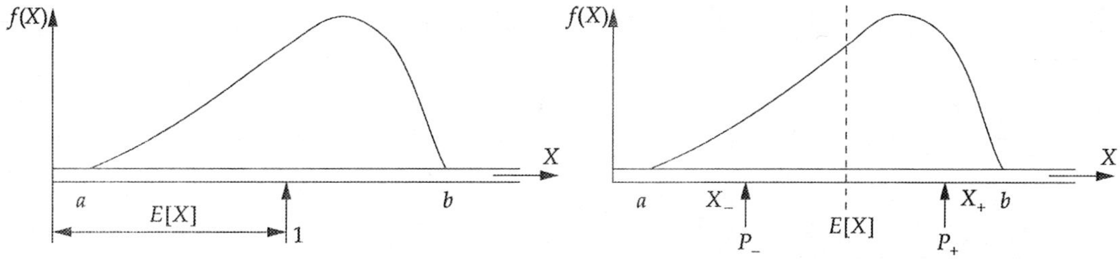

(a) Single support

(b) Two supports

Fig. 5.1 Continuously distributed vertical force system on rigid beam

The reactions P_- and P_+ are said to be two-point estimates of the distribution of $F_X(x)$. Moments, interpreted in common probabilistic symbolism, specify four equations with four unknowns. The first reaction and three additional side constraints, namely; first, second and third moments of the original random variable X must be preserved.

Sum of verticals

$$P_+ + P_- = 1 \qquad \text{...(5.1a)}$$

First moment, i.e., centre of gravity

$$P_+(X_+) + P_-(X_-) = E[f(x)] = \bar{x} \qquad \text{... (5.1b)}$$

Second moment, i.e., moment of inertia

$$P_+(X_+ - \bar{x})^2 + P_-(X_- - \bar{x})^2 = \sigma[f(x)]^2 \qquad \text{...(5.1c)}$$

Third moment, i.e., skewness

$$P_+ (X_+ - \bar{x})^3 + P(X_- - \bar{x})^3 = \beta(1)\, \sigma^3[x] \qquad \text{... (5.1d)}$$

where $\beta(1)$ is the skewness coefficient, when symmetric, $\beta(1) = 0.0$

Solving Eqs. 5.1(a) to 5.1(d), we get

$$P_+ = \frac{1}{2}\left[1 \pm \sqrt{1 - \frac{1}{1 + [\beta(1)/2]^2}}\right] \qquad \text{...(5.2a)}$$

$$P_- = 1 - P_+ \qquad \text{... (5.2b)}$$

$$X_+ = \bar{x} + \sigma[x]\sqrt{\frac{P_-}{P_+}} \qquad \text{... (5.2c)}$$

$$X_- = \bar{x} - \sigma[x]\sqrt{\frac{P_+}{P_-}} \qquad \text{...(5.2d)}$$

The sign preceding the radical in the first of Eqs. (5.2a to 5.2d) is that of $\beta(1)$.

If $\beta(1) = 0$, these reduce to

$$P_+ = P_- = \frac{1}{2} \quad \text{...(5.3a)}$$

$$X_+ = \bar{x} + \sigma[x] \quad \text{...(5.3b)}$$

$$X_- = \bar{x} - \sigma[x] \quad \text{...(5.3c)}$$

It should be emphasized that point estimates and their points of application, in concept, serve to transfer information about the distribution of the variable. This will become apparent with further application of the method.

Given information concerning the first three moments of a random variable x, what can be said about the dependent random variable $y = y(x)$? With only $\bar{x}$, $\sigma[x]$, and $\beta(1)$ known, using Eqs. (5.2a to 5.2d), X_-, X_+, P_-, and P_+ are obtained, shown schematically in Fig. 5.2, this information about the unknown distribution of x produces the two estimates of the y variable, y_- and y_+, i.e., information about the x variable is transferred through the functional relationship between x and $y(x)$ to provide the two values of y_+ and y_-. The two weighting functions P_- and P_+ will then scale these estimates.

$$E[y] = \bar{y} = ((p_- y_-) + (p_+ y_+)) \quad (5.4a)$$

$$E[y^2] = (p_- y^2_-) + (p_+ y^2_+) \quad (5.4b)$$

In general,

$$E[y^M] = (P_- y^M_-) + (P_+ y^M_+) \quad (5.4c)$$

Of course, M cannot be greater than the number of known moments of x.

If we have only information regarding the expected value and the coefficient of variation, (or the standard deviation) of x, there is no reason (information) to assume $y(x)$ to be skewed, as we know only the first two moments. The principle of maximum entropy specifies that $\beta(1) = 0$ and Eqs. (5.2a and 5.2b) are applicable.

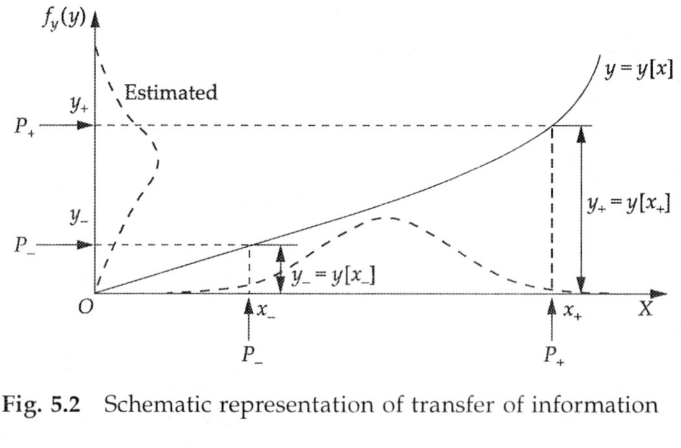

Fig. 5.2 Schematic representation of transfer of information

5.2 BI-VARIATE POINT ESTIMATE

For two random variables, Rosenblueth considered the probability distribution to be analogous to a distributed vertical load acting over a rigid plate, supported at four points

$P_{++}, P_{+-}, P_{-+}, P_{--}$ (Fig. 5.3). For this case the Mth moment of the functional relationship, say, $y = y(x_1, x_2)$, is

$$E\left[y^M\right] = P_{++}y_{++}^M + P_{+-}y_{+-}^M + P_{-+}y_{-+}^M + P_{--}y_{--}^M \qquad \ldots(5.5a)$$

where
$$y_{\pm\pm} = y\left(\bar{x}_1 \pm \sigma[x_1], \bar{x}_2 \pm \sigma[x_2]\right) \qquad \ldots(5.5b)$$

$$P_{++} = P_{--} = \frac{1+\rho}{4} \qquad \ldots(5.5c)$$

$$P_{+-} = P_{-+} = \frac{1-\rho}{4} \qquad \ldots(5.5d)$$

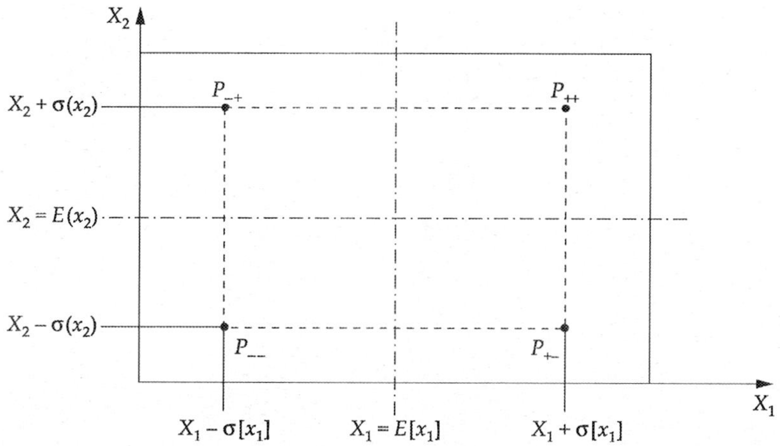

Fig. 5.3 Representation of two variables as a loading on rigid plate

In these equations ρ is the correlation coefficient between x_1 and x_2 and $\bar{x}_1 = E[x_1], \bar{x}_2 = E[x_2]$. The influence of correlation coefficient on weighting factors is shown for a few cases in Fig. 5.4.

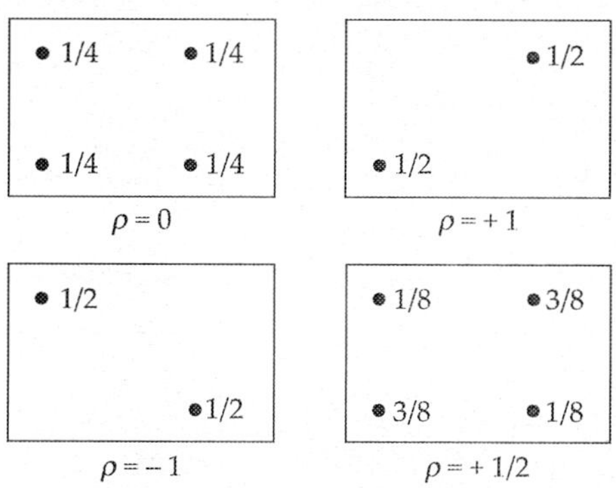

Fig. 5.4 Influence of correlation coefficient on weighting factors

The values of correlation factor are given below.

Correlation value	**Low**	**Medium**	**Important**	**Strong**	**Very Strong**
	0.0 to 0.3	0.3 to 0.5	0.5 to 0.7	0.7 to 0.9	0.9 to 1.0

5.3 POINT ESTIMATE FOR TWO RANDOM VARIABLES

The performance function and its margin of safety (*MS*), of two random variables *X* and *Y*, will take the form $MS = g(X, Y)$.

Where the uncertainty in the random variables is described by their standard deviation, σ_X, σ_Y (or coefficient of variance), and their mean values taken as μ_X, μ_Y. For the general case, we can also presume that the variables are dependent, with a correlation coefficient of ρ_{XY}, and that the PDF's are all symmetric.

For the two random variables, the first step requires the evaluation of *MS* at four unique combinations of the random variables, as shown below

$$MS_{++} = g[(\mu_X + \sigma_X), (\mu_Y + \sigma_Y)] \quad \text{...(5.6a)}$$

$$MS_{+-} = g[(\mu_X + \sigma_X), (\mu_Y - \sigma_Y)] \quad \text{...(5.6b)}$$

$$MS_{-+} = g[(\mu_X - \sigma_X), (\mu_Y + \sigma_Y)] \quad \text{...(5.6c)}$$

$$MS_{--} = g[(\mu_X - \sigma_X), (\mu_Y - \sigma_Y)] \quad \text{...(5.6d)}$$

The MS_{++}, represents the value of the function $g(X, Y)$, as evaluated for $X = (\mu_X + \sigma_X)$ and $Y = (\mu_Y + \sigma_Y)$ and MS_{-+} represents the value of the function $g(X, Y)$, as evaluated for $X = (\mu_X - \sigma_X)$ and $Y = (\mu_Y + \sigma_Y)$.

The expected *MS* will be given by their sum as

$$E(MS) = \mu_{MS} = P_{++}\, MS_{++} + P_{+-}\, MS_{+-} + P_{-+}\, MS_{-+} + P_{--}\, MS_{--} \quad \text{... (5.7)}$$

where

$$P_{++} = P_{--} = \frac{1}{4}\,[(1 + \rho_{XY})] \quad \text{...(5.8a)}$$

$$P_{+-} = P_{+-} = \frac{1}{4}\,[(1 - \rho_{XY})] \quad \text{...(5.8b)}$$

The variance of *MS* will be given by

$$\text{Var}\,(MS) = E[MS^2] - (E\,[MS])^2 \quad \text{...(5.9)}$$

$$E[MS^2] = P_{++}\, MS^2_{++} + P_{+-}\, MS^2_{+-} + P_{-+}\, MS^2_{-+} + P_{--}\, MS^2_{--} \quad \text{...(5.10)}$$

Hence, for two variables, the *MS* values are evaluated for four conditions using the adopted performance function.

5.4 POINT ESTIMATE FOR THREE RANDOM VARIABLES

The method can be extended to three random variables *X*, *Y*, *Z*, which will lead to an expected *MS*, given by summation of eight products.

$$E\,[MS] = P_{+++}\, MS_{+++} + P_{++-}\, MS_{++-} + P_{+-+}\, MS_{+-+} + P_{+--}\, MS_{+--}$$
$$+ P_{---}\, MS_{---} + P_{--+}\, MS_{--+} + P_{-+-}\, M_{-+-} + P_{--+}\, MS_{-++} \quad \text{...(5.11)}$$

$$P_{+++} = P_{---} = 1/8(1 + \rho_{XY} + \rho_{YZ} + \rho_{ZX}) \quad \text{...(5.12a)}$$

$$P_{++-} = P_{--+} = 1/8(1 + \rho_{XY} - \rho_{YZ} - \rho_{ZX}) \quad \text{...(5.12b)}$$

$$P_{+-+} = P_{-+-} = 1/8(1 - \rho_{XY} - \rho_{YZ} + \rho_{ZX}) \quad \text{...(5.12c)}$$

$$P_{+--} = P_{-++} = 1/8(1 - \rho_{XY} + \rho_{YZ} - \rho_{ZX}) \quad \text{...(5.12d)}$$

The variance is obtained from the expression Var (MS) = $E[MS^2] - (E[MS])^2$.

In general, the expected value, $E[MS]$, is given by the product of the P coefficients and the MS values, and the variance is computed using Eq. 5.9. So for even more number of random variables, the new P coefficients must be derived to follow the sequence presented for two and three random variables or use contingency Table 5.1. For each case, the numeric coefficients (in Eqs 5.8 and 5.12) is the reciprocal of 2^n, where n is the number of random variables being considered. The signs in front of the correlation coefficients are related to the subscripts of P, with + + or – – correlation leading to a positive sign. Rosenblueth method requires more evaluations of the performance function. Use cyclic notations for any number of variables with appropriate signs.

5.5 GENERALIZED POINT ESTIMATE METHOD

Rosenblueth (1975) generalized PEM for any number of random variables. For a function of three random variables $y = y(x_1, x_2, x_3)$,

$$y_{\pm\pm\pm} = y(\bar{x}_1 \pm \sigma[x_1], \bar{x}_2 \pm \sigma[x_2], \bar{x}_3 \pm \sigma[x_3]) \quad \text{...(5.13)}$$

$$P_{+++} = P_{---} = \frac{1}{2^3}(1 + \rho_{12} + \rho_{23} + \rho_{31})$$

$$P_{++-} = P_{--+} = \frac{1}{2^3}(1 + \rho_{12} - \rho_{23} - \rho_{31})$$

$$P_{+-+} = P_{-+-} = \frac{1}{2^3}(1 - \rho_{12} - \rho_{23} + \rho_{31})$$

$$P_{+--} = P_{-++} = \frac{1}{2^3}(1 - \rho_{12} + \rho_{23} - \rho_{31}) \quad \text{...(5.14)}$$

where ρ_{ij} is the correlation coefficient of the random variables x_i and x_j. The sign of ρ_{ij} is determined by the sign of the multiplication of ij, that is i = (–), j = (+) yields

$$ij = (-)(+) = (-)$$

The Mth expectation is obtained from

$$E[y^M] = P_{+++}y^M_{+++} + P_{++-}y^M_{++-} + \ldots + P_{---}y^M_{---} \quad (5.15)$$

Equation 5.14 has $2^3 = 8$ terms, all permutations of the three +'s and –'s. In general, for N variables, there are 2^N terms and $N(N-1)/2$ correlation coefficients, the latter being the number of combinations of N objects taken two at a time. The coefficient on the right-hand side of Eq. 5.14, in general, is $(1/2)^N$. However, Table 5.1 is a useful generator of the + and – computation and assessment of functions.

Example 5.1 The recommendation of the American Concrete Institute (Galambos et al., 1982) for the design of reinforce concrete structures is, on simplified form,

$$R \geq 1.6\ W_g + 1.9\ W_L$$

where R is the strength of the element W_g is the dead load and W_L is the lifetime live load.

(a) If $E[W_g] = 19$ kN/m, $E[W_L] = 24$ kN/m, CoV $(W_g) = 0.10$, CoV $(W_L) = 0.25$ and $\rho_{W_g, W_L} = 0.75$. Find the expected value and the standard deviation of MS for the case $R = 1.6\ W_g + 1.9\ W_L$.

Solution:

(a) $R_{ij} = R_{W_g, W_L} = 1.6\ W_g + 1.9\ W_L,$

Variable	E[X]	CoV[X]	$\sigma[X] = E[X] \times \text{CoV}[X]$	$X_+ = E[X] + \sigma[X]$	$X_- = E[X] - \sigma[X]$
W_g	19	0.10	1.9	20.9	17.1
W_L	24	0.25	6	30	18

Here $\rho_{W_g, W_L} = 0.75$

Resistance	R_{ij}	R^2_{ij}
R_{++}	1.6(20.9) + 1.9(30) = 90.44	8179.3936
R_{+-}	1.6(20.9) + 1.9(18) = 67.64	4600.9039
R_{-+}	1.6(17.1) + 1.9(30) = 84.36	7116.6096
R_{--}	1.6(17.1) + 1.9(18) = 61.56	3789.6336

$$P_{++} = P_{--} = (1/4)\ (1 + \rho_{W_g, W_L}) = (1/4)(1 + 0.75) = 0.4375$$
$$P_{+-} = P_{-+} = (1/4)\ (1 - \rho_{W_g, W_L}) = (1/4)(1 - 0.75) = 0.0625$$

Therefore,

$$E[R] = \sum R_{ij} P_{ij} = 0.4375 \times (90.44 + 61.56) + 0.0625(67.83 + 84.36) = 76.0119$$

$$E[R^2] =$$
$$\sum R_{ij}^2 P_{ij} = 0.4375(8179.3936 + 3789.6336) + 0.0625(4600.9089 + 7116.6096) = 5968.7943$$

$$\text{Var}[R] = E[R^2] - (E[R])^2 = (5968.7943) - (76.0119)^2 = 190.9853$$
$$\sigma[R] = \sqrt{\text{Var}[R]} = \sqrt{190.9853} = 13.8197$$

Hence, $\text{CoV}(R) = \dfrac{\sigma[R]}{E[R]} = \dfrac{13.8197}{76.0119} = 0.1818$

The exact solution from mathematical expectation principles,

$$E[R] = 1.6\ E[W_R] + 1.9[W_S] = (1.6 \times 19) + (1.9 \times 24) = 76$$

$$\text{Var}[R] = 1.6^2\ \text{Var}[W_g] + 1.9^2\ \text{Var}[W_L] + 2 \times 1.6 \times 1.9\ \sigma_{W_g}\sigma_{W_L}\rho_{W_gW_L}$$

$$\text{Var}[R] = [1.6^2 \times 9^2] + [1.9^2 \times 6^2] + [2 \times 1.6 \times 1.9 \times 1.9 \times 6 \times 0.75]$$

$$\boxed{\text{Var}[R] = 191.1856}$$

$$\sigma_R = \sqrt{\text{Var}[R]} = \sqrt{191.1856} = 13.8270$$

Hence, $$\text{CoV}[R] = \frac{\sigma_R}{E[R]} = \frac{13.8270}{76.0119} = 0.1819$$

Example 5.2 A simply supported beam, fully restrained laterally along its length, carries uniformly distributed dead and live loads of 3 kN/m and 4 kN/m, respectively, over an effective span of 5 m. In addition, concentrated dead and live loads of 40 kN and 50 kN, respectively are transmitted from secondary beam to the main beam at midpoint. The beam is designed as per IS: 800–2007. The details of the section are ISMB350@524 N/m, $Z = 889.57 \times 10^3$ mm^3, $f_y = 250$ N/mm^2, $M = 201.56 \times 10^6$ N-mm.

The coefficient of variation is as follows: $\text{CoV}_{f_y} = 0.05$, $\text{CoV}_z = 0.05$, $\text{CoV}_M = 0.20$, the variables are uncorrelated. The performance function is $g(x) = f_y\ Z - M$. Find the reliability index, β: (1) when the variables are uncorrelated; and (2) when correlation between the random variables f_y and Z is 0.4.

Solution: Basic variables, mean, standard deviation, upper and lower bound values

Sl. No.	**Variables**	***E*[*x*]**	**CoV[*X*]**	**σ[*X*] = *E*[*X*] × CoV[*X*]**	**Upper bound $X_+ = E[X] + \sigma[X]$**	**Lower bound $X_- = E[X] - \sigma[X]$**
1	$f_y = 250$ N/mm^2	272.4058	0.05	13.6203	286.2610	258.7855
2	$Z = 889570$ mm^3	9.6929×10^5	0.05	4.8465×10^4	1.0178×10^6	9.2083×10^5
3	$M = 201560000$ N-mm	1.5166×10^8	0.20	3.0333×10^7	1.8200×10^8	1.2133×10^8

Case 1: The number of basic variables is three in number, and the samples generated is $2^3 = 8$ (eight).

The variables are uncorrelated, i.e. $\rho_{(fy,\,Z)} = \rho_{(1,\,2)} = 0.0$, $\rho_{(Z,\,M)} = \rho_{(2,\,3)} = 0.0$, $\rho_{(M,fy)} = \rho_{(3,\,1)} = 0.0$

$$P_{(---)} = 1/8 \times [1 + \rho_{(1,\,2)} + \rho_{(2,\,3)} + \rho_{(3,\,1)}]$$
$$P_{(+--)} = 1/8 \times [1 - \rho_{(1,\,2)} + \rho_{(2,\,3)} - \rho_{(3,\,1)}]$$
$$P_{(-+-)} = 1/8 \times [1 - \rho_{(1,\,2)} - \rho_{(2,\,3)} + \rho_{(3,\,1)}]$$
$$P_{(++-)} = 1/8 \times [1 + \rho_{(1,\,2)} - \rho_{(2,\,3)} - \rho_{(3,\,1)}]$$
$$P_{(--+)} = 1/8 \times [1 + \rho_{(1,\,2)} - \rho_{(2,\,3)} - \rho_{(3,\,1)}]$$
$$P_{(+-+)} = 1/8 \times [1 - \rho_{(1,\,2)} - \rho_{(2,\,3)} + \rho_{(3,\,1)}]$$
$$P_{(-++)} = 1/8 \times [1 - \rho_{(1,\,2)} + \rho_{(2,\,3)} - \rho_{(3,\,1)}]$$
$$P_{(+++)} = 1/8 \times [1 + \rho_{(1,\,2)} + \rho_{(2,\,3)} + \rho_{(3,\,1)}]$$

Sl. No.	$g_{(1,2,3)} \times 10^8$ $g(x) = f_y Z - M$	$P_{(1,2,3)}$	$g_{(1,2,3)} \times P_{(1,2,3)}$ $\times 10^7$	$g^2_{(1,2,3)} \times P_{(1,2,3)}$ $\times 10^{15}$
$g_{(---)}$	1.1697	0.1250	1.4621	1.7102
$g_{(+--)}$	1.4205	0.1250	1.7756	2.5223
$g_{(-+-)}$	1.4205	0.1250	1.7756	2.5223
$g_{(++-)}$	1.6977	0.1250	2.1222	3.6029
$g_{(--+)}$	0.5630	0.1250	0.7038	0.3962
$g_{(+-+)}$	0.8139	0.1250	1.0173	0.8279
$g_{(-++)}$	0.8139	0.1250	1.0173	0.8279
$g_{(+++)}$	1.0911	0.1250	1.3639	1.4881
Sum	$= 8.9902 \times 10^8$	**= 1.0000**	$E(g) = 1.1238 \times 10^8$	$E(g^2) = 1.3898 \times 10^{16}$

$$\sigma_g = \sqrt{[g^2(1,2,3) \times P(1,2,3)] - [g(1,2,3) \times P(1,2,3)]^2}$$

$$\sigma_g = \sqrt{1.3898 \times 10^{16} - [1.1238 \times 10^8]^2} = 3.5619 \times 10^9$$

$$\beta = E(g)/\sigma_g = 1.1238 \times 10^8 / 3.5624 \times 10^7 = 3.1550$$

$$P_f = 8.3 \times 10^{-4}$$

Case 2: If the variables are correlated $\rho_{fy,z} = 0.4$ and other variables are uncorrelated, for the performance function $g(x) = f_y Z - M$.

Basic variables, mean, standard deviation, upper and lower bound values

Sl. No.	Variable	$E[x]$	CoV[X]	$\sigma[X] = E[X] \times$ CoV[X]	$X_+ = E[X] + \sigma[X]$	$X_- = E[X] - \sigma[X]$
1	$f_y = 250$ N/mm^2	272.4058	0.05	13.6203	286.2610	258.7855
2	$Z = 889570$ mm^3	9.6929×10^5	0.05	4.8465×10^4	1.0178×10^6	9.2083×10^5
3	$M = 201560000$ N-mm	1.5166×10^8	**0.20**	3.0333×10^7	1.8200×10^8	1.2133×10^8

The number of basic variables is three and the samples generated is $2^3 = 8$.

The variables are uncorrelated, i.e. $\rho_{(fy,Z)} = \rho_{(1,2)} = 0.4$, $\rho_{(Z,M)} = \rho_{(2,3)} = 0.0$, $\rho_{(M,fy)} = \rho_{(3,1)} = 0.0$

$$P_{(---)} = 1/8 \times [1 + \rho_{(1,2)} + \rho_{(2,3)} + \rho_{(3,1)}]$$

$$P_{(+--)} = 1/8 \times [1 - \rho_{(1,2)} + \rho_{(2,3)} - \rho_{(3,1)}]$$

$$P_{(-+-)} = 1/8 \times [1 - \rho_{(1,2)} - \rho_{(2,3)} + \rho_{(3,1)}]$$

$$P_{(++-)} = 1/8 \times [1 + \rho_{(1,2)} - \rho_{(2,3)} - \rho_{(3,1)}]$$

$$P_{(--+)} = 1/8 \times [1 + \rho_{(1,2)} - \rho_{(2,3)} - \rho_{(3,1)}]$$

$$P_{(+-+)} = 1/8 \times [1 - \rho_{(1,2)} - \rho_{(2,3)} + \rho_{(3,1)}]$$
$$P_{(-++)} = 1/8 \times [1 - \rho_{(1,2)} + \rho_{(2,3)} - \rho_{(3,1)}]$$
$$P_{(+++)} = 1/8 \times [1 + \rho_{(1,2)} + \rho_{(2,3)} + \rho_{(3,1)}]$$

Sl. No.	$g_{(1,2,3)}$ $\times 10^8$ $g(x) = f_y Z - M$	$P_{(1,2,3)}$	$g_{(1,2,3)} \times P_{(1,2,3)}$ $\times 10^7$	$g^2_{(1,2,3)} \times P_{(1,2,3)}$ $\times 10^{15}$
$g_{(---)}$	1.1697	0.1750	2.0469	2.3942
$g_{(+--)}$	1.4205	0.0750	1.0654	1.5134
$g_{(-+-)}$	1.4205	0.0750	1.0654	1.5134
$g_{(++-)}$	1.6977	0.1750	2.9711	5.0441
$g_{(--+)}$	0.5630	0.1750	0.9853	0.5547
$g_{(+-+)}$	0.8139	0.0750	0.6104	0.4968
$g_{(-++)}$	0.8139	0.0750	0.6104	0.4968
$g_{(+++)}$	1.0911	0.1750	1.9094	2.0834
Sum	**8.9902 × 10^8**	**1.0000**	$E(g)$ **= 1.1264 × 10^8**	$E(g^2)$ **= 1.4097 × 10^{16}**

$$\sigma_g = \sqrt{[g^2(1,2,3) \times P(1,2,3)] - [g(1,2,3) \times P(1,2,3)]^2}$$

$$\sigma_g = \sqrt{1.4097 \times 10^{16} - [1.1264 \times 10^8]^2} = 3.7539 \times 10^7$$

$$\beta = E(g)/\sigma_g = 1.1264 \times 10^8 / 3.729 \times 10^7 = 3.006$$

Example 5.3 A simply supported beam, fully restrained laterally along its length, carries a uniformly distributed load of 45 kN/m, over an effective span of 5 m. The beam is designed as per IS: 800–2007 for limit state of strength $g(x) = f_y Z - (WL^2/8)$. The details of the section are ISMB350@524 N/m, $Z = 889.57 \times 10^3$ mm^3, $f_y = 250$ N/mm^2, $M = 201.56 \times 10^6$ N/mm.

The coefficient of variation is as follows: $\text{CoV}_{fy} = 0.05$, $\text{CoV}_Z = 0.05$, $\text{CoV}_w = 0.20$, $\text{CoV}_L = 0.05$ and the variables are uncorrelated. Find the reliability index, β.

Solution: The performance function $g(x) = f_y Z - WL^2/8$

The variables are uncorrelated, i.e. $\rho_{(1,2)} = 0.4$; $\rho_{(2,3)} = 0.0$; $\rho_{(3,4)} = 0.0$; $\rho_{(4,1)} = 0$

The number of basic variables $N = 4$, the sample size is $2^4 = 16$

$P(1) = [1/16] \times [1 + \rho_{(1,2)} + \rho_{(2,3)} + \rho_{(3,4)} + \rho_{(4,1)}]$

Table showing the basic variables, mean, standard deviation, upper and lower bound values.

Sl. No.	Variable	$E[x]$	CoV[X]	$\sigma[X] = E[X] \times$ CoV[X]	$X_+ = E[X] + \sigma[X]$	$X_- = E[X] - \sigma[X]$
1	$f_y = 250$ N/mm^2	272.4058	0.05	13.6203	286.0256	258.7851
2	$Z = 889570$ mm^3	9.6929×10^5	0.05	4.8465×10^4	1.0178×10^6	9.2083×10^5
3	$M = 45$ N/mm	33.8600	**0.20**	6.7720	27.0880	40.6321
4	$L = 5000$ mm	5.4481×10^3	0.05	272.4053	5.7205×10^3	5.1757×10^3

Sl. No.	$g_{(1,2,3)} \times 10^8$ $g(x) = f_y Z - (WL^2/8)$	$P_{(1,2,3)}$	$g_{(1,2,3)} \times P_{(1,2,3)} \times 10^7$	$g^2_{(1,2,3)} \times P_{(1,2,3)} \times 10^{15}$
$g_{(++++)}$	1.2490	0.0625	0.7806	0.9750
$g_{(-+++)}$	0.9717	0.0625	0.6073	0.5902
$g_{(+-++)}$	0.9717	0.0625	0.6073	0.5902
$g_{(--++)}$	0.7209	0.0625	0.4506	0.3248
$g_{(++-+)}$	1.8030	0.0625	1.1269	2.0318
$g_{(-+-+)}$	1.5258	0.0625	0.9536	1.4550
$g_{(+--+)}$	1.5258	0.0625	0.9536	1.4550
$g_{(---+)}$	1.2749	0.0625	0.7968	1.0159
$g_{(+++-)}$	1.5505	0.0625	0.9691	1.5025
$g_{(-++-)}$	1.2733	0.0625	0.7958	1.0132
$g_{(+-+-)}$	1.2733	0.0625	0.7958	1.0132
$g_{(--+-)}$	1.0224	0.0625	0.6390	0.6533
$g_{(++--)}$	2.0040	0.0625	1.2525	2.5100
$g_{(-+--)}$	1.7268	0.0625	1.0792	1.8636
$g_{(+---)}$	1.7268	0.0625	1.0792	1.8636
$g_{(----)}$	1.4759	0.0625	0.9225	1.3615
Sum	**2.2096×10^9**	**$P = 1.0000$**	**$E(g) = 1.3810 \times 10^8$**	**$E(g^2) = 2.0219 \times 10^{16}$**

$$\sigma_g = \sqrt{[g^2_{(1,2,3)} \times P_{(1,2,3)}] - [g_{(1,2,3)} \times P_{(1,2,3)}]^2}$$

$$\sigma_g = \sqrt{[2.0219 \times 10^{16}] - [1.3810 \times 10^8]^2} = 3.3873 \times 10^7$$

$$\beta = E(g)/\sigma_g = 1.3810 \times 10^8 / 3.729 \times 10^7 = 4.0769$$

Example 5.4 Determine the reliability index of a simply supported beam shown in Fig. 5.5, at the limit state of flexure. The failure surface is given by $g(x) = f_y Z - (W_g + W_L) L^2/8$. The characteristic values for the beam section ISMB 350 @ 52.4 kg/m are given in the Table below.

Sl. No.	Characteristic value	CoV
1	$f_y = 250$ N/mm^2	0.05
2	$Z = 778.9 \times 10^3$ mm^3	0.05
3	$W_g = 0.524$ N/mm	0.10
4	$W_L = 20$ N/mm	0.15
5	$L = 6500$ mm	0.05

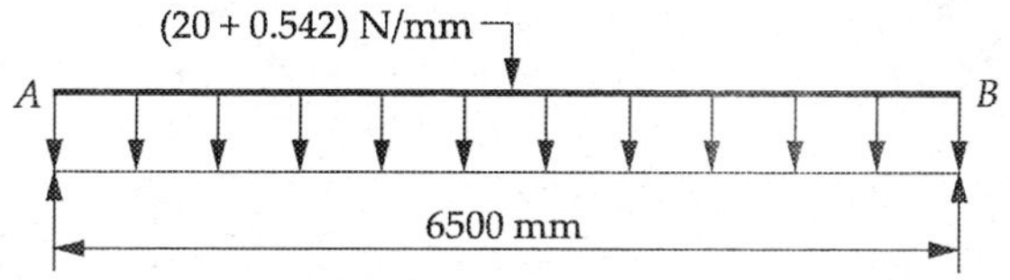

Fig. 5.5 Simply supported steel-beam

Solution: The number of basic variables is 3 and the samples generated is $2^5 = 32$. All the basic variables are uncorrelated.

$P_{(+++++)} = P_{(-----)} = 1/32$ and all correlation coefficients are $\rho = 0.0$

The performance function $MS = f_y Z - (W_g + W_L) \dfrac{L^2}{8}$

Sl. No.	Characteristic Value	CoV	Mean Value (μ)	S.D (σ)	Upper bound X_+	Lower bound X_+
1	$f_y = 250$ N/mm^2	0.05	2.7240×10^2	1.3620×10^1	2.8602×10^2	2.5878×10^2
2	$Z = 778.9 \times 10^3$ mm^3	0.05	8.4860×10^5	4.2430×10^4	8.9103×10^5	8.0617×10^5
3	$W_g = 0.524$ N/mm	0.10	0.0449×10^1	0.0044×10^1	0.4998×10^1	0.0404×10^1
4	$W_L = 20$ N/mm	0.15	1.6041×10^1	0.2406×10^1	1.8448×10^1	1.3635×10^1
5	$L = 6500$ mm	0.05	7.0825×10^3	3.5413×10^2	7.4366×10^3	6.7284×10^3

$$MS_1 = f_y{}^{+} Z^{+} - \frac{(W_g^+ + W_L^+)}{8}(L^+)^2$$

$$= 286.025 \times 891.026 \times 10^3 - \frac{(0.4998 + 18.448)}{8} \times 7436.63^2$$

$$= 123.87 \times 10^6 \text{ N-mm}$$

$$MS_2 = f_y^- Z^+ - \frac{(W_g^+ + W_L^+)}{8}(L^+)^2$$

$$= 258.78 \times 891.026 \times 10^3 - \frac{(0.4998 + 18.448)}{8} \times 7436.63^2$$

$$= 99.599 \times 10^6 \text{ N-mm}$$

$$MS_3 = f_y^+ Z^- - \frac{(W_g^+ + W_L^+)}{8}(L^+)^2$$

$$= 286.025 \times 806.167 \times 10^3 - \frac{(0.4998 + 18.448)}{8} \times 7436.63^2$$

$$= 99.599 \times 10^6 \text{ N-mm}$$

$$MS_4 = f_y^- Z^- - \frac{(W_g^+ + W_L^+)}{8}(L^+)^2$$

$$= 258.78 \times 806.167 \times 10^3 - \frac{(0.4998 + 18.448)}{8} \times 7436.63^2$$

$$= 77.635 \times 10^6 \text{ N-mm}$$

$$MS_5 = f_y^+ Z^+ - \frac{(W_g^- + W_L^+)}{8}(L^+)^2$$

$$= 286.025 \times 891.026 \times 10^3 - \frac{(0.4 + 18.448)}{8} \times 7436.63^2$$

$$= 124.56 \times 10^6 \text{ N-mm}$$

$$MS_6 = f_y^- Z^+ - \frac{(W_g^- + W_L^+)}{8}(L^+)^2$$

$$= 258.78 \times 891.026 \times 10^3 - \frac{(0.4 + 18.448)}{8} \times 7436.63^2$$

$$= 100.284 \times 10^6 \text{ N-mm}$$

$$MS_7 = f_y^+ Z^- - \frac{(W_g^- + W_L^+)}{8}(L^+)^2$$

$$= 286.025 \times 806.167 \times 10^3 - \frac{(0.0404 + 18.448)}{8} \times 7436.63^2$$

$$= 100.288 \times 10^6 \text{ N-mm}$$

$$MS_8 = f_y^- Z^- - \frac{(W_g^- + W_L^+)}{8}(L^+)^2$$

$$= 258.78 \times 806.167 \times 10^3 - \frac{(0.4 + 18.448)}{8} \times 7436.63^2$$

$$= 78.325 \times 10^6 \text{ N-mm}$$

$$MS_9 = f_y^+ Z^+ - \frac{(W_g^+ + W_L^-)}{8}(L^+)^2$$

$$= 286.025 \times 891.026 \times 10^3 - \frac{(0.4998 + 13.635)}{8} \times 7436.63^2$$

$$= 157.143 \times 10^6 \text{ N-mm}$$

$$MS_{10} = f_y^- Z^+ - \frac{(W_g^+ + W_L^-)}{8} (L^+)^2$$

$$= 258.78 \times 891.026 \times 10^3 - \frac{(0.4998 + 13.635)}{8} \times 7436.63^2$$

$$= 132.866 \times 10^6 \text{ N-mm}$$

$$MS_{11} = f_y{}^+ Z^- - \frac{(W_g^+ + W_L^-)}{8} \times (L^+)^2$$

$$= 286.025 \times 806.167 \times 10^3 - \frac{(0.4998 + 13.635)}{8} \times 7436.63^2$$

$$= 132.018 \times 10^6 \text{ N-mm}$$

$$MS_{12} = f_y^- Z^- - \frac{(W_g^+ + W_L^-)}{8} (L^+)^2$$

$$= 258.78 \times 806.167 \times 10^3 - \frac{(0.4998 + 13.635)}{8} \times 7436.63^2$$

$$= 110.906 \times 10^6 \text{ N-mm}$$

$$MS_{13} = f_y^+ Z^+ - \frac{(W_g^- + W_L^-)}{8} (L^+)^2$$

$$= 286.025 \times 891.026 \times 10^3 - \frac{(0.4 + 13.635)}{8} \times 7436.63^2$$

$$= 157.832 \times 10^6 \text{ N-mm}$$

$$MS_{14} = f_y^- Z^+ - \frac{(W_g^- + W_L^-)}{8} (L^+)^2$$

$$= 258.78 \times 891.026 \times 10^3 - \frac{(0.4 + 13.635)}{8} \times 7436.63^2$$

$$= 133.550 \times 10^6 \text{ N-mm}$$

$$MS_{15} = f_y^+ Z^- - \frac{(W_g^- + W_L^-)}{8} (L^+)^2$$

$$= 286.025 \times 806.167 \times 10^3 - \frac{(0.4 + 13.635)}{8} \times 7436.63^2$$

$$= 133.56 \times 10^6 \text{ N-mm}$$

$$MS_{16} = f_y^- Z^- - \frac{(W_g^- + W_L^-)}{8} (L^+)^2$$

$$= 258.78 \times 806.167 \times 10^3 - \frac{(0.4 + 13.635)}{8} \times 7436.63^2$$

$$= 111.596 \times 10^6 \text{ N-mm}$$

$$MS_{17} = f_y^+ Z^+ - \frac{(W_g^+ + W_L^+)}{8} (L^-)^2$$

$$= 286.025 \times 891.026 \times 10^3 - \frac{(0.4998 + 18.448)}{8} \times 6728.37^2$$

$$= 147.632 \times 10^6 \text{ N-mm}$$

$$MS_{18} = f_y^- Z^+ - \frac{(W_g^+ + W_L^+)}{8} (L^-)^2$$

$$= 258.78 \times 891.026 \times 10^3 - \frac{(0.4998 + 18.448)}{8} \times 6728.37^2$$

$$= 123.356 \times 10^6 \text{ N-mm}$$

$$MS_{19} = f_y^+ Z^- - \frac{(W_g^+ + W_L^+)}{8} \times (L^-)^2$$

$$= 286.025 \times 806.167 \times 10^3 - \frac{(0.4998 + 18.448)}{8} \times 6728.37^2$$

$$= 123.36 \times 10^6 \text{ N-mm}$$

$$MS_{20} = f_y^- Z^- - \frac{(W_g^+ + W_L^+)}{8} (L^-)^2$$

$$= 258.78 \times 806.167 \times 10^3 - \frac{(0.4998 + 18.448)}{8} \times 6728.37^2$$

$$= 101.396 \times 10^6 \text{ N-mm}$$

$$MS_{21} = f_y^{+} Z^{+} - \frac{(W_g^{-} + W_L^{+})}{8} (L^{-})^2$$

$$= 286.025 \times 891.026 \times 10^3 - \frac{(0.4 + 18.448)}{8} \times 6728.37^2$$

$$= 148.197 \times 10^6 \text{ N-mm}$$

$$MS_{22} = f_y^{-} Z^{+} - \frac{(W_g^{-} + W_L^{+})}{8} (L^{-})^2$$

$$= 258.78 \times 891.026 \times 10^3 - \frac{(0.4 + 18.448)}{8} \times 6728.37^2$$

$$= 123.921 \times 10^6 \text{ N-mm}$$

$$MS_{23} = f_y^{+} Z^{-} - \frac{(W_g^{-} + W_L^{+})}{8} (L^{-})^2$$

$$= 286.025 \times 806.167 \times 10^3 - \frac{(0.4 + 18.448)}{8} \times 6728.37^2$$

$$= 123.925 \times 10^6 \text{ N-mm}$$

$$MS_{24} = f_y^{-} Z^{-} - \frac{(W_g^{-} + W_L^{+})}{8} (L^{-})^2$$

$$= 258.78 \times 806.167 \times 10^3 - \frac{(0.4 + 18.448)}{8} \times 6728.37^2$$

$$= 101.961 \times 10^6 \text{ N-mm}$$

$$MS_{25} = f_y^{+} Z^{+} - \frac{(W_g^{+} + W_L^{-})}{8} (L^{-})^2$$

$$= 286.025 \times 891.026 \times 10^3 - \frac{(0.4998 + 13.635)}{8} \times 6728.37^2$$

$$= 174.868 \times 10^6 \text{ N-mm}$$

$$MS_{26} = f_y^{-} Z^{+} - \frac{(W_g^{+} + W_L^{-})}{8} (L^{-})^2$$

$$= 258.78 \times 891.026 \times 10^3 - \frac{(0.4998 + 13.635)}{8} \times 6728.37^2$$

$$= 150.592 \times 10^6 \text{ N-mm}$$

$$MS_{27} = f_y^{+} Z^{-} - \frac{(W_g^{+} + W_L^{-})}{8} (L^{-})^2$$

$$= 286.025 \times 806.167 \times 10^3 - \frac{(0.4998 + 13.635)}{8} \times 6728.37^2$$

$$= 150.596 \times 10^6 \text{ N-mm}$$

$$MS_{28} = f_y^{-} Z^{-} - \frac{(W_g^{+} + W_L^{-})}{8} (L^{-})^2$$

$$= 258.78 \times 806.167 \times 10^3 - \frac{(0.4998 + 13.635)}{8} \times 6728.37^2$$

$$= 128.632 \times 10^6 \text{ N-mm}$$

$$MS_{29} = f_y^{+} Z^{+} - \frac{(W_g^{-} + W_L^{-})}{8} (L^{-})^2$$

$$= 286.025 \times 891.026 \times 10^3 - \frac{(0.4 + 13.635)}{8} \times 6728.37^2$$

$$= 175.433 \times 10^6 \text{ N-mm}$$

$$MS_{30} = f_y^{-} Z^{+} - \frac{(W_g^{-} + W_L^{-})}{8} (L^{-})^2$$

$$= 258.78 \times 891.02 \times 10^3 - \frac{(0.4 + 13.635)}{8} \times 6728.37^2$$

$$= 151.157 \times 10^6 \text{ N-mm}$$

$$MS_{31} = f_y^{+} Z^{-} - \frac{(W_g^{-} + W_L^{-})}{8} (L^{-})^2$$

$$= 286.025 \times 806.167 \times 10^3 - \frac{(0.4 + 13.635)}{8} \times 6728.37^2$$

$$= 151.161 \times 10^6 \text{ N-mm}$$

$$MS_{32} = f_y^{-} Z^{-} - \frac{(W_g^{-} + W_L^{-})}{8} (L^{-})^2$$

$$= 258.78 \times 806.167 \times 10^3 - \frac{(0.4 + 13.635)}{8} \times 6728.37^2$$

$$= 129.197 \times 10^6 \text{ N-mm}$$

Note: The number of basic variables can be reduced to 4 by combining W_g and W_L to W resulting in $g(x) = f_y Z - (WL^2)/8$.

Therefore,

$$E[R] = \sum R_{ij} P_{ij} = 127.755 \times 10^6 \text{N-mm}$$

$$E[R^2] = \sum R_{ij}^2 P_{ij} = 16889.0 \times 10^{12} \text{N-mm}$$

$$\text{Var}[R] = E[R^2] - (E[R])^2 = [(16889.0 - 127.755 \times 127.755) \times 10^{12}] = 567.6599 \times 10^{12}$$

$$\sigma_{MS} = 23.8256 \times 10^6 \text{ N-mm}$$

Reliability index

$$\beta = \frac{E[R]}{\sigma_{MS}} = \frac{127.755 \times 10^6}{23.8256} = 5.3620$$

Example 5.5 Determine the reliability index of a simply supported beam shown in Fig. 5.6, at the limit state of flexure. The failure surface is given by $g(x) = R–S$ and the margin of safety is given by MS and the resistance and loads are evaluated individually to compute the reliability index of the RCC beam. $MS = \left[f_y A_{st} d - \dfrac{0.77 f_y^2 A_{st}^2}{b f_{ck}} \right] - WL^2/8$, the characteristic values for the beam section, are given in the Fig. 5.6.

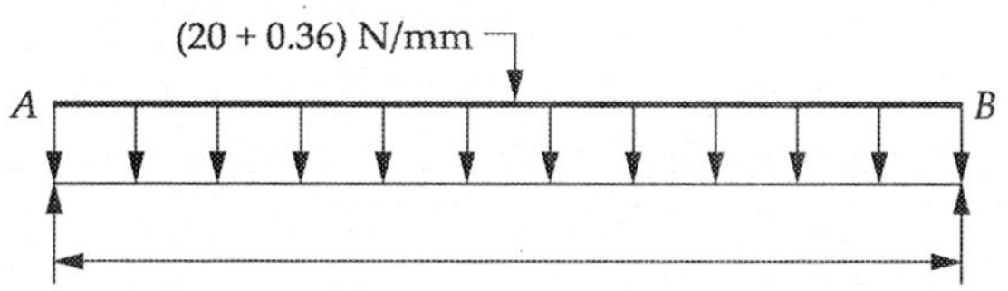

Fig. 5.6 Simply supported RCC-beam

Solution: The number of basic variables are 7 in number, and the samples generated is $2^7 = 128$. All the basic variables are uncorrelated. Hence, the resistance has 5 basic variables, and load has 2 basic variables, therefore the resistance has 32 samples and load has 4 samples. To circumvent the above difficulty resistance and loads are computed separately, to simplify the problem.

$P_{(+++++)} = P_{(-----)} = 1/32$ and all the correlation coefficients are $\rho = 0.0$.

The probability of each sample is 1/32 for resistance and ¼ for load variables.

Characteristic value	Coefficient of variation CoV	Mean value μ	Standard deviation σ	Upper bound X_+	Lower bound X_-
$f_y = 415$ N/mm^2	5%	452.1920	22.6090	474.8010	429.5830
$A_{st} = 1162$ mm^2	5%	1266.1400	63.3070	1329.4470	1202.8330
$d = 485$ mm	5%	528.4700	26.4000	554.8700	502.0700
$b = 250$ mm	5%	272.4030	13.6200	286.0230	258.7830
$f_{ck} = 20$ N/mm^2	5%	21.7920	1.0890	22.8810	20.7030
$W = 20.36$ N/mm	25%	14.4300	3.6070	18.0370	10.8230
$L = 6500$ mm	5%	7082.5388	354.1269	7436.6657	6728.4119

Resistance, $R = \left[f_y A_{st} d - \dfrac{0.77 f_y^2 A_{st}^2}{b f_{ck}} \right]$; Load, $S = WL^2/8$

The variables have the following indexes, namely $\begin{cases} f_{ck} = 1 \\ b = 2 \\ d = 3 \\ A_{st} = 4 \\ f_y = 5 \end{cases}$

Resistance:

$$R_1 = \left[f_y^+ A_{st}^+ d^+ - \frac{0.77 f_y^{2+} A_{st}^{2+}}{b^+ f_{ck}^+} \right] = 303.367 \times 10^6 \text{ N-mm}$$

$$R_2 = \left[f_y^+ A_{st}^+ d^+ - \frac{0.77 f_y^{2+} A_{st}^{2+}}{b^+ f_{ck}^-} \right] = 298.435 \times 10^6 \text{ N-mm}$$

$$R_3 = \left[f_y^+ A_{st}^+ d^+ - \frac{0.77 f_y^{2+} A_{st}^{2+}}{b^- f_{ck}^+} \right] = 298.433 \times 10^6 \text{ N-mm}$$

$$R_4 = \left[f_y^+ A_{st}^+ d^+ - \frac{0.77 f_y^{2+} A_{st}^{2+}}{b^- f_{ck}^-} \right] = 292.982 \times 10^6 \text{ N-mm}$$

$$R_5 = \left[f_y^+ A_{st}^+ d^- - \frac{0.77 f_y^{2+} A_{st}^{2+}}{b^+ f_{ck}^+} \right] = 270.038 \times 10^6 \text{ N-mm}$$

$$R_6 = \left[f_y^+ A_{st}^+ d^- - \frac{0.77 f_y^{2+} A_{st}^{2+}}{b^+ f_{ck}^-} \right] = 265.107 \times 10^6 \text{ N-mm}$$

$$R_7 = \left[f_y^+ A_{st}^+ d^- - \frac{0.77 f_y^{2+} A_{st}^{2+}}{b^- f_{ck}^+} \right] = 265.104 \times 10^6 \text{ N-mm}$$

$$R_8 = \left[f_y^+ A_{st}^+ d^- - \frac{0.77 f_y^{2+} A_{st}^{2+}}{b^- f_{ck}^-} \right] = 259.653 \times 10^6 \text{ N-mm}$$

$$R_9 = \left[f_y^+ A_{st}^- d^+ - \frac{0.77 f_y^{2+} A_{st}^{2-}}{b^+ f_{ck}^+} \right] = 278.515 \times 10^6 \text{ N-mm}$$

$$R_{10} = \left[f_y^+ A_{st}^- d^+ - \frac{0.77 f_y^{2+} A_{st}^{2-}}{b^+ f_{ck}^-} \right] = 274.478 \times 10^6 \text{ N-mm}$$

$$R_{11} = \left[f_y^+ A_{st}^- d^+ - \frac{0.77 f_y^{2+} A_{st}^{2-}}{b^- f_{ck}^+} \right] = 274.475 \times 10^6 \text{ N-mm}$$

$$R_{12} = \left[f_y^+ A_{st}^- d^+ - \frac{0.77 f_y^{2+} A_{st}^{2-}}{b^- f_{ck}^-} \right] = 270.013 \times 10^6 \text{ N-mm}$$

$$R_{13} = \left[f_y^+ A_{st}^- d^- - \frac{0.77 f_y^{2+} A_{st}^{2-}}{b^+ f_{ck}^+} \right] = 248.360 \times 10^6 \text{ N-mm}$$

$$R_{14} = \left[f_y^+ A_{st}^- d^- - \frac{0.77 f_y^{2+} A_{st}^{2-}}{b^+ f_{ck}^-} \right] = 244.323 \times 10^6 \text{ N-mm}$$

$$R_{15} = \left[f_y^+ A_{st}^- d^- - \frac{0.77 f_y^{2+} A_{st}^{2-}}{b^- f_{ck}^+} \right] = 244.321 \times 10^6 \text{ N-mm}$$

$$R_{16} = \left[f_y^+ A_{st}^- d^- - \frac{0.77 f_y^{2+} A_{st}^{2-}}{b^- f_{ck}^-} \right] = 239.859 \times 10^6 \text{ N-mm}$$

$$R_{17} = \left[f_y^- A_{st}^+ d^+ - \frac{0.77 f_y^{2-} A_{st}^{2+}}{b^+ f_{ck}^+} \right] = 278.515 \times 10^6 \text{ N-mm}$$

$$R_{18} = \left[f_y^- A_{st}^+ d^+ - \frac{0.77 f_y^{2-} A_{st}^{2+}}{b^+ f_{ck}^-} \right] = 274.478 \times 10^6 \text{ N-mm}$$

$$R_{19} = \left[f_y^- A_{st}^+ d^+ - \frac{0.77 f_y^{2-} A_{st}^{2+}}{b^- f_{ck}^+} \right] = 274.476 \times 10^6 \text{ N-mm}$$

$$R_{20} = \left[f_y^- A_{st}^+ d^+ - \frac{0.77 f_y^{2-} A_{st}^{2+}}{b^- f_{ck}^-} \right] = 270.014 \times 10^6 \text{ N-mm}$$

$$R_{21} = \left[f_y^- A_{st}^+ d^- - \frac{0.77 f_y^{2-} A_{st}^{2+}}{b^+ f_{ck}^+} \right] = 248.361 \times 10^6 \text{ N-mm}$$

$$R_{22} = \left[f_y^- A_{st}^+ d^- - \frac{0.77 f_y^{2-} A_{st}^{2+}}{b^+ f_{ck}^-} \right] = 244.324 \times 10^6 \text{ N-mm}$$

$$R_{23} = \left[f_y^- A_{st}^+ d^- - \frac{0.77 f_y^{2-} A_{st}^{2+}}{b^- f_{ck}^+} \right] = 244.321 \times 10^6 \text{ N-mm}$$

$$R_{24} = \left[f_y^- A_{st}^+ d^- - \frac{0.77 f_y^{2-} A_{st}^{2+}}{b^- f_{ck}^-} \right] = 239.859 \times 10^6 \text{ N-mm}$$

$$R_{25} = \left[f_y^- A_{st}^- d^+ - \frac{0.77 f_y^{2-} A_{st}^{2-}}{b^+ f_{ck}^+} \right] = 255.297 \times 10^6 \text{ N-mm}$$

$$R_{26} = \left[f_y^- A_{st}^- d^+ - \frac{0.77 f_y^{2-} A_{st}^{2-}}{b^+ f_{ck}^-} \right] = 251.992 \times 10^6 \text{ N-mm}$$

$$R_{27} = \left[f_y^- A_{st}^- d^+ - \frac{0.77 f_y^{2-} A_{st}^{2-}}{b^- f_{ck}^+} \right] = 251.990 \times 10^6 \text{ N-mm}$$

$$R_{28} = \left[f_y^- A_{st}^- d^+ - \frac{0.77 f_y^{2-} A_{st}^{2-}}{b^- f_{ck}^-} \right] = 248.337 \times 10^6 \text{ N-mm}$$

$$R_{29} = \left[f_y^- A_{st}^- d^- - \frac{0.77 f_y^{2-} A_{st}^{2-}}{b^+ f_{ck}^+} \right] = 228.014 \times 10^6 \text{ N-mm}$$

$$R_{30} = \left[f_y^- A_{st}^- d^- - \frac{0.77 f_y^{2-} A_{st}^{2-}}{b^+ f_{ck}^-} \right] = 224.709 \times 10^6 \text{ N-mm}$$

$$R_{31} = \left[f_y^- A_{st}^- d^- - \frac{0.77 f_y^{2-} A_{st}^{2-}}{b^- f_{ck}^+} \right] = 224.707 \times 10^6 \text{ N-mm}$$

$$R_{32} = \left[f_y^- A_{st}^- d^- - \frac{0.77 f_y^{2-} A_{st}^{2-}}{b^- f_{ck}^-} \right] = 221.055 \times 10^6 \text{ N-mm}$$

$$E(R) = 259.622 \times 10^6 \text{ N-mm}$$
$$E(R) = 67878 \times 10^{12} \text{ N-mm}$$
$$\text{Var} = E(R^2) - E(R)^2 = (67878 - 259.622 \times 259.622) \times 10^{12} = 474.330 \times 10^{12}$$
$$\sigma_R = 21.779 \times 10^6 \text{ N-mm}$$

Load:

$$S_1 = \frac{W^+ l^{+2}}{8} = 124.6898 \times 10^6 \text{ N-mm}$$

$$S_2 = \frac{W^+ l^{-2}}{8} = 102.0703 \times 10^6 \text{ N-mm}$$

$$S_3 = \frac{W^- l^{+2}}{8} = 74.8193 \times 10^6 \text{ N-mm}$$

$$S_4 = \frac{W^- l^{-2}}{8} = 61.2467 \times 10^6 \text{ N-mm}$$

$$E(S) = \frac{WL^2}{8} = \frac{14.43 \times 7082.5388^2}{8} = 90.4803 \times 10^6$$

$$E(S) = 8.825 \times 10^{15} = \text{Var}(S) = E(S^2) - E(S)^2 = (8825 - 90.68^2) \times 10^{12} = 600.95 \times 10^{12}$$
$$\sigma_S = 24.51 \times 10^6 \text{ N-mm}$$

Reliability index

$$\beta = \frac{\mu_R - \mu_S}{\sqrt{\sigma_R^2 + \sigma_S^2}}$$

$$= \frac{259.622 \times 10^6 - 90.68 \times 10^6}{\sqrt{(21.779 \times 10^6)^2 + (24.5100 \times 10^6)^2}} = 5.15$$

Comments: It can be seen that the number of basic variables in a given problem is more than 3, the number of unknown exceeding the number of available conditions (first moment, variance, skewness). The problem determining the coordinates and the corresponding probability masses for more than 3 is indeterminate and *the solution is not unique.*

Table 5.1 Generator of + and – for random variables

Sl. No.	No.	Basic variables				
	2^n	1	2	3	4	5
1		+	+	+	+	+
2	$n = 1, 2^1 = 2$	–	+	+	+	+
3		+	–	+	+	+
4	$n = 2, 2^2 = 4$	–	–	+	+	+
5		+	+	–	+	+
6		–	+	–	+	+
7		+	–	–	+	+
8	$n = 3, 2^3 = 8$	–	–	–	+	+
9		+	+	+	–	+
10		–	+	+	–	+
11		+	–	+	–	+
12		–	–	+	–	+
13		+	+	–	–	+
14		–	+	–	–	+
15		+	–	–	–	+
16	$n = 4, 2^4 = 16$	–	–	–	–	+
17		+	+	+	+	–
18		–	+	+	+	–
19		+	–	+	+	–
20		–	–	+	+	–
21		+	+	–	+	–
22		–	+	–	+	–
23		+	–	–	+	–
24		–	–	–	+	–
25		+	+	+	–	–
26		–	+	+	–	–
27		+	–	+	–	–
28		–	–	+	–	–
29		+	+	–	–	–
30		–	+	–	–	–
31		+	–	–	–	–
32	$n = 5, 2^5 = 32$	–	–	–	–	–

Gaining Information from Experimental Data

In practice, rarely one can perform prototype test on a civil engineering system prior to construction. In fact, full-scale test is costly and rarely conducted. Under these circumstances, generalization must be made from the limited observations of experiments conducted on scale down models. The knowledge base represents only a small sample of large population about which we wish to discover the truth.

How to extract information from experiments and this domain falls into statistics of sampling theory?

In this section an analytical procedure is employed using simple matrix operations to obtain variance, covariance, coefficient of variation and correlation coefficients for a number of random variables.

Example 5.6 Thirty-six samples for three groups of concrete cube of size 150 mm × 150 mm, cylinder size 150 mm × 300 mm, and cube size of 100 mm × 100 mm were tested in the laboratory. Find the variances, covariance, coefficient of variation and correlation coefficients for a number of random variables.

The experimental values of 36 samples tested in the laboratory are tabulated in Table 5.2, in column 1, 2, 3 for cube 100 mm, cylinder 150 mm, and cube 100 mm respectively.

Table 5.2 Experimental results of concrete strength in N/mm^2

Sl. No.	Cube 150 mm (1)	Cylinder 150 mm (2)	Cube 100 mm (3)	Cube 150 – mean 150 (4) = (1) – mean (1)	Cylinder 150 –mean 150 (5) = (2) – mean (2)	Cube 100 –mean 100 (6) = (3) – mean (3)
1	44.7	39.59	49.70	2.4531	3.9878	2.0944
2	50.58	41.81	58.86	8.3331	6.2078	11.2544
3	48.25	40.70	53.63	6.0031	5.0978	6.0244
4	43.75	37.00	51.01	1.5031	1.3978	3.4044
5	42.73	35.52	47.74	0.4831	–0.0822	0.1344
6	47.59	38.78	54.90	5.3431	3.1778	7.2944
7	44.57	36.45	48.51	2.3231	0.8478	0.9044
8	38.66	33.67	47.09	–3.5869	–1.9322	–0.5156
9	41.86	34.41	46.43	–0.3869	–1.1922	–1.1756
10	45.05	37.74	50.36	2.8031	2.1378	2.7544
11	42.44	35.52	48.40	0.1931	–0.0822	0.7944
12	37.79	31.08	44.15	–4.4569	–4.5222	–3.4556
13	43.02	35.15	51.01	0.7731	–0.4522	3.4044
14	45.93	39.22	58.21	3.6831	3.6178	10.6044
15	44.76	37.37	53.18	2.5131	1.7678	5.5744
16	42.15	33.67	48.05	–0.0969	–1.9322	0.4444

Contd...

17	42.73	35.89	46.28	0.4831	0.2878	–1.3256
18	47.09	41.44	50.45	4.8431	5.8378	2.8444
19	43.89	39.22	48.75	1.6431	3.6178	1.1444
20	40.11	34.40	44.45	–2.1369	–1.2022	–3.1556
21	41.27	34.04	43.45	–0.9769	–1.5622	–4.1556
22	43.60	38.08	47.78	1.3531	2.4778	0.1744
23	40.98	35.58	45.19	–1.2669	–0.0222	–2.4156
24	37.21	32.63	41.25	–5.0369	–2.9722	–6.3556
25	40.40	34.41	45.13	–1.8469	–1.1922	–2.4756
26	46.22	38.50	50.36	3.9731	2.8978	2.7544
27	43.60	35.97	47.74	1.3531	0.3678	0.1344
28	37.50	32.99	43.82	–4.7469	–2.6122	–3.7856
29	40.11	32.19	45.78	–2.1369	–3.4122	–1.8256
30	41.27	36.02	48.40	–0.9769	0.4178	0.7944
31	40.69	33.65	45.78	–1.5569	–1.9522	–1.8256
32	37.79	30.86	39.24	–4.4569	–4.7422	–8.3656
33	36.62	31.08	39.89	–5.6269	–4.5222	–7.7156
34	41.57	34.77	45.78	–0.6769	–0.8322	–1.8256
35	38.66	32.49	43.16	–3.5869	–3.1122	–4.4456
36	35.75	29.79	39.89	–6.4969	–5.8122	–7.7156
Mean	42.2469	35.6022	47.6056			
Standard deviation	**3.4989**	**3.0667**	**3.4989**			

Matrix A is formed by subtracting the mean or expected values from column 1, 2, and 3 of the cube 100, cylinder 150, cube 100 respectively is tabulated in column 4, 5, and 6. The A matrix size is (36 × 1) and A^T size is (1 × 36). Performing the matrix multiplication of A^T and A and divide by 35 ($N - 1 = 36 - 1 = 35$), we obtain the covariance matrix.

$$C_{XX} = (1/35) \times A^T \times A \qquad \text{...(5.16)}$$

The general covariance matrix is given below for three variables.

$$C_{XX} = \begin{bmatrix} \sigma_1^2 & \rho_{12}\sigma_1\sigma_2 & \rho_{13}\sigma_1\sigma_3 \\ \rho_{21}\sigma_2\sigma_1 & \sigma_2^2 & \rho_{23}\sigma_2\sigma_3 \\ \rho_{31}\sigma_3\sigma_1 & \rho_{32}\sigma_3\sigma_2 & \sigma_3^2 \end{bmatrix} \qquad \text{...(5.17)}$$

$$C_{XX} = \begin{bmatrix} 12.2422 & 10.1046 & 14.6100 \\ 10.1046 & 9.4044 & 12.0879 \\ 14.6100 & 12.0879 & 21.2069 \end{bmatrix}$$

Obtaining the square root of the diagonal elements in the covariance matrix produces the corresponding standard deviations.

$\sigma_1^2 = 12.2422; \sigma_2^2 = 9.04044; \sigma_3^2 = 21.2069;$

$\sigma_1 = 3.4989; \sigma_2 = 3.0667; \sigma_3 = 4.6051$

The corresponding matrix with standard deviation along the principal diagonal and correlation coefficients for the off-diagonal elements is shown in matrix S.

$$\rho_{12} = \rho_{21} = \frac{\text{COV}(1,2)}{\sigma_1 \sigma_2} = \frac{10.1046}{3.4989 \times 3.0667} = 0.9417;$$

$$\rho_{13} = \rho_{31} = \frac{\text{COV}(1,3)}{\sigma_1 \sigma_3} = \frac{14.6100}{3.4989 \times 4.6051} = 0.9067;$$

$$\rho_{23} = \rho_{32} = \frac{\text{COV}(2,3)}{\sigma_3 \sigma_2} = \frac{12.0879}{3.0667 \times 4.6051} = 0.8559$$

$$S = \begin{bmatrix} 3.4989 & 0.9417 & 0.9067 \\ 0.9417 & 3.0667 & 0.8559 \\ 0.9067 & 0.8559 & 4.6051 \end{bmatrix} \qquad \ldots(5.18)$$

Another representation commonly used in the analysis is correlation matrix K, which substitute's unity for the standard deviations along the principal diagonal. (There is a perfect positive correlation between a variable and itself.) In principal component analysis, it is customary to use the correlation matrix K, wherein all the variances (the diagonal elements) are unity and the off-diagonal elements are the respective correlation coefficients. Hence, the eigenvectors can be obtained by substituting the eigenvalues in to the matrix equation.

$(K - \lambda_i I)\, X_i = 0$, where X_i is the column of eigenvectors corresponding to the eigenvalues λ_i. We note that as each of the elements of the principal diagonal of K is unity, for this representation, the sum of the eigenvalues will be equal to the number of variables,

$$K \text{ or } \rho = \begin{bmatrix} 1.0000 & 0.9417 & 0.9067 \\ 0.9417 & 1.0000 & 0.8559 \\ 0.9067 & 0.8559 & 1.0000 \end{bmatrix} \qquad \ldots(5.19)$$

The eigenvalues and eigenvectors for the correlation matrix is obtained

$$\sum_{i=1}^{N} \lambda_i = N \qquad \ldots (5.20)$$

The eigenvector:

$$\varphi(\lambda) = \lambda^3 - 3\lambda^2 + 0.5585\,\lambda - 0.0201 = 0.0$$

$$\phi = \begin{bmatrix} 0.7931 & 0.1630 & 0.5869 \\ -0.5506 & 0.6038 & 0.5764 \\ -0.2604 & -0.7803 & 0.5686 \end{bmatrix} \quad \ldots(5.21)$$

Eigenvalues:

$$\lambda^2 = \begin{bmatrix} 0.0484 & 0 & 0 \\ 0 & 0.1482 & 0 \\ 0 & 0 & 2.8034 \end{bmatrix} \quad \ldots(5.22)$$

Eigenvector	**Eigenvalues**		
	λ_1	λ_2	λ_3
	2.8034	0.1482	0.0484
1	0.5869	0.1630	0.7931
2	0.5764	0.6038	–0.5506
3	0.5686	–0.7803	–0.2604
Cumulative proportion	93.5%	98.4%	100%

The eigenvalues are seen to range from a maximum of 2.8034 for variable cube 150 mm to 0.0484 for cube 100 mm. The first variable cube 150 mm accounts for 93.5% of the total variance.

UNSOLVED PROBLEMS

1. A multivariate function $Y = X_1 + 2X_2 + 3X_3$ in which X_1, X_2, and X_3 are normal random variables. The statistical properties of the three random variables are the following.
   Derive the exact values of mean and variance of Y.

Sl. No.	Variables	$E(X)$	Var (X)
1	X_1	7.25	7.45
2	X_2	7.67	12.96
3	X_3	6.92	14.59

   $\rho(X_1, X_2) = 0.89$, $\rho(X_1, X_3) = 0.75$, and $\rho(X_2, X_3) = 0.89$.

2. A multivariate function $Y = \dfrac{P}{A} + \dfrac{M}{I}C$, in which P, A, M, I, and C are normal random variables. The statistical properties of the random variables are the following:

Derive the exact values of mean and variance of Y. Assume random variables as uncorrelated.

Sl. No.	Variables	$E(X)$	CoV (X)
1	P	3	0.1
2	A	1	0.05
3	M	10	0.1
4	I	4	0.05
5	C	2	0.0

3. Estimate the expected value and coefficient of variation for the active earth pressure $K_a = \tan^2(45 - \varphi/2)$, if $E[\varphi] = 30^\circ$ and passive earth pressure $K_p = \tan^2(45 + \varphi/2)$.
4. The theoretical quantity of flow through a triangular-notched weir is

$$Q = \frac{8}{15} c_d \sqrt{2g} H^{\frac{5}{2}\tan\left(\frac{\theta}{2}\right)}$$

where c_d is an experimental coefficient and g is the gravitational acceleration. A number of tests were conducted to obtain a measure of parameter c_d; however, only the expected values are as follows:

Sl. No.	Variables	$E(X)$
1	Q	0.750
2	H	0.885
3	θ	45°

Estimate the expected value and the coefficient of variation of the parameter.
5. The steady water flow per unit time at a constant depth in a prismatic open channel can be calculated by using the Manning formula as $Q = \frac{c_m}{n} AR^{2/3}S^{1/2}$ where c_m is a constant value 1 in SI units; n is the Manning roughness factor, which depends on the boundary materials used; R is the hydraulic radius; and S is the slope of the bottom of the channel. Consider a trapezoidal cross section made with gravel. Assume C_m is a constant. However, n, A, R, and S are statistically independent random variables with means of 0.025, 8.0 m^2, 1.1 m, and 0.003, respectively, and the corresponding CoV are 0.30, 0.1, 0.1 and 0.1 respectively.
6. The ultimate bearing capacity Q per unit length of a long footing of width B, founded at a depth D below the ground surface is $Q = \frac{\gamma B^2}{2} N_\gamma + \gamma D B N_c + c D B N_q$, where γ is the unit weight of the soil and c is the soil strength (cohesion). The

dimensionless factors $N\gamma$, N_q, and N_c, are bearing capacity factors, are functions of the ϕ, parameter of strength (angle of internal friction or friction angle) as given in the following Table. A very long footing, 2.4384 m wide, is founded at a depth of 1.2192 m in a soil with the following parameters. Assuming that correlation coefficient ρ (ϕ, c) = −0.50: (a) estimate the expected value and the standard deviation of the bearing Q; (b) if a central factor of safety (CSF) of 4 is used and it is assumed that the coefficient of variation of the load on the footing, the demand, is 50%, $V(D)$ = 50%, estimate the reliability of the design.

Parameter x	$E[x]$	$\sigma[x]$	x_+	x_-
γ, N/m^3	1570	0	1570	1570
ϕ, degrees	35	5	40	30
c, N/m^2	6580	3840	10420	2740

$N_{\gamma+} = 109.41$, $N_{q+} = 64.20$, $N_{c+} = 75.31$
$N_{\gamma}- = 22.40$, $N_{q-} = 18.40$, $N_{c-} = 30.14$

7. Determine the reliability index of a simply supported beam in Example 5.4, shown in Fig. 5.5. The limit state of stress, serviceability, and shear formulation are given below, with the respective failure surface as $g(x)$. For the beam section ISMB 350 @ 524 N/m the characteristic values are given in Example 5.4. Solve the three problems by point estimate method.

   (i) Stress formulation

$$g(x) = f_y - \frac{M}{I} y$$

   (ii) Serviceability formulation

$$g(x) = \Delta_{\text{lim}} - \Delta_{\text{act}}$$

$$\Delta_{\text{lim}} = \frac{L}{360};\ \Delta_{\text{act}} = \frac{5}{384}\frac{WL^4}{EI}$$

   (iii) Shear formulation

$$MS = R - S$$
$$R = \tau_s\, t_W\, d$$
$$S = 0.5\,(W_g + W_L)\,L$$

   Find (a) reliability index, β; (b) resistance factor, ϕ; (c) load factor, γ.

6

CHAPTER

Reliability Analysis by Taylor Series

6.1 INTRODUCTION

A Taylor series is commonly used as a basis of approximation in engineering analysis. The objective of this section is to provide a review of the Taylor series. A Taylor series is the sum of functions based on continually increasing derivatives. For a function $f(x)$ that depends on only one independent variable x the value of the function at point $(x_o + h)$ can be approximated by the following Taylor series.

$$f(x_o + h) = f(x_o) + hf^{(1)}(x_o) + \frac{h^2}{2!}f^{(2)}(x_o) + \frac{h^3}{3!}f^{(3)}(x_o) + \ldots + \frac{h^n}{n!}f^{(n)}(x_o) + R_{n+1} \qquad \ldots(6.1)$$

In which x_o is some base value (or starting value) of the independent variable x; h is the distance between x_o and the point x at which the value of the function is needed, that is, $h = (x - x_o)$, $n!$ is the factorial of n, i.e., $n! = n(n-1)(n-2)\ldots 1$; R_{n+1} is the remainder of the Taylor series expansion and the superscript (n) on the function $f^{(n)}$ indicates the nth derivative of the function $f(x)$. The derivatives $f(x)$ in Eq. (6.1) is evaluated at x_o. Equation 6.1 can also be written in the following form:

$$f(x_o + h) = \sum_{k=0}^{\infty} \frac{h^k}{k!} f^{(k)}(x_o) \qquad \ldots(6.2)$$

where $0! = 1$ by convention. Equations 6.1 and 6.2 are based on the assumption that continuous derivatives exists in an interval that includes the points x and x_o. Equations 6.1 and 6.2 can be used with finite number of terms to produce approximations for the evaluated function $f(x_o + h)$. The order of the approximation is defined by the order of the highest derivative that is included in the approximation. For example, the first-order approximation is defined by the terms of the Taylor series expansion up to the first derivative as follows:

$$f(x_o + h) \approx f(x_o) + hf^{(1)}(x_o) \qquad \ldots(6.2a)$$

Similarly, the second and third order approximations are given respectively by,

$$f(x_o + h) \approx f(x_o) + hf^{(1)}(x_o) + \frac{h^2}{2!}f^{(2)}(x_o) \qquad \ldots(6.2b)$$

and

$$f(x_o+h) \approx f(x_o) + hf^{(1)}(x_o) + \frac{h^2}{2!}f^{(2)}(x_o) + \frac{h^3}{3!}f^{(3)}(x_o) \quad \text{... (6.2c)}$$

The accuracy of the approximation improves as the order of the approximation is increased. The importance of the individual terms of the Taylor series depends on the nature of the function and the distance, h. The higher order terms become more important as the non-linearity of the function increases and the difference $(x - x_o)$ increases. If the function is linear, then only the term with the first derivative is necessary. As the separation distance h increases, the importance of the non-linear terms usually increases. Thus, the error increases, as fewer terms of the Taylor series are included.

We shall next consider an approximate method that permits evaluation of more complicated formulations. Recall from elementary calculus that Taylor's formula for the expansion of function $f(x)$, about a point $x = \mu_x$, is

$$f(x) = f(\mu_x) + f'(\mu_x)(x-\mu_x) + \frac{f''(\bar{x})}{2!}(x-\mu_x)^2 + \ldots + \frac{f^{(N-1)}(\mu_x)}{(N-1)}(x-\mu_x)^{N-1} + R_N \quad \text{...(6.3)}$$

where $f^m(\mu_x)$ is the mth derivative evaluated at $x = \mu_x$ and R_N is the remainder (which may or may not be zero). The special case with $\mu_x = 0$ is Maclaurin's series.

Truncating the series in Eq. 6.1 after the quadratic term (a second order approximation) and taking the expected value of both sides, we have, with $E(x) = \mu_x$

$$E[f(x)] = E\left[f(\mu_x) + f'(\mu_x)(x-\mu_x) + \frac{f''(\mu_x)}{2}(x-\mu_x)^2\right]$$

This produces,

$$E[f(x)] = f(\mu_x) + \frac{f''(\mu_x)}{2}V[x] \quad \text{...(6.4)}$$

Recall $E[(x-\mu_x)] = 0$

In a similar manner, the second-order approximation gives the variance.

$$V[f(x) = [f'(\mu_x)]^2 V[x] - 1/4[f''(\mu_x)]^2 V^2[x] + [\beta(2)-1] + \beta(1)\sigma^3[x][f'(\mu_x)][f''(\mu_x)] \quad \text{... (6.5)}$$

where $\beta(1)$, $\beta(2)$ are the coefficients of skewness and Kurtosis. If $f(x)$ is symmetrical (or unless there is a reason to assume it to be otherwise), $\beta(1) = 0$, and Eq. 6.3 reduces to

$$E[f(x)] = f(\mu_x) + \frac{f''(\mu_x)}{2}V[x] \quad \text{...(6.6a)}$$

$$V[f(x)] = f'[(\mu_x)]^2 V[x] + 1/4[f''(\mu_x)]^2 V^2[x]\beta(2) \quad \text{... (6.6b)}$$

The principle of maximum entropy suggests that the normal distribution be assumed if only the first two moments are known. For the normal distribution $\beta(2) = 3$, hence, Eq. 6.3 for this case becomes

$$E[f(x)] = f(\mu_x) + \frac{f''(\mu_x)}{2} V[x] \quad \text{...(6.7a)}$$

$$V[f(x)] = f'(\mu_x)^2 V[x] - (1/4)[f''(\mu_x)]^2 V^2[x] \quad \text{...(6.7b)}$$

In a nut shell if Y is an univariate function given by $Y = g(x)$, then the function using Taylor series can be expressed as

$$Y = g(x)$$

$$Y = g(\mu_x) + \frac{(x-\mu_x)}{1!}\frac{dg}{dx} + \frac{(x-\mu_x)^2}{2!}\frac{d^2g}{dx^2} + \ldots \quad \text{...(6.8a)}$$

$|\leftarrow$ I[st] order terms $\rightarrow|$

$$E(Y) = E[g(\mu_x) + \frac{(x-\mu_x)}{1!}\frac{dg}{dx} + \frac{(x-\mu_x)^2}{2!}\frac{d^2g}{dx^2} + \ldots]$$

$$\cong g(\mu_x) + 0 + \frac{1}{2}\frac{d^2g}{dx^2}\text{Var}(x) \quad \text{...(6.8b)}$$

$$\text{Var}(Y) = \text{Var}[g(\mu_x)] + \text{Var}\left[\frac{(x-\mu_x)}{1!}\frac{dg}{dx}\right] + \text{Var}\left[\frac{(x-\mu_x)^2}{2!}\frac{d^2g}{dx^2}\right] + \ldots$$

$$\cong 0 + \left(\frac{dg}{dx}\right)^2 \text{Var}(x) - \frac{1}{4}\left(\frac{d^2g}{dx^2}\right)^2 \text{Var}^2(x) \quad \text{...(6.8c)}$$

For a function of two variables $f(x_1, x_2)$, Taylor's expansion at the point $f(\mu_{x_1}, \mu_{x_2})$, is

$$f(x_1, x_2) = f(\mu_{x_1}, \mu_{x_2}) + \frac{\partial f}{\partial x_1}(x_1 - \mu_{x_1}) + \frac{\partial f}{\partial x_2}(x_2 - \mu_{x_2})$$

$$+\frac{1}{2}\left[\frac{\partial^2 f(x_1-\mu_{x_1})}{\partial x_1} + \frac{\partial^2 f(x_2-\mu_{x_2})}{\partial x_2} + 2\frac{\partial^2 f}{\partial x_1 \partial x_2}(x_1-\mu_{x_1})(x_2-\mu_{x_2})\right] + R \quad \text{...(6.8d)}$$

where, R is the remainder.

Taylor's expansion can be written using the summation notation

$$f(x_1, x_2) = f(\mu_{x_1}, \mu_{x_2}) + \sum_{i=1}^{2}\frac{\partial f}{\partial x_i}(x_i - \mu_{x_i}) + \frac{1}{2}\sum_{i=1}^{2}\sum_{j=1}^{2}\frac{\partial^2 f}{\partial x_i \partial x_j}(x_i - \mu_{x_i})(x_j - \mu_{x_j}) + R \quad \text{...(6.8e)}$$

If the function Y is multivariate then

$$Y = g(x_1, x_2, x_3, \ldots x_n)$$

$$Y = g(\mu_{x_1}, \mu_{x_2}, \mu_{x_3}, \ldots \mu_{x_n}) + \sum_{i=1}^{n}(x_i - \mu_{x_i})\frac{\partial g}{\partial x_i} + \frac{1}{2}\sum_{i=1}^{n}\sum_{j=1}^{n}(x_i - \mu_{x_i})(x_j - \mu_{x_j})\frac{\partial^2 g}{\partial x_i \cdot \partial x_j} + \ldots \quad \text{...(6.9a)}$$

$|\leftarrow$ I[st] order terms $\rightarrow|$

$$E(Y) = E\left\{ g(\mu_{x1}, \mu_{x2}, \mu_{x3}, \ldots \mu_{xn}) + \sum_{i=1}^{n}(x_i - \mu_{xi})\frac{\partial g}{\partial x_i} + \frac{1}{2}\sum_{i=1}^{n}\sum_{j=1}^{n}(x_i - \mu_{xi})(x_j - \mu_{xj})\frac{\partial^2 g}{\partial x_i \cdot \partial x_j} + \ldots \right\}$$

$$E(Y) \cong g(\mu_{x1}, \mu_{x2}, \mu_{x3}, \ldots \mu_{xn}) + 0 + \frac{1}{2}\sum_{i=1}^{n}\sum_{j=1}^{n}\text{COV}(x_i, x_j)\frac{\partial^2 g}{\partial x_i \cdot \partial x_j} \qquad \ldots(6.9b)$$

If $x_i, x_2, x_3, \ldots x_j$ are independent then the Eq. (6.9b) becomes

$$E(Y) \cong g(\mu_{x_1}, \mu_{x_2}, \mu_{x_3}, \ldots \mu_{x_n}) + 0 + \frac{1}{2}\sum_{i=1}^{n}\left(\frac{\partial^2 g}{\partial x_i^2}\right)\cdot \text{Var}(x_i) \qquad \ldots(6.9c)$$

$$\text{Var}(Y) = \text{Var}\left\{ g(\mu_{x_1}, \mu_{x_2}, \mu_{x_3}, \ldots \mu_{x_n}) + \sum_{i=1}^{n}(x_i - \mu_{x_i})\frac{\partial g}{\partial x_i} + \frac{1}{2}\sum_{i=1}^{n}\sum_{j=1}^{n}(x_i - \mu_{x_i})(x_j - \mu_{x_j})\frac{\partial^2 g}{\partial x_i \cdot \partial x_j} + \ldots \right\}$$

$$\cong 0 + \sum_{i=1}^{n}\left(\frac{\partial g}{\partial x_i}\right)^2 \cdot \text{Var}(x_i) + \sum_{i\neq j}^{n}\sum_{j\neq i}^{n}\left(\frac{\partial g}{\partial x_i}\right)\left(\frac{\partial g}{\partial x_j}\right)\cdot \text{COV}(x_i, x_j) \qquad \ldots(6.9d)$$

Equation (6.9d) can also be expressed as

$$\text{Var}(Y) \cong \sum_{i=1}^{n}\sum_{j=1}^{n}\left(\frac{\partial g}{\partial x_i}\right)\left(\frac{\partial g}{\partial x_j}\right)\cdot \text{COV}(x_i, x_j) \qquad \ldots(6.9e)$$

It may be noted that all the derivatives in the above expressions from Eqs 6.8 and 6.9 are evaluated at the mean values.

Taylor's series can be written in matrix notation as

$$f(x) = f(\mu_x) + \nabla f^T\{x - \mu_x\} + \{x - \mu_x\}^T[H]\{x - \mu_x\} + R \qquad \ldots(6.9f)$$

6.2 GRADIENT VECTOR

Consider a function $f(x)$ of n variables $x_1\ x_2 \ldots x_n$ the partial derivatives of the function with respect to x_1 at any point μ_{x1} is $\frac{\partial f}{\partial x_1}$, with respect to x_2 at any point μ_{x2} is $\frac{\partial f}{\partial x_2}, \ldots,$ in general $\frac{\partial f}{\partial x_i}$ $i = 1$ to n, into a column vector, called gradient vector and represent it by any of the following ways.

Therefore, for convenience and compactness of notation, the partial derivative are arranged as column called **gradient vector** and represented by ∇_f.

The gradient of a function $f(x)$ of n variables $x_1, x_2 \ldots x_n$ at a point $\mu_x = \mu_{x_1}, \mu_{x_2}, \mu_{x_n}$.

$$\nabla_{f(x)} = \begin{Bmatrix} \dfrac{\partial f}{\partial x_1} \\ \dfrac{\partial f}{\partial x_2} \\ \vdots \\ \dfrac{\partial f}{\partial x_n} \end{Bmatrix} \qquad \ldots (6.10a)$$

$$\nabla_{f(x)^T} = \left\{ \frac{\partial f}{\partial x_1}, \frac{\partial f}{\partial x_2}, \ldots \frac{\partial f}{\partial x_n} \right\}^T \qquad \ldots (6.10b)$$

6.3 HESSIAN MATRIX

Differentiating the gradient vector $\nabla f(x)$ once again (each component of the gradient vector is differentiable with respect to x_i) then we obtain a matrix of 2nd order this second (order) partial derivative of a function $f(x)$ is said to be "Hessian matrix", or simply the Hessian. That is differentiating each component of the gradient vector given with respect to $x_1, x_2 \ldots x_n$, we obtain

$$\frac{\partial^2 f}{\partial x_i \partial x_j} = \begin{bmatrix} \dfrac{\partial^2 f}{\partial x_1} & \dfrac{\partial^2 f}{\partial x_1 \partial x_2} & \cdots & \dfrac{\partial^2 f}{\partial x_1 \partial x_n} \\ \dfrac{\partial^2 f}{\partial x_2 \partial x_1} & \dfrac{\partial^2 f}{\partial x_2^2} & \cdots & \dfrac{\partial^2 f}{\partial x_1 \partial x_n} \\ \dfrac{\partial^2 f}{\partial x_n \partial x_1} & \dfrac{\partial^2 f}{\partial x_n \partial x_1} & \cdots & \dfrac{\partial^2 f}{\partial x_n^2} \end{bmatrix} \qquad \ldots (6.11)$$

$$\nabla^2 f(x) = [H]$$

where all derivatives are calculated, at given point μ_x. Hessian is an $(n \times n)$ matrix usually denoted as H or $\nabla^2 = H$ is $(n \times n)$ matrix, since $f(x)$ is assumed to be twice continuous differentiable and cross partial derivatives are equal. It is important to note that each element of the Hessian is a function in itself which is evaluated at the given point μ_x. Also, since $f(x)$ is assumed to be twice continuously differentiable, the cross partial derivatives are equal.

$$\frac{\partial^2 f}{\partial x_i \partial x_j} = \frac{\partial^2 f}{\partial x_j \partial x_i}, \text{ i.e. } \nabla^2 = H \text{ is } (n \times n) \text{ matrix}$$

$$H = \left[\frac{\partial^2 f}{\partial x_1 \partial x_n} \right], i = 1 \text{ to } n, j = 1 \text{ to } n$$

A function can be approximated by polynomials in a neighbourhood of any point in terms of its value and derivatives using Taylor series expansion.

Example 6.1 The cantilever beam contains two concentrated forces L_1 and L_2 with expected values as shown in Fig. 6.1, and their coefficient of variation is 10%. Find the mean and the standard deviation of the bending moment, M_A at point A, if the moment-resisting capacity of the beam is a normal variate, $MR = N(600, 100)$. What is the probability that the beam will fail (assume the moments are not correlated).

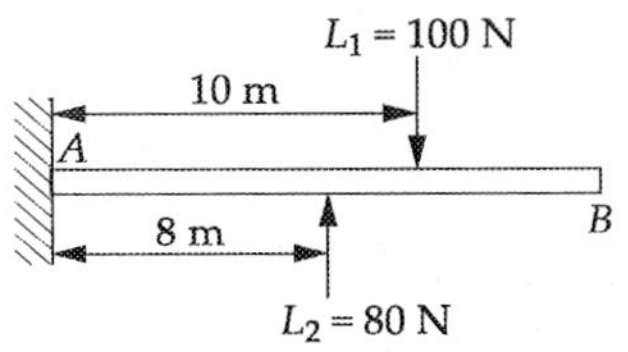

Fig. 6.1 Cantilever beam

Solution:

(a) The bending moment at A, with $E[L_1] = 100$, $E[L_2] = 80$, $\sigma[L_1] = 10$, and $\sigma[L_2] = 8$.

$V[L_1] = 100$ and $V[L_2] = 64$

$$M_A = 10\, L_1 - 8 L_2$$
$$E[M_A] = 10 \times E[L_1] - 8 \times E[L_2] = 360$$
$$V[M_A] = 100\, V[L_1] + 64\, V[L_2] = 14{,}096\ (\text{Nm})^2$$
$$\sigma(M_A) = \sqrt{VEM_A} = \sqrt{14096}$$
$$\sigma[M_A] = 118.72 \text{ N-m.}$$

$$\beta = M_R - M_S / \sqrt{(\sigma_R)^2 + (\sigma_S)^2} = M_R - M_A / \sqrt{(\sigma_R)^2 + (\sigma_A)^2}$$

$$\beta = (600 - 360)/\sqrt{(118)^2 + (100)^2} = 1.5461$$

Hence, the probability of failure, or $P[M_A \geq M_R] = P[M_R - M_A \leq 0]$

$$p_f = \Phi\left(-\left(\frac{240}{155}\right)\right) = 0.0618$$

Its reliability is $P_s = 1 - P_f = 1 - 0.0618 = 0.9382$

Example 6.2 Compute the gradient and Hessian and obtain the second order expansion using Taylor expansion for function of two variable $f(x) = 3x_1^3\, x_2$ at point $x = (1, 1)$.

Solution: The given function is $f(x) = 3x_1^3\, x_2$ at point $x = (1, 1)$.

Taylor expansion of a function on two variables is

$$f(x_1, x_2) = f(\mu_{x_1}, \mu_{x_2}) + \sum_{i=1}^{2} \frac{\partial f}{\partial x_1}(x_i - \mu_{x_i}) + \frac{1}{2}\sum_{i=1}^{2}\sum_{j=1}^{2} \frac{\partial^2 f}{\partial x_i\, \partial x_j}(x_i - \mu_{x_i})(x_j - \mu_{x_j}) + R$$

$$\nabla f^T = \begin{Bmatrix} \dfrac{\partial f}{\partial x_1} \\ \dfrac{\partial f}{\partial x_2} \\ \vdots \\ \dfrac{\partial f}{\partial x_n} \end{Bmatrix}; \qquad [H] = \begin{bmatrix} \dfrac{\partial^2 f}{\partial x_1^2} & \dfrac{\partial^2 f}{\partial x_1 \partial x_2} \\ \dfrac{\partial^2 f}{\partial x_2 \partial x_1} & \dfrac{\partial^2 f}{\partial x_2^2} \end{bmatrix}$$

$$f(x) = 3x_1^3 x_2$$

$$\frac{\partial f}{\partial x_1} = 3 \times 3 \times x_1^2 = 9x_1^2\, x_2$$

$$\frac{\partial f}{\partial x_2} = 3 \times x_1^2 = 3x_1^3$$

$$\nabla_{f(1,\,2)} = \begin{bmatrix} 9 \\ 3 \end{bmatrix}$$

$$\frac{\partial^2 f}{\partial x_1^2} = 18x_1x_2 = 18, \; \frac{\partial^2 f}{\partial x_1 \partial x_2} = 9x_1^2$$

$$\frac{\partial^2 f}{\partial x_2^2} = 0, \; \frac{\partial^2 f}{\partial x_2 \partial x_1} = 9x_1^2$$

$$H = \begin{bmatrix} 18 & 9 \\ 9 & 0 \end{bmatrix}$$

Taylor expansion in matrix notation, simplifying the expression and expanding the vector products, we obtain the second order approximation of the function.

$$f(x) = f(\mu_x) + \nabla f^T \{x - \mu_x\} + \{x - \mu_x\}^T [H] \{x - \mu_x\} + R$$

$$= 3 + \begin{bmatrix} 9 \\ 3 \end{bmatrix}^T \begin{bmatrix} (x_1 - 1) \\ (x_2 - 1) \end{bmatrix} + \frac{1}{2} \begin{bmatrix} (x_1 - 1) \\ (x_2 - 1) \end{bmatrix} \begin{bmatrix} 18 & 9 \\ 9 & 0 \end{bmatrix} \begin{bmatrix} (x_1 - 1) \\ (x_2 - 1) \end{bmatrix}$$

$$= 3 + [9 \quad 3] \begin{bmatrix} (x_1 - 1) \\ (x_2 - 1) \end{bmatrix} + \frac{1}{2} [(x_1 - 1), (x_2 - 1)] \begin{bmatrix} 18 & 9 \\ 9 & 0 \end{bmatrix} \begin{bmatrix} (x_1 - 1) \\ (x_2 - 1) \end{bmatrix}$$

On simplifying

$$f(x_1, x_2)_{(1,\,1)} = 9x_1^2 + 9x_1x_2 - 18x_1 - 9x_2 + 9$$

Example 6.3 A wooden beam with rectangular section is subjected to a bending moment M and allowable bending stress f_b. The margin of safety of the beam is given by $MS = f_b - \dfrac{6M}{b.d^2}$. Using Taylor series linearization the equation for margin of safety by considering first order terms.

Sl. No.	Variables	Mean value (μ)	CoV	Standard deviation (σ)	Distribution
1	f_b	1600 N/mm^2	0.32	512 N/mm^2	Normal
2	M	10^5 N-mm	0.12	12000 N-mm	Normal
3	b	5.6 mm	0.04	0.224 mm	Normal
4	d	11.4 mm	0.03	0.342 mm	Normal

Solution: Using Eq. 6.9(a) we have

$$Y = g(\mu_{x_1}, \mu_{x_2}, \mu_{x_3}, \ldots \mu_{x_n}) + \sum_{i=1}^{n}(x_i - \mu_{x_i})\frac{\partial g}{\partial x_i} + \frac{1}{2}\sum_{i=1}^{n}\sum_{j=1}^{n}(x_i - \mu_{x_i})(x_j - \mu_{x_j})\frac{\partial^2 g}{\partial x_i \cdot \partial x_j} + \ldots$$

Considering only the first order terms the equation for margin of safety is

$$MS = \left(f_b - \frac{6M}{bd^2}\right)_\mu + \left[(f_b - \mu_{f_b})\left(\frac{\partial MS}{\partial f_b}\right)_\mu\right] + \left[(M - \mu_M)\left(\frac{\partial MS}{\partial M}\right)_\mu\right] + \left[(b - \mu_b)\left(\frac{\partial MS}{\partial b}\right)_\mu\right] + \left[(d - \mu_d)\left(\frac{\partial MS}{\partial d}\right)_\mu\right]$$

where the derivatives are evaluated at the mean point it is also called sensitivity or elasticity of the variable.

$$\frac{\partial MS}{\partial f_b} = 1$$

$$\frac{\partial MS}{\partial M} = -\frac{6}{b.d^2} = -\frac{6}{5.6 \times 11.4^2} = -0.008244$$

$$\frac{\partial MS}{\partial b} = \frac{6M}{b^2 d^2} = \frac{6 \times 10^5}{5.6^2 \times .11.4^2} = 147.2$$

$$\frac{\partial MS}{\partial d} = \frac{12M}{bd^3} = \frac{12 \times 10^5}{5.6 \times 11.4^3} = 144.6$$

Substitute the mean values of the random variables

$g(x) = MS = \mu_{fb} - 0.008244\ \mu_M + 147.2\ \mu_b + 144.601\ \mu_d - 2473.0$

$\mu_{MS} = 1600 - 0.008244 \times 100000 + 147.2 \times 5.6 + 144.6 \times 11.4 - 2473$

$\mu_{MS} = 775.4$

$\sigma_{MS}{}^2 = (1)^2 \times \sigma_{fb}{}^2 + (-0.008244)^2 \times \sigma_M{}^2 + (147.2)^2 \times \sigma_b{}^2 + (144.6)^2 \times \sigma_d{}^2$

$\sigma_{MS}{}^2 = (1)^2 \times 512^2 + (-0.008244)^2 \times 12000^2 + (147.2)^2 \times 0.224^2 + (144.6)^2 \times 0.342^2 = 275464$

$\sigma_{MS}{}^2 = 275463.5737$

$\sigma_{MS} = 524.8462$

Reliability index,

$$\beta = \frac{\mu_{MS}}{\sigma_{MS}} = \frac{775.4}{524.8462} = 1.4775$$

Example 6.4 Analyze by Taylor series, a tie member of roof truss subjected to 300 kN for the limit state of yield of $g(x) = f_y A - S$ of the steel member for strength formulation. Analyze for the following data. Determine the reliability index, by converting the non-linear function into a linear polynomial function considering only: (i) first order terms (ii) second order terms.

Sl. No.	Variables	Characteristic value	CoV	Distribution
1	f_y	250 N/mm^2	0.05	Normal
2	A	1432 mm^2	0.05	Normal
3	S	300×10^3 N	0.20	Normal

Solution: **Analysis by Taylor series method**

Sl. No.	Variables	Characteristic value (x)	CoV	Mean value (μ)	Standard deviation (σ)	Distribution
1	f_y	250 N/mm^2	0.05	272.405	13.620	Normal
2	S	300 kN	0.20	225.734×10^3	45.147	Normal
3	A	1432 mm^2	0.05	1560.34	78.017	Normal

Case 1: Compute the mean value of the basic variable from characteristic values.

Here, f_y, A and S are random variables

The performance function $g(x) = R - S$ (The failure mode in yielding)

The limit state function, $g(x) = f_y A - S$

The given function is $g(x) = f_y A - S$

The Taylor expansion is $g(x) = g(x_i) + \nabla g\{x_i - \mu_x\}^T + \frac{1}{2} g\{x_i - \mu_x\}^T [H] g\{x_i - \mu_x\}$

where $\nabla g(x) =$ Gradient vector

$g(\mu_x) = 272.405 \times 1560.34 - 225.734 \times 1000$

$g(\mu_x) = 199310.4177$ N

Gradient vector

$$\begin{Bmatrix} \dfrac{\partial}{\partial f_y} = A \\ \dfrac{\partial}{\partial A} = f_y \\ \dfrac{\partial}{\partial S} = -1 \end{Bmatrix} = \begin{Bmatrix} 1560.34 \\ 272.405 \\ -1 \end{Bmatrix}$$

$$\nabla g\{x_i - \mu_x\}^T = \{(f_y - \mu_{fy})(A - \mu_A)(S - \mu_S)\}^T \begin{Bmatrix} \dfrac{\partial}{\partial f_y} \\ \dfrac{\partial}{\partial A} \\ \dfrac{\partial}{\partial S} \end{Bmatrix}$$

$$= \{(f_y - 272.405)(A - 1560.34)(S - 225734)\}^T \begin{Bmatrix} 1560.34 \\ 272.405 \\ -1 \end{Bmatrix}$$

$$= 1560.34(f_y - 272.405) + 272.405(A - 1560.34 - 1(S - 225734))$$

$$= 1560.34 f_y - 425044.4177 + 272.405A - 425044.4177 - S + 225734.00$$

$$\nabla g\{x_i - \mu_x\}^T = 1560.34 f_y + 272.405A - S - 624354.8354$$

$$V[g(x)] = \sigma_{g(x)} = \sqrt{\left(\frac{\partial}{\partial f_y}\right)^2 (\sigma f_y)^2 + \left(\frac{\partial}{\partial A}\right)^2 (\sigma A)^2 + \left(\frac{\partial}{\partial S}\right)^2 \sigma_S^2}$$

$$= \sqrt{(1560.34)^2 (13.62)^2 + (272.405)^2 (78.017)^2 + (-1)^2 (45.147)^2}$$

$$= \sqrt{451640312.40 + 451656892.50 + 2038251603}$$

$$\sigma_{g(x)} = \sqrt{903299243.1} = 30054.9371 \text{ N}$$

The nonlinear equation is converted into a simply polynomial

$$g(x) = g(\mu_x) + \nabla g\{x_i - \mu_x\}^T$$

$$= 199310.4177 + 1560.34 f_y + 272.405A - S - 624354.8358$$

$$g(x) = 1560.34 f_y + 272.405 A - S - 425044.8345$$

$$\therefore \quad \beta = \frac{g(x)}{\sigma_{g(x)}}$$

$$= \frac{1560.34\ (250) + 272.405\ (1432) - 300{,}000 - 425044.8345}{30054.9371}$$

$$\beta = \frac{55124.1255}{30054.9371} = 1.8341$$

Case 2: Second order terms are considered to solve the problem

$$[H] = \begin{bmatrix} \frac{\partial^2}{\partial f_y^2} = 0 & \frac{\partial^2}{\partial f_y\, \partial A} = 1 & \frac{\partial^2}{\partial f_y\, \partial S} = 0 \\ \frac{\partial^2}{\partial A \partial f_y} = 1 & \frac{\partial^2}{\partial A^2} = 0 & \frac{\partial^2}{\partial A \partial S} = 0 \\ \frac{\partial^2}{\partial S \partial f_y} = 0 & \frac{\partial^2}{\partial S \partial A} = 0 & \frac{\partial^2}{\partial S^2} = 0 \end{bmatrix} = \begin{bmatrix} 0 & 1 & 0 \\ 1 & 0 & 0 \\ 0 & 0 & 0 \end{bmatrix}$$

$$\{x_i - \mu_x\}^T [H] g\{x_i - \mu_x\} = [(f_y - 272.405)(A - 1560.34)](S - 225734)]$$

$$= \begin{bmatrix} 0 & 1 & 0 \\ 1 & 0 & 0 \\ 0 & 0 & 0 \end{bmatrix} g\{x_i - \mu_x\}$$

$$= [(f_y - 272.405)(A - 1560.34)0] \begin{bmatrix} f_y - 272.405 \\ A - 1560.340 \\ S - 225734 \end{bmatrix}$$

$$= (f_y - 272.405)^2 + (A - 1560.34)^2 + 0$$

$$= f_y^2 + 272.405^2 - 544.81 f_y + A^2 + 1560.34^2 - 3120.68A$$

$$= \{x_i - \mu_x\}^T [H] g\{x_i - \mu_x\} = f_y^2 + A^2 - 544.81 f_y - 3120.68A + 2508865.40$$

$$g(x) = g(\mu_x) + \nabla g\{x_i - \mu_x\}^T + \frac{1}{2}\{x_i - \mu_x\}^T [H] g\{x_i - \mu_x\}$$

$$= 199310.4177 + 1560.34 f_y + 272.405A - S - 624354.8354$$

$$+ (f_y^2 + A^2 - 544.81 f_y - 3120.68A + 2508865.40)\frac{1}{2}$$

The second order polynomial expansion of the nonlinear function is given by the following expression.

$$g(x) = \frac{1}{2} f_y^2 + \frac{A^2}{S} + 1287.935 f_y - 1287.935A - S + 829388.2823$$

The values of the function $g(x)$ are computed and reliability index is computed.

$$= \frac{1}{2}(250)^2 + \frac{(1432)^2}{2} + 1287.935(250) - 1287.935(1432) - 300000 + 829388.2823$$

$$= 31250 + 1025312.0 + 321983.75 - 1844322.92 + 829388.2823 - 300000$$

$$g(x) = 63611.1123$$

$$\beta = \frac{g(x)}{(\sigma_x)} = \frac{63611.112}{30054.9371} = 2.1165$$

Example 6.5 Euler's classical (buckling) formula for critical axial unit stress at which a slender rod of length L damped at base will buckle is $\sigma_{cr} = \frac{\pi^2 E}{4 \times (L/r)^2}$; where, E is the modulus of elasticity, r is the radius of gyration. The ratio L/r is called slenderness ratio. What is the probability that an applied stress will reduce buckling (assume un-correlated variables)

Sl. No.	Mean value (μ)	Standard deviation (σ)	Distribution
E	2.0685×10^{11} N/mm^2	0	Deterministic
λ	150	20	Normal
σ_{actual}	17.2375×10^6 N/mm^2	4.3094×10^6 N/mm^2	Normal

Solution:

Given: $E = 2.0685 \times 10^{11}$ N/m^2, $\lambda = 150$, $\sigma_\lambda = 20$

$\sigma_{actual} = 17.2375 \times 10^6$ N/m^2

$\sigma_f = 4.309375 \times 10^6$ N/m^2

$$\sigma_{cr} = \frac{\pi^2 E}{4 \times (L/r)^2}$$

Let,

$$\lambda = \frac{L}{r}$$

$$\sigma_{cr} = \frac{\pi^2 E}{4(\lambda)^2}$$

Let,

$$\frac{\pi^2 E}{4} = C$$

$$C = \frac{\pi^2 \times 2.0685 \times 10^{11}}{4}$$

$$C = 5.1038 \times 10^{11} \text{ N/m}^2$$

$$g(x) = \mu_R - \mu_S$$

$$\mu_R = \sigma_{cr} = \frac{\pi^2 E}{4(\lambda)^2} = \frac{C}{(\lambda)^2} = \frac{5.1038 \times 10^{11}}{152^2} = 22.6836 \times 10^6 \text{ N/m}^2$$

$$\mu_S = \sigma_{actual} = 17.2375 \times 10^6 \text{ N/m}^2$$

$$\sigma_{cr} = 6.286 \times 10^6 \text{ N/m}^2$$

$$\sigma_{actual} = \sigma_s = 4.309375 \times 10^6 \text{ N/m}^2$$

$$\beta = \frac{\mu_R - \mu_S}{\sqrt{(\sigma_{Rm}^2 + \sigma_{Sm}^2)}}$$

$$\beta = \frac{(22.68364078 \times 10^6) - (17.2375 \times 10^6)}{\sqrt{(6.286 \times 10^6)^2 + (4.309375 \times 10^6)^2}} = \frac{22682982.76}{7621276.959} = 2.97$$

$$P_{(Buckling)} = 1 - \Phi[\beta]$$
$$= 1 - \Phi[2.97]$$
$$= 1 - 0.9985$$
$$= 0.0015 \text{ or } 0.15\%$$

UNSOLVED PROBLEMS

1. Determine the reliability index for a simply supported beam for the limit state of shear $g(x) = \tau_s t_w d - 0.5P$. The beam is subjected to point load P at the mid-span; where d is the depth of the beam, t_w is the thickness of the web, τ_s is the shear strength of the material. $\mu_P = 4000$ N, $\sigma_P = 1000$ N, $\mu_{\tau s} = 95$ N/mm², $\sigma_{\tau s} = 10$ N/mm², $\mu_d = 50$ mm, $\sigma_d = 2.5$ mm and $d/t_w = 40$.
2. Determine the mean, standard deviation for the concrete member, and the variables are as given below.

$$M_R = [(A_{st} f_y d - (0.77 A_{st}^2 f_y^2 / b f_{ck})]$$

Sl. No.	Characteristic value	CoV	Distribution
1	$A_{st} = 7390$ mm²	0.05	Normal
2	$f_y = 415$ N/mm²	0.03	Normal
3	$d = 1450$ mm	0.05	Normal
4	$b = 300$ mm	0.05	Normal
5	$f_{ck} = 20$ N/mm²	0.05	Normal

3. The limit state of failure of $g(x) = Zf_y - M_{ext}$ of the steel beam for strength formulation is given. Determine the section modulus of the section for a reliability index of 3.09. Given the following data:

Sl. No.	Variables	Characteristic value	CoV	Distribution
1	f_y	25 kN/cm^2	0.05	Normal
2	M_{ext}	200 kN/cm^2	0.20	Normal
3	Z	?	0.05	Normal

4. The limit state of serviceability $g(x) = \delta_{act} - L/360$; $g(x) = \dfrac{5WL^4}{384EI} - L/360$ for the steel beam is given. Determine the second moment of inertia of the section for a reliability index of 3.09 given the following data:

Sl. No.	Variables	Characteristic value	CoV	Distribution
1	W	150 kN	0.25	Normal
2	L	6.5 m	0.05	Normal
3	E	200 GPa	0.05	Normal
4	I	?	0.05	Normal

7

CHAPTER

Reliability Analysis by Hasofer-Lind Method

7.1 HASOFER-LIND METHOD

The Hasofer-Lind method is applicable for normal random variables. It first defines the reduced variables as

$$X_i' = \frac{X_i - \mu_{X_i}}{\sigma_{X_i}} \; (i = 1, 2, ..., n) \qquad ...(7.1)$$

where, X_i' is a random variable with zero mean and unit standard deviation. Equation 7.1 is used to transform the original limit state $g(X) = 0$ to the reduced limit state, $g(X') = 0$. The X coordinate system is referred to as the original coordinate system. The X' coordinate system is referred to as the transformed or reduced coordinate system. Note that X_i' is standard normal. These notations will be used throughout this chapter to denote different coordinate systems. The safety index β is defined as the minimum distance from the origin of the axes in the reduced coordinate system to the limit state surface (failure surface). It can be expressed as

$$\beta = \sqrt{(X'^*)^T (X'^*)} \qquad ...(7.2)$$

The minimum distance point on the limit state surface is called the design point or checking point. It is denoted by vector X^* in the original coordinate system and by vector X'^* in the reduced coordinate system. These vectors represent the values of all the random variables, that is, X_1, X_2... X_n, at the design point corresponding to the coordinate system being used.

This method can be explained with the help of Fig. 7.1. Consider the linear limit state equation in two variables,

$$Z = R - S = 0 \qquad ...(7.3)$$

Note that R and S need not be normal variables. A set of reduced variables is introduced as

$$S' = \frac{S - \mu_S}{\sigma_S} \qquad ...(7.4)$$

$$R' = \frac{R - \mu_R}{\sigma_R} \qquad \ldots(7.5)$$

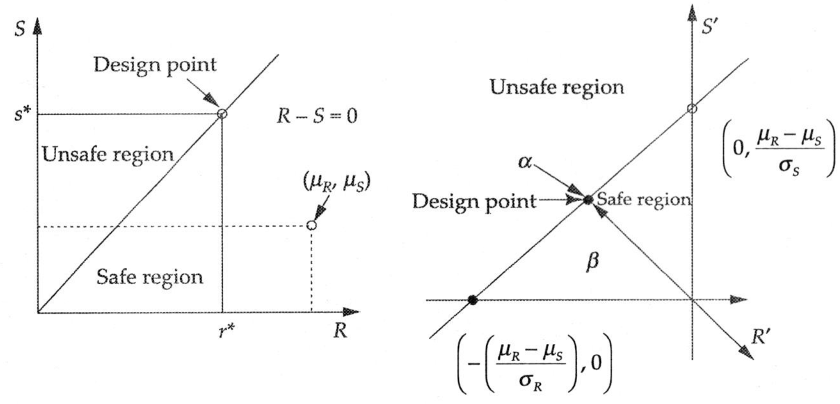

(a) Original coordinates (b) Reduced coordinates

Fig. 7.1 Hasofer-Lind reliability index: Linear performance function

If we substitute these into Eq. 7.3, the limit state equation in the reduced coordinate system becomes

$$g(X') = \sigma_R R' - \sigma_S S' + \mu_R - \mu_S = 0 \qquad \ldots(7.6)$$

The transformation of the limit state equation from the original to the reduced coordinate system is shown in Fig. 7.1(b). The safe and failure regions are also shown. From Fig. 7.1(b) it is apparent that if the failure line (limit state line) is closer to the origin in the reduced coordinate system, the failure region is larger, and if it is farther away from the origin, the failure region is smaller. Thus, the position of the limit state surface relative to the origin in the reduced coordinate system is a measure of the reliability of the system. The coordinates of the intercepts of Eq. 7.6 on the R' and S' axes can be shown to be $[-(\mu_R - \mu_S)/\sigma_R, 0]$, and $[0, (\mu_R - \mu_S)/\sigma_S]$, respectively.

Using simple trigonometry, we can calculate the distance of the limit state line (Eq. 7.6) from the origin as

$$\beta = \frac{\mu_R - \mu_S}{\sqrt{\sigma_R^2 + \sigma_S^2}} \qquad \ldots(7.7)$$

The distance is referred to as the reliability index or safety index. It is the same as the reliability index defined by the MVFOSM method, if both R and S are normal variables. However, it is obtained in a completely different way based on geometry. It indicates that if the limit state is linear and if the random variables R and S are normal, both the methods will give an identical reliability or safety index.

Let random variables be represented by the vector $X = (x_1, x_2 \ldots x_n)$ in the original coordinate system and $X' = (x_1', x_2' \ldots x_n')$ in the reduced coordinate system. The limit state

$g(X') = 0$ is a nonlinear function as shown in the reduced coordinates for two variables in Fig. 7.2. At this stage, X'_i is assumed to be uncorrelated.

Here, $g(X') > 0$ denotes the safe state and $g(X') < 0$ denotes the failure state. Again, the Hasofer-Lind reliability index β is defined as the minimum distance from the origin to the design point on the limit state in the reduced coordinates and can be expressed by Eq. 7.2, where X'^* represents the coordinate of the design point or the point of minimum distance from the origin to the limit state. In this definition the reliability index is invariant, because regardless of the form in which the limit state equation is written, its geometric shape and the distance from the origin remains constant. For the limit state surface where the failure region is away from the origin, it is easy to see from Fig. 7.2 that X'^* is the most probable failure point. The Hasofer-Lind reliability index can be used to calculate a first-order approximation of the failure probability as $p_f = \Phi(-\beta)$. This is the integral of the standard normal density function along the line joining the origin and X'^*. It is obvious that the nearer X'^* is to the origin, the larger is the failure probability. Thus, the minimum distance point on the limit state surface is also the most probable failure point. The point of minimum distance from the origin to the limit state surface, X'^*, represents the worst combination of the stochastic variables and is appropriately named as the *design point* or *the most probable point* (MPP) of failure.

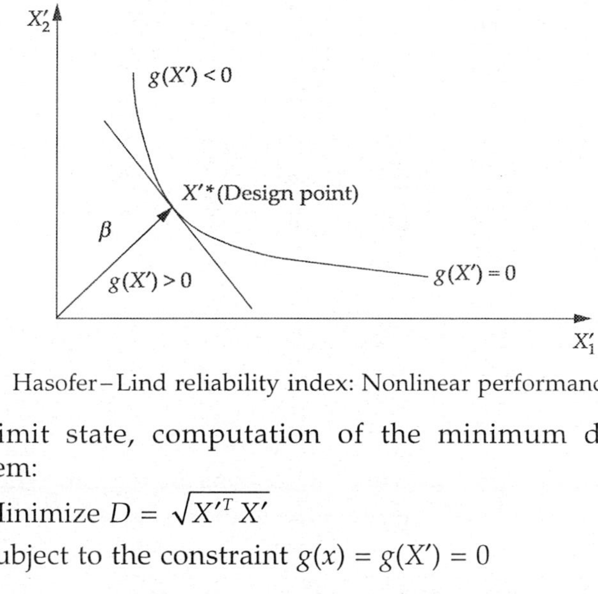

Fig. 7.2 Hasofer–Lind reliability index: Nonlinear performance function

For nonlinear limit state, computation of the minimum distance becomes an optimization problem:

$$\text{Minimize } D = \sqrt{X'^T X'} \qquad \ldots(7.8)$$

$$\text{Subject to the constraint } g(x) = g(X') = 0$$

where x' represents the coordinates of the checking point on the limit state equation in the reduced coordinates to be estimated. Using the method of Lagrange multipliers, we can obtain the minimum distance as

$$\beta = -\frac{\sum_{i=1}^{n} x_i'^* \left(\frac{\partial g}{\partial X_i'}\right)^*}{\sqrt{\sum_{i=1}^{n}\left[\left(\frac{\partial g}{\partial X_i'}\right)^*\right]^2}} \qquad \ldots(7.9)$$

where $(\partial g/\partial X'_i)^*$ is the ith partial derivative evaluated at the design point with coordinates $(x'^*_1, x'^*_2, \ldots x'^*_n)$. The asterisk after the derivative indicates that it is evaluated at $(x'^*_1, x'^*_2, \ldots x'^*_n)$. The design point in the reduced coordinates is given by

$$x'^*_1 = -\alpha_i \beta \quad (i = 1, 2, \ldots, n) \qquad \ldots(7.10)$$

where,

$$\alpha_i = \frac{\left(\dfrac{\partial g}{\partial X'_i}\right)^*}{\sqrt{\sum_{i=1}^{n}\left(\dfrac{\partial g}{\partial X'_i}\right)^{2*}}} \qquad \ldots(7.11)$$

are the direction cosines along the coordinate axes x'_i. In the space of the original coordinates and using Eq. 7.10, we find the design point to be

$$x''_i = \mu_{x_i} - \alpha_i \sigma_{x_i} \beta \qquad \ldots(7.12)$$

An algorithm to compute β and x''_i as follows:

Step 1: Define the appropriate limit state equation.

Step 2: Assume initial values of the design point x^*_i, $i = 1, 2, \ldots, n$. Typically, the initial design point may be assumed to be at the mean values of the random variables. Obtain the reduced $x'^*_i = \dfrac{(x^*_i - \mu_{x_i})}{\sigma_{x_i}}$.

Step 3: Evaluate $\left(\dfrac{\partial g}{\partial X'_i}\right)^*$ and α_i at x''_i.

Step 4: Obtain the new design point x''_i in terms of β, as in Eq. 7.10.

Step 5: Substitute the new x''_i in the limit state equation $g(x''_i) = 0$ and solve for β.

Step 6: Using the β value obtained in step 5, evaluate $\beta_{HL} x''_i = -\alpha_i \beta$.

Step 7: Repeat steps 3 through 6 until β converges.

This algorithm is geometrically shown in Fig. 7.3. The algorithm constructs a linear approximation to the limit state at every search point and finds the distance from the origin to the limit state. In Fig. 7.3, point B represents the initial design point, usually assumed to be at the mean values of the random variables, as noted in step 2. Note that B is not on the limit state equation $g(X') = 0$.

The tangent to the limit state at B is represented by the line BC. Then AD will give an estimate of β in the first iteration, as noted in step 5. As the iteration continues, β value converges.

It can be shown that for a nonlinear limit state surface, β lacks invariance (comparability); the ordering of β values may not be consistent with the ordering of actual reliabilities. An example of this is shown in Fig. 7.2 with two limit state surfaces: one flat and other curved. The shaded region to the right of each limit state represents the corresponding failure region. Clearly, the structure with the flat limit state surface has a different reliability than the one with the curved limit state surface; however, the β_{HL} values are identical for both surfaces and suggest equal reliability. To overcome this inconsistency, the generalized reliability index, β_g, can be defined as

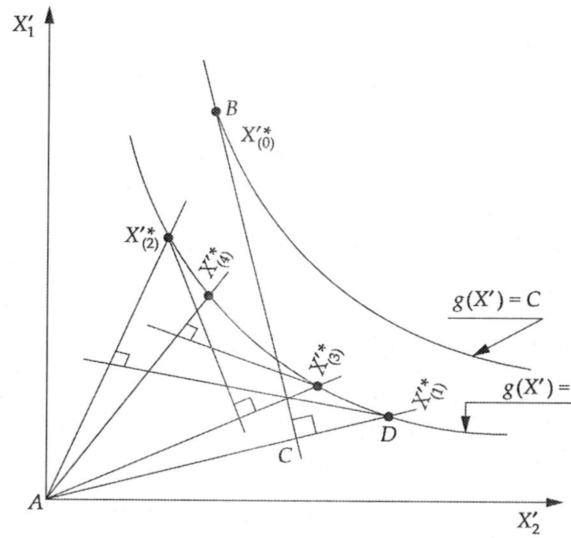

Fig. 7.3 Geometric illustration of algorithm to compute reliability index, β

$$\beta_g = \Phi^{-1}\left[\int_{g(x')>0} ....\int \phi(x_1')\ \phi(x_2')\ ....\phi(x_n')\ d(x_1')\ dx_2'\ ....dx_n'\right] \quad ...(7.13)$$

where Φ and ϕ are the cumulative distribution function and the probability density function of a standard normal variable, respectively. The reliability index includes the entire safe region and provides a consistent ordering of second-moment reliability. The integral in Eq. 7.13 and is difficult to compute directly. Hence, by approximating the nonlinear limit state by a polyhedral surface consisting of tangent hyper planes at selected points on the surface solution can be obtained by numerical procedures.

Example 7.1 Denoting R and S as the random variables representing the resistance and the applied load on a structure, assume that the limit state equation is represented by $Z = (R - S)$ in the original coordinate system and by the reduced coordinate system $g(X') = \sigma_R R' - \sigma_S S' + \mu_R - \mu_S = 0$. These are shown in Fig. 7.1(a) and 7.1(b), respectively. Using Eq. 7.11 we can evaluate the direction cosines α_R and α_S as

$$\alpha_R = \frac{(\partial g/\partial R')}{\sqrt{(\partial g/\partial R')^2 + (\partial g/\partial S')^2}} = \frac{\sigma_R}{\sqrt{\sigma_R^2 + \sigma_S^2}} \quad ...(7.14a)$$

Similarly, $$\alpha_S = -\frac{\sigma_S}{\sqrt{\sigma_R^2 + \sigma_S^2}} \qquad \text{...(7.14b)}$$

Notice that in Fig. 7.1(b), $\alpha_R = \cos\theta_R$ and $\alpha_S = \cos\theta_S$. The angles θ_R and θ_S are defined as counterclockwise angles of rotation from the positive direction of R' and S' axes to the positive direction of α vector (i.e., vector of direction cosines). Using Eq. 7.10, we can show the coordinates of the checking point in the reduced coordinate to be

$$r'^* = -\alpha_R\beta = -\frac{\sigma_R}{\sqrt{\sigma_R^2 + \sigma_S^2}}\beta \qquad \text{...(7.15)}$$

$$s'^* = -\alpha_S\beta = -\frac{\sigma_S}{\sqrt{\sigma_R^2 + \sigma_S^2}}\beta \qquad \text{...(7.16)}$$

Substituting these new checking points in Eq. 7.6, we can calculate the reliability index β as

$$\beta = \frac{\mu_R - \mu_S}{\sqrt{\sigma_R^2 + \sigma_S^2}} \qquad \text{...(7.17)}$$

Since, it is a linear limit state equation, iteration is not required. This is the same result as obtained by Eq. 7.7, indicating that the algorithm works correctly. Using Eq. 7.12, we find the new checking point in the original coordinates to be

$$r^* = \mu_R - \left(\frac{\sigma_R}{\sqrt{\sigma_R^2 + \sigma_S^2}}\right)\sigma_R\left(\frac{\mu_R - \mu_S}{\sqrt{\sigma_R^2 + \sigma_S^2}}\right) = \frac{\mu_R\sigma_S^2 + \mu_S\sigma_R^2}{\sigma_R^2 + \sigma_S^2} \qquad \text{...(7.18)}$$

$$s^* = \mu_S - \left(\frac{\sigma_S}{\sqrt{\sigma_R^2 + \sigma_S^2}}\right)\sigma_S\left(\frac{\mu_R - \mu_S}{\sqrt{\sigma_R^2 + \sigma_S^2}}\right) = \frac{\mu_R\sigma_S^2 + \mu_S\sigma_R^2}{\sigma_R^2 + \sigma_S^2} \qquad \text{...(7.19)}$$

In this case, $r^* = s^*$, indicating that the checking point is on the limit state line, which is at a 45° angle to both coordinate axes. This method can also be used for reliability design. The value of the random variable in the limit state equation can be evaluated for a specified value of reliability index.

Several important observations can be made by comparing the safety indices calculated by the MVFOSM (Eq. 7.7) and the AFSOM proposed by Hasofer and Lind (Eq. 7.9). As long as the limit state equation of resistance and load is linear and all the variables are normal, the safety indices calculated by the two methods will be the same. However, strictly speaking, the MVFOSM does not use any information on the distribution of the resistance and load, whereas the AFOSM proposed by Hasofer and Lind is applicable when they are normal. The most important difference is that in the MVFOSM method, the design point is at the mean values of R and S, indicating they are not on the limit state line. The AFOSM (Hasofer–Lind) method indicates that the design point is on the limit state line.

Example 7.2 A cable of resistance R needs to carry a weight S. Assume both R and S are normal random variables with mean of 120 kN and 50 kN, respectively and corresponding standard deviation of 18 kN and 12 kN, respectively.

Solution: The limit state equation can be represented by Eq. 7.3. Then, the safety index according to the MVFOSM and the Hasofer-Lind methods will be the same, that is,

$$\beta = \frac{120 - 50}{\sqrt{18^2 + 12^2}} = 3.236$$

The design point according to the MVFOSM is (120, 50) as shown in Fig. 7.1(a). The coordinates of the design point according to the Hasofer -Lind method can be estimated as

$$r^* = s^* = \frac{\mu_R \sigma_S^2 + \mu_S \sigma_R^2}{\sigma_R^2 + \sigma_S^2} = \frac{120 \times 12^2 + 50 \times 18^2}{18^2 + 12^2} = 71.54$$

The coordinates for the checking point are (71.54, 71.54), indicating that it is on the limit state equation as shown in Fig. 7.1(b).

Example 7.3 A statically indeterminate beam shown in Fig. 7.4 is loaded with a concentrated force P and the following deflection failure criterion is given by

$$\Delta_{\max} = \frac{5}{48}\frac{PL^3}{EI} \geq \frac{1}{30}L$$

where $\Delta_{\max}$ is the maximum deflection, E the modulus of elasticity and I the relevant moment of inertia. Further, let $\Delta_{\max}$, P, L, E and I be the realizations of uncorrelated random variables $P \cdot L$, E and I with the following data

Sl. No.	Variables	Mean value (μ)	Coefficient of variation	Standard deviation (σ)
1	μ_p	4 kN	0.25	1
2	μ_L	5 m	0.00	
3	μ_E	2×10^7 kN/m^2	0.25	0.5×10^7
4	μ_I	10^{-4} m^4	0.10	10^{-5}

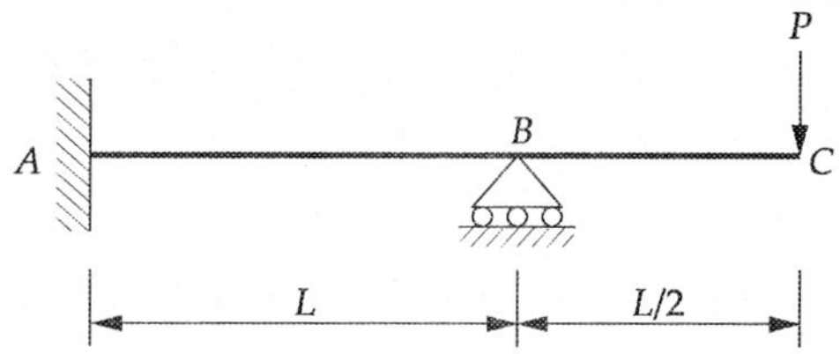

Fig. 7.4 Statically indeterminate beam

Solution: Substituting, $L = 5$ m into the failure function, we have

$$E.I - 78.125\,P = 0$$

The basic variables *I*, *E* and *P* are then normalized $Z_1 = \frac{I - \mu_I}{\sigma_I}$, $Z_2 = \frac{E - \mu_E}{\sigma_E}$, and $Z_3 = \frac{P - \mu_P}{\sigma_P}$, in the normalized coordinate system the failure surface is given by

$$(10^{-4} + 0.2 \times 10^{-4}\, Z_1) \times (2 \times 10^7 + 0.5 \times 10^7\, Z_2) - 78.125 \times (4 + Z_3) = 0$$

$$2000 + 400\,Z_1 + 500\,Z_2 + 100\,Z_1\,Z_2 - 312.5 - 78.125\,Z_3 = 0$$

$$400\,Z_1 + 500\,Z_2 + 100\,Z_1\,Z_2 - 78.125\,Z_3 + 1687.5 = 0$$

The reliability index β and the design point are then determined by the following equations

$$\beta = \frac{-1687.5}{400\alpha_1 + 500\alpha_2 + 100\beta \times \alpha_1\alpha_2 - 78.125\alpha_3}$$

$$\alpha_1 = -\frac{1}{k}(400 + 100\beta\alpha_2)$$

$$\alpha_2 = -\frac{1}{k}(500 + 100\beta\alpha_1)$$

$$\alpha_3 = \frac{1}{k}(78.125)$$

$$\alpha_i = \frac{-\frac{\partial f}{\partial Z_j}(\beta\bar{\alpha})}{\left[\sum_{k=1}^{n}\left(\frac{\partial f}{\partial Z_k}(\beta\bar{\alpha})\right)^2\right]^{\frac{1}{2}}}, \quad i = 1, 2, \ldots n$$

$$f(\beta\alpha_1, \beta\alpha_2 \ldots \beta\alpha_n) = 0$$

where the same notation as in (Eq. 7.11) is used and where *k* is determined by putting

$$\sum_{i=1}^{3} \alpha_i^2 = \alpha_1^2 + \alpha_2^2 + \alpha_3^2 = 1$$

The first equation for β is written in a form suitable for iteration. The iteration is now performed by choosing starting values for β, α_1, α_2 and α_3 and calculating new values until only small modifications are obtained. This is shown in the following Table.

Start		**Iteration No.**				
		1	2	3	4	5
β	3	3.62	3.51	3.36	3.30	3.29
α_1	−0.58	−0.53	−0.33	−0.23	−0.19	−0.18
α_2	−0.58	−0.82	−0.91	−0.95	−0.97	−0.97
α_3	+0.58	0.22	0.23	0.19	0.18	0.17

The reliability index is $\beta = 3.29$ and the design point $A = 3.29(-0.18, -0.97, 0.17) = (-0.59, -3.19, 0.56)$. Note that the correct sign for α_i, $i = 1, 2, 3$ was chosen for the starting values in the Table. This can usually be done in the following way. A positive sign can be recommended, when the basic variables correspond to load effect, R (e.g., P) and a negative when variables correspond to resistance effect, S (e.g., E and I).

Example 7.4 Determine the reliability index for a simply supported beam for the limit state of shear $g(x) = \tau_s t_w d - 0.5P$. The beam is subjected to point load P at the mid span, where d is the depth of the beam, t_w is the thickness of the web, τ_s is the shear strength of the material. Assume the mean and standard deviation and coefficient of variations for different variables as $\mu_P = 4000$ N, $\sigma_P = 1000$ N, $(\text{CoV})_P = 0.25$ $\mu_{\tau_s} = 95$ N/mm^2, $\sigma_{\tau_s} = 10$ N/mm^2, $(\text{CoV})_{\tau_s} = 0.1053$, $\mu_d = 50$ mm, $\sigma_d = 2.5$ mm, $(\text{CoV})_d = 0.05$ and $d/t_w = 40$.

Solution:

$$g(x) = \tau_s t_w d - 0.5P \qquad \frac{d}{t_w} = 40 \qquad \therefore\ t_w = \frac{d}{40}$$

$$g(x) = \tau_s \frac{d^2}{40} - 0.5P$$

Normalize the variables

Here, $Z_1 = \dfrac{\tau_S - \mu_{\tau_S}}{\sigma_{\tau_S}} \qquad Z_2 = \dfrac{d - \mu_d}{\sigma_d} \qquad Z_3 = \dfrac{P - \mu_P}{\sigma_P}$

$$\Rightarrow \quad \tau_S = Z_1\sigma_{\tau_S} + \mu_{\tau_S} = (10Z_1 + 95)\,\text{N/mm}^2$$

$$\Rightarrow \quad d = Z_2\sigma_d + \mu_d = (2.5Z_2 + 50)\,\text{mm}$$

$$\Rightarrow \quad P = Z_3\sigma_P + \mu_P = (1000Z_3 + 4000)\,\text{N}$$

$$\therefore \quad g(x) = \tau_S t_w d - 0.5P$$

$$\Rightarrow \quad g(x) = \tau_S \frac{d^2}{40} - 0.5P$$

$$g(x) = \frac{1}{40}[(95 + 10Z_1)(50 + 2.5Z_2)^2] - 0.5(4000 + 1000Z_3)$$

$$\Rightarrow g(x) = \frac{1}{40}\{(95 + 10Z_1)(6.25Z_2^2 + 250Z_2 + 2500)\} - 0.5(4000 + 1000Z_3)$$

$$\Rightarrow g(x) = \frac{1}{40}[593.75Z_2^2 + 62.5Z_1Z_2^2 + 23750Z_2 + 2500Z_1Z_2 + 237500 + 25000Z_1] - (2000 + 500Z_3)$$

$$g(x) = [14.84375Z_2^2 + 1.5625Z_1Z_2^2 + 593.75Z_2 + 62.5Z_1Z_2 + 5937.5 + 625Z_1] - (2000 + 500Z_3)$$

$$\therefore g(x) = 5937.5 + 625Z_1 + 14.84375Z_2^2 + 1.5625Z_1Z_2^2 + 593.75Z_2 + 62.5Z_1Z_2 - 2000 - 500Z_3 \qquad \ldots(1)$$

At limit state condition, $g(x) = 0$

$$0 = 5937.5 + 625Z_1 + 14.84375Z_2^2 + 1.5625Z_1Z_2^2 + 593.75Z_2 + 62.5Z_1Z_2 - 2000 - 500Z_3$$

$$\Rightarrow \quad 1.5625Z_1Z_2^2 + 14.84375Z_2^2 + 625Z_1 + 593.75Z_2 + 62.5Z_1Z_2 - 500Z_3 = -3937.5$$

Put $Z_1 = \alpha_1\beta$, $Z_2 = \alpha_2\beta$, $Z_3 = \alpha_3\beta$ in the above Eq. (1)

$$\Rightarrow \quad \beta[1.5625\alpha_1\alpha_2^2\beta^2 + 14.8375\alpha_2^2\beta + 625\alpha_1 + 593.75\alpha_2 + 62.5\alpha_1\alpha_2\beta - 500\alpha_3] = -3937.5$$

$$\beta = \frac{-3937.5}{[1.5625\alpha_1\alpha_2^2\beta^2 + 14.8375\alpha_2^2\beta + 625\alpha_1 + 593.75\alpha_2 + 62.5\alpha_1\alpha_2\beta - 500\alpha_3]} \quad \ldots(2)$$

Differentiating Eq. (1) w.r.t Z_1, Z_2 and Z_3 respectively

$$\frac{\partial g(x)}{\partial Z_1} = 1.5625Z_2^2 + 625 + 62.5Z_2$$

$$\Rightarrow \quad \frac{\partial g(x)}{\partial Z_1} = 1.5625\,a_2^2\beta^2 + 625 + 62.5a_1\beta \quad \ldots(3)$$

$$\frac{\partial g(x)}{\partial Z_2} = 3.125Z_1Z_2 + 29.6875Z_2 + 62.5Z_1 + 593.75$$

$$\Rightarrow \quad \frac{\partial g(x)}{\partial Z_2} = 3.125\alpha_1\alpha_2\beta^2 + 29.6875\,\alpha_2\beta + 62.5\alpha_1\beta + 593.75 \quad \ldots(4)$$

$$\frac{\partial g(x)}{\partial Z_3} = -500 \quad \ldots(5)$$

$$\Rightarrow \quad \left[\frac{\partial g(x)}{\partial Z_1}\right]^2 + \left[\frac{\partial g(x)}{\partial Z_2}\right]^2 + \left[\frac{\partial g(x)}{\partial Z_3}\right]^2 = k^2 \quad \ldots(6)$$

Trial 1:

Put $\beta = 3$, $\alpha_1 = \alpha_2 = -\frac{1}{\sqrt{3}}, \alpha_3 = \frac{1}{\sqrt{3}}$

Now, Eq. (2) becomes

$$\beta_{\text{new}}^{\text{I}} = \frac{-3937.5}{\begin{bmatrix}1.5625\times(-1/\sqrt{3})(-1/\sqrt{3})^2(3)^2 + 14.8375(-1/\sqrt{3})^2\times 3 + 625(-1/\sqrt{3}) + 593.75(-1/\sqrt{3})\\ +62.5(-1/\sqrt{3})\times(-1/\sqrt{3})\times 3 - 500\times(1/\sqrt{3})\end{bmatrix}}$$

$\beta_{\text{new}}^{\text{I}} = 4.29066$

Here,

$$\frac{\partial g(x)}{\partial Z_1} = 1.5625\,(-1/\sqrt{3})^2(3)^2 + 625 + 62.5(-1/\sqrt{3})(3) = 521.43$$

$$\frac{\partial g(x)}{\partial Z_2} = 3.125(-1/\sqrt{3})(-1/\sqrt{3})(3)^2 + 29.6875(-1/\sqrt{3})(3) + 62.5(-1/\sqrt{3})(3) + 593.75 = 443.45$$

$$\frac{\partial g(x)}{\partial Z_3} = -500$$

Now, Eq. (6) becomes

$$\Rightarrow \qquad K^2 = (521.43)^2 + (443.45)^2 + (-500)^2 = 718543$$

$$\therefore \qquad K = \sqrt{718543} = 847.66$$

Now,

$$\alpha^I_{1\,\text{new}} = -\frac{\dfrac{\partial g(x)}{\partial Z_1}}{K} = -\frac{521.43}{847.66} = -0.6151$$

$$\alpha^I_{2\,\text{new}} = -\frac{\dfrac{\partial g(x)}{\partial Z_2}}{K} = -\frac{443.45}{847.66} = -0.5231$$

$$\alpha^I_{3\,\text{new}} = \frac{\dfrac{\partial g(x)}{\partial Z_3}}{K} = -\frac{500}{847.66} = -0.5898$$

Check:

$$\alpha_1^2 + \alpha_2^2 + \alpha_3^2 = [(-0.6151)^2 + (-0.5231)^2 + (-0.5898)^2] = 0.999845 \approx 1$$

Trial 2:

Put $\beta = 4.29066 \quad \alpha_1 = -0.6151, \quad \alpha_2 = -0.5231, \quad \alpha_3 = 0.5898$

$$\beta^{II} = \frac{-3937.5}{\begin{bmatrix} 1.5625\times(-0.6151)(-0.5231)^2(4.29066)^2 + 14.8375(-0.5231)^2 \times 4.29066 + 625(-0.6151) \\ +593.75(-0.5231) + 62.5(-0.6151)\times(-0.5231)\times 4.29066 - 500\times 0.5898 \end{bmatrix}}$$

$$\therefore \qquad \beta^{II} = 4.4185$$

Here,

$$\frac{\partial g(x)}{\partial Z_1} = 492.58$$

$$\frac{\partial g(x)}{\partial Z_2} = 380.66$$

$$\frac{\partial g(x)}{\partial Z_3} = -500$$

$$\Rightarrow \qquad K^2 = (492.58)^2 + (380.66)^2 + (-500)^2 = 637545.1$$

$$\therefore \qquad K = \sqrt{637545.1} = 798.46$$

Now,

$$\alpha_1^{II} = -\frac{\dfrac{\partial g(x)}{\partial Z_1}}{K} = -\frac{492.58}{798.46} = -0.6169$$

$$\alpha_2^{II} = -\frac{\frac{\partial g(x)}{\partial Z_2}}{K} = -\frac{380.66}{798.46} = -0.4767$$

$$\alpha_3^{II} = \frac{\frac{\partial g(x)}{\partial Z_3}}{K} = \frac{500}{798.46} = 0.6262$$

Check:

$$\alpha_1^2 + \alpha_2^2 + \alpha_3^2 = [(-0.6169)^2 + (-0.4767)^2 + (0.6262)^2] = 0.999934 \approx 1$$

Trial 3:

Put $\beta = 4.4185 \quad \alpha_1 = -0.6169, \quad \alpha_2 = -0.4767, \quad \alpha_3 = 0.6262$

$$\beta^{III} = \frac{-3937.5}{\begin{bmatrix} 1.5625 \times (-0.6169)(-0.4767)^2(4.4185)^2 + 14.8375(-0.4767)^2 \times 4.4185 + 625(-0.6169) \\ +593.75(-0.4767) + 62.5(-0.6169) \times (-0.4767) \times 4.4185 - 500 \times 0.6262 \end{bmatrix}}$$

$\therefore \qquad \beta^{III} = 4.4246$

Here,

$$\frac{\partial g(x)}{\partial Z_1} = 500.27$$

$$\frac{\partial g(x)}{\partial Z_2} = 378.78$$

$$\frac{\partial g(x)}{\partial Z_3} = -500$$

$\Rightarrow \qquad K^2 = (500.27)^2 + (378.78)^2 + (-500)^2 = 643755.97$

$\therefore \qquad K = \sqrt{643755.97} = 802.34$

Now,

$$\alpha_1^{III} = -\frac{\frac{\partial g(x)}{\partial Z_1}}{K} = -\frac{500.27}{802.34} = -0.6235$$

$$\alpha_2^{III} = -\frac{\frac{\partial g(x)}{\partial Z_2}}{K} = -\frac{378.78}{802.34} = -0.4721$$

$$\alpha_3^{III} = \frac{\frac{\partial g(x)}{\partial Z_3}}{K} = \frac{500}{802.34} = 0.6231$$

Check:

$$\alpha_1^2 + \alpha_2^2 + \alpha_3^2 = [(-0.6235)^2 + (-0.4721)^2 + (0.6231)^2] = 0.999884 \approx 1$$

$\therefore$ Reliability index, $\beta = 4.42$

Example 7.5 Determine the reliability index of a simply supported beam in Fig. 7.5 at the limit state of flexure. The failure surface is given by $g(x) = f_y Z - WL^2/8$. Given characteristic strength, $f_y = 250$ N/mm^2, section modulus, $Z = 3060 \times 10^3$ mm^3, load on beam, $W = 25$ kN/m, span, $L = 6.5$ m, and the coefficient of variation, $(\text{CoV})_{f_y} = 0.05$, $(\text{CoV})_Z = 0.10$, $(\text{CoV})_W = 0.15$ and $(\text{CoV})_L = 0.05$. Solve the problem by the Hasofer-Lind method.

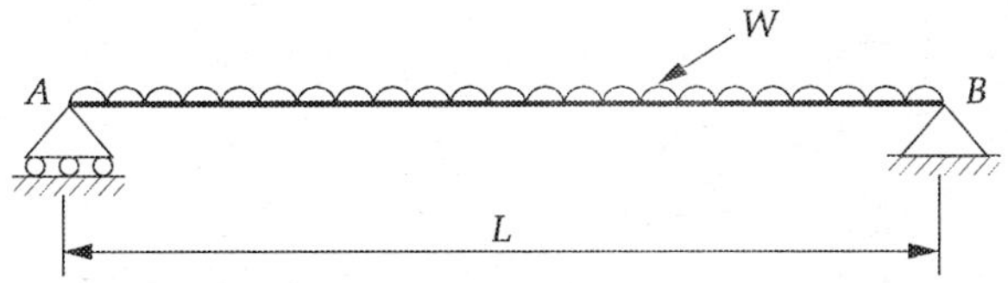

Fig. 7.5 Simply supported beam carrying UDL

Solution: The strengths are underestimated and loads are overestimated, the mean values are computed from the characteristic values and are tabulated as given below:

Sl. No.	Variables	Characteristic value	Mean value (μ)	CoV	Standard deviation (σ)
1	f_y	250 N/mm^2	272.41	0.05	13.62
2	Z	3060×10^3 mm^3	3662.48×10^3	0.10	366248
3	W	25 N/mm	20.052	0.15	3.008
4	L	6500 mm	7082.54	0.05	354.13

The failure surface is given by $g(x) = f_y Z - WL^2/8$

Normalize the variables

Here, $Z_1 = \dfrac{f_y - \mu_{f_y}}{\sigma_{f_y}}$ $\quad Z_2 = \dfrac{Z - \mu_Z}{\sigma_Z}$ $\quad Z_3 = \dfrac{W - \mu_W}{\sigma_W}$ $\quad Z_4 = \dfrac{L - \mu_L}{\sigma_L}$

$\Rightarrow \quad f_y = \mu_{f_y} + \sigma_{f_y} Z_1 = (272.41 + 13.62Z_1)$ N/mm^2

$\Rightarrow \quad Z = \mu_Z + \sigma_Z Z_2 = (3662.48 \times 10^3 + 366.248Z_2)$ mm^3

$\Rightarrow \quad W = \mu_W + \sigma_W Z_3 = (20.052 + 3.008Z_3)$ N/mm

$\Rightarrow \quad L = \mu_L + \sigma_L Z_4 = (7082.54 + 354.13Z_4)$ mm

$g(x) = R - S$, where, $R = f_y Z \quad S = WL^2/8$

$$\Rightarrow \; g(x) = f_y Z - WL^2/8$$

$$g(x) = \left(\mu_{f_y} + \sigma_{f_y} Z_1\right)\left(\mu_Z + \sigma_Z Z_2\right) - \frac{(\mu_W + \sigma_W Z_3)(\mu_L + \sigma_L Z_4)^2}{8}$$

$$\Rightarrow \; g(x) = (272.41 + 13.62Z_1)\,(3662.48 \times 10^3 + 366248Z_2) - \frac{1}{8}\,[(20.052 + 3.008Z_3)\,(7082.54 + 354.13Z_4)^2]$$

$$\Rightarrow \; g(x) = (9.977 \times 10^8 + 4.988 \times 10^7 Z_1 + 9.977 \times 10^7 Z_2 + 4.988 \times 10^6 Z_1 Z_2) - \frac{1}{8}\,[(100.58 \times 10^7 + 15.088 \times 10^7 Z_3 + 25.148 \times 10^5 Z_4^{\,2} + 3.76 \times 10^5 Z_3 Z_4^{\,2} + 100.58 \times 10^6 Z_4 + 15.088 \times 10^6 Z_3 Z_4)]$$

$$\Rightarrow \; g(x) = (9.977 \times 10^8 + 4.988 \times 10^7 Z_1 + 9.977 \times 10^7 Z_2 + 4.9885 \times 10^6 Z_1 Z_2) - [(12.5725 \times 10^7 + 1.886 \times 10^7 Z_3 + 3.1435 \times 10^5 Z_4^{\,2} + 0.47 \times 10^5 Z_3 Z_4^{\,2} + 12.5725 \times 10^6 Z_4 + 1.886 \times 10^6 Z_3 Z_4)]$$

$$\Rightarrow \; g(x) = [8.7192 \times 10^8 + 4.988 \times 10^7 Z_1 + 9.977 \times 10^7 Z_2 + 4.988 \times 10^6 Z_1 Z_2 - 1.886 \times 10^7 Z_3 - 3.1435 \times 10^5 Z_4^{\,2} - 0.47 \times 10^5 Z_3 Z_4^2 - 12.5725 \times 10^6 Z_4 - 1.886 \times 10^6 Z_3 Z_4] = 0 \quad \ldots(1)$$

Put $Z_1 = \alpha_1\beta \quad Z_2 = \alpha_2\beta \quad Z_3 = \alpha_3\beta Z_4 = \alpha_4\beta$ as in Eq. (1)

$$\Rightarrow \; [8.7192 \times 10^8 + 4.988 \times 10^7\,\alpha_1\beta + 9.977 \times 10^7\,\alpha_2\beta + 4.988 \times 10^6\alpha_1\alpha_2\beta^2 - 1.886 \times 10^7\alpha_3\beta - 3.1435 \times 10^5\alpha_4^2\beta^2 - 0.47 \times 10^5\;\alpha_3\alpha_4^2\beta^3 - 12.5725 \times 10^6\alpha_4\beta - 1.886 \times 10^6\alpha_3\alpha_4\beta^2] = 0$$

$$\Rightarrow \; \beta\,[4.988 \times 10^7\;\alpha_1 + 9.977 \times 10^7\alpha_2 + 4.988 \times 10^6\alpha_1\alpha_2\beta - 1.886 \times 10^7\alpha_3 - 3.1435 \times 10^5\;\alpha_4^{\,2}\beta - 0.47 \times 10^5\,\alpha_3\alpha_4^{\,2}\beta^2 - 12.5725 \times 10^6\alpha_4 - 1.886 \times 10^6\alpha_3\alpha_4\beta] = -8.7192 \times 10^8$$

$$\beta = \frac{-8.7192 \times 10^8}{\begin{bmatrix} 4.988 \times 10^7\alpha_1 + 9.977 \times 10^7\alpha_2 + 4.988 \times 10^6\alpha_1\alpha_2\beta - 1.886 \times 10^7\alpha_3 - 3.1435 \times 10^5\alpha_4^{\,2}\beta \\ -0.47 \times 10^5\alpha_3\alpha_4^2\beta^2 - 12.5725 \times 10^6\alpha_4 - 1.886 \times 10^6\alpha_3\alpha_4\beta \end{bmatrix}} \quad \ldots(2)$$

Differentiating Eq. (1) w.r.t Z_1, Z_2, Z_3 and Z_4 respectively

$$\frac{\partial g(x)}{\partial Z_1} = 4.988 \times 10^7 + 4.988 \times 10^6\,Z_2$$

$$\Rightarrow \; \frac{\partial g(x)}{\partial Z_1} = 4.988 \times 10^7 + 4.988 \times 10^6\,\alpha_2\beta \quad \ldots(3)$$

$$\frac{\partial g(x)}{\partial Z_2} = 9.977 \times 10^7 + 4.988 \times 10^6 Z_1$$

$$\Rightarrow \quad \frac{\partial g(x)}{\partial Z_2} = 9.977 \times 10^7 + 4.988 \times 10^6 \alpha_1 \beta \quad \ldots(4)$$

$$\frac{\partial g(x)}{\partial Z_3} = -1.886 \times 10^7 - 0.47 \times 10^5 Z_4^2 - 1.886 \times 10^6 Z_4$$

$$\Rightarrow \quad \frac{\partial g(x)}{\partial Z_3} = -1.886 \times 10^7 - 0.47 \times 10^5 \alpha_4^2 \beta^2 - 1.886 \times 10^6 \alpha_4 \beta \quad \ldots(5)$$

$$\frac{\partial g(x)}{\partial Z_4} = -6.287 \times 10^5 Z_4 - 0.94 \times 10^5 Z_3 Z_4 - 12.5725 \times 10^6 - 1.886 \times 10^6 Z_3$$

$$\frac{\partial g(x)}{\partial Z_4} = -6.287 \times 10^5 \alpha_4 \beta - 0.94 \times 10^5 \alpha_3 \alpha_4 \beta^2 - 12.5725 \times 10^6 - 1.886 \times 10^6 \alpha_3 \beta \quad \ldots(6)$$

$$\Rightarrow \quad \left[\frac{\partial g(x)}{\partial Z_1}\right]^2 + \left[\frac{\partial g(x)}{\partial Z_2}\right]^2 + \left[\frac{\partial g(x)}{\partial Z_3}\right]^2 + \left[\frac{\partial g(x)}{\partial Z_4}\right]^2 = k^2 \quad \ldots(7)$$

Trial 1:

Put $\beta = 3$, $\alpha_1 = \alpha_2 = -\frac{1}{\sqrt{4}} = -\frac{1}{2} = -0.5$ $\alpha_3 = \alpha_4 = \frac{1}{\sqrt{4}} = \frac{1}{2} = 0.5$

$$\beta^I = \frac{-8.7192 \times 10^8}{\left[\begin{array}{r} 4.988 \times 10^7 \times (-0.5) + 9.977 \times 10^7 \times (-0.5) + 4.988 \times 10^6 \times (-0.5) \times (-0.5) \times (3) \\ -1.886 \times 10^7 \times (0.5) - 3.1435 \times 10^5 \times (0.5)^2 \times (3) - 0.47 \times 10^5 \times (0.5) \times (0.5)^2 \times (3)^2 \\ -12.5725 \times 10^6 \times (0.5) - 1.886 \times 10^6 \times (0.5) \times (0.5) \times (3) \end{array}\right]}$$

$\beta^I{}_{new} = 9.8518$

Here,

$$\therefore \quad \frac{\partial g(x)}{\partial Z_1} = 4.988 \times 10^7 + 4.988 \times 10^6 \times (-0.5) \times 3 = 42.397 \times 10^6$$

$$\therefore \quad \frac{\partial g(x)}{\partial Z_2} = 9.977 \times 10^7 + 4.988 \times 10^6 \times (-0.5) \times 3 = 92.28 \times 10^6$$

$$\therefore \quad \frac{\partial g(x)}{\partial Z_3} = -1.886 \times 10^7 - 0.47 \times 10^5 \times (0.5)^2 \times (3)^2 - 1.886 \times 10^6 \times 0.5 \times 3 = -21.79 \times 10^6$$

$$\therefore \quad \frac{\partial g(x)}{\partial Z_4} = -6.287 \times 10^5 \times 0.5 \times 3 - 0.94 \times 10^5 \times 0.5 \times 0.5 \times (3)^2 - 12.5725 \times 10^6 - 1.886 \times 10^6 \times 0.5 \times 3$$

$$= -16.55 \times 10^6$$

Now, Eq. (7) becomes

$\Rightarrow \quad K^2 = (42.397 \times 10^6)^2 + (92.28 \times 10^6)^2 + (-21.79 \times 10^6)^2 + (-16.55 \times 10^6)^2 = 1.106 \times 10^{16}$

$\therefore \quad K = \sqrt{1.106 \times 10^{16}} = 105.18 \times 10^6$

Now,

$$\alpha_1{}^I{}_{\text{new}} = -\frac{\frac{\partial g(x)}{\partial Z_1}}{K} = -\frac{42.397 \times 10^6}{105.18 \times 10^6} = -0.4030$$

$$\alpha_2{}^I{}_{\text{new}} = -\frac{\frac{\partial g(x)}{\partial Z_2}}{K} = -\frac{92.28 \times 10^6}{105.18 \times 10^6} = -0.8773$$

$$\alpha_3{}^I{}_{\text{new}} = \frac{\frac{\partial g(x)}{\partial Z_3}}{K} = \frac{21.79 \times 10^6}{105.18 \times 10^6} = 0.2072$$

$$\alpha_4{}^I{}_{\text{new}} = \frac{\frac{\partial g(x)}{\partial Z_4}}{K} = \frac{16.55 \times 10^6}{105.18 \times 10^6} = 0.1574$$

Check:

$$\alpha_1^2 + \alpha_2^2 + \alpha_3^2 + \alpha_4^2 = \left[(-0.4030)^2 + (-0.8773)^2 + (0.2072)^2 + (0.1574)^2\right] = 0.99977 \approx 1$$

Trial 2:

$\beta = 9.8518,\ \alpha_1 = -0.4030,\ \alpha_2 = -0.8773,\ \alpha_3 = 0.2072,\ \alpha_4 = 0.1574$

$$\beta^{II} = \frac{-8.7192 \times 10^8}{\left[\begin{array}{r} 4.988 \times 10^7 \times (-0.4030) + 9.977 \times 10^7 \times (-0.8773) + 4.988 \times 10^6 \times (-0.4030) \times (-0.8773) \times 9.8518 \\ -1.886 \times 10^7 \times (0.2072) - 3.1435 \times 10^5 \times (0.1574)^2 \times (9.8518) - 0.47 \times 10^5 \times (0.2072) \times (0.1574)^2 \\ \times (9.8518)^2 - 12.5725 \times 10^6 \times (0.1574) - 1.886 \times 10^6 \times (0.2072) \times (0.1574) \times (9.8518) \end{array}\right]}$$

$\therefore \quad \beta^{II} = 9.0023$

Here,

$$\therefore \quad \frac{\partial g(x)}{\partial Z_1} = 6.759 \times 10^6$$

$$\therefore \quad \frac{\partial g(x)}{\partial Z_2} = 79.96 \times 10^6$$

$$\therefore \qquad \frac{\partial g(x)}{\partial Z_3} = -21.89 \times 10^6$$

$$\therefore \qquad \frac{\partial g(x)}{\partial Z_4} = -17.69 \times 10^6$$

$$\Rightarrow \quad K^2 = (6.759 \times 10^6)^2 + (79.96 \times 10^6)^2 + (-21.89 \times 10^6)^2 + (-17.69 \times 10^6)^2 = 7.23 \times 10^{15}$$

$$\therefore \quad K = \sqrt{7.23 \times 10^{15}} = 85.04 \times 10^6$$

Now,

$$\alpha_1^{II} = -\frac{\frac{\partial g(x)}{\partial Z_1}}{K} = -\frac{6.759 \times 10^6}{85.04 \times 10^6} = -0.0794$$

$$\alpha_2^{II} = -\frac{\frac{\partial g(x)}{\partial Z_2}}{K} = -\frac{79.96 \times 10^6}{85.04 \times 10^6} = -0.9402$$

$$\alpha_3^{II} = \frac{\frac{\partial g(x)}{\partial Z_3}}{K} = \frac{21.89 \times 10^6}{85.04 \times 10^6} = 0.2574$$

$$\alpha_4^{II} = \frac{\frac{\partial g(x)}{\partial Z_4}}{K} = \frac{17.69 \times 10^6}{85.04 \times 10^6} = 0.2080$$

Check:

$$\alpha_1^2 + \alpha_2^2 + \alpha_3^2 + \alpha_4^2 = \left[(-0.0794)^2 + (-0.9402)^2 + (0.2574)^2 + (0.2080)^2\right] = 0.99979 \approx 1$$

Trial 3:

$\beta = 9.0023$, $\alpha_1 = -0.0794$, $\alpha_2 = -0.9402$, $\alpha_3 = 0.2574$, $\alpha_4 = 0.2080$

$$\beta^{III} = \frac{-8.7192 \times 10^8}{\left[\begin{array}{r} 4.988 \times 10^7 \times (-0.0794) + 9.977 \times 10^7 \times (-0.9402) + 4.988 \times 10^6 \times (-0.0794) \times (-0.9402) \times 9.0023 \\ -1.886 \times 10^7 \times (0.2574) - 3.1435 \times 10^5 \times (0.2080)^2 \times (9.0023) - 0.47 \times 10^5 \times (0.2574) \times (0.2080)^2 \\ \times (9.0023)^2 - 12.5725 \times 10^6 \times (0.2080) - 1.886 \times 10^6 \times (0.2574) \times (0.2080) \times (9.0023) \end{array}\right]}$$

$$\therefore \quad \beta^{III} = 8.4681$$

Here,

$$\therefore \qquad \frac{\partial g(x)}{\partial Z_1} = 7.65 \times 10^6$$

$$\therefore \qquad \frac{\partial g(x)}{\partial Z_2} = 96.20 \times 10^6$$

$$\therefore \quad \frac{\partial g(x)}{\partial Z_3} = -22.55 \times 10^6$$

$$\therefore \quad \frac{\partial g(x)}{\partial Z_4} = -18.53 \times 10^6$$

$$\Rightarrow K^2 = (7.65 \times 10^6)^2 + (96.20 \times 10^6)^2 + (-22.55 \times 10^6)^2 + (-18.53 \times 10^6)^2 = 1.016 \times 10^{16}$$

$$\therefore \quad K = \sqrt{1.016 \times 10^{16}} = 100.82 \times 10^6$$

Now,

$$\alpha_1^{III} = -\frac{\frac{\partial g(x)}{\partial Z_1}}{K} = -\frac{7.65 \times 10^6}{100.82 \times 10^6} = -0.0759$$

$$\alpha_2^{III} = -\frac{\frac{\partial g(x)}{\partial Z_2}}{K} = -\frac{96.20 \times 10^6}{100.82 \times 10^6} = -0.9541$$

$$\alpha_3^{III} = \frac{\frac{\partial g(x)}{\partial Z_3}}{K} = \frac{22.55 \times 10^6}{100.82 \times 10^6} = 0.2237$$

$$\alpha_4^{III} = \frac{\frac{\partial g(x)}{\partial Z_4}}{K} = \frac{18.53 \times 10^6}{100.82 \times 10^6} = 0.1837$$

Check:

$$\alpha_1^2 + \alpha_2^2 + \alpha_3^2 + \alpha_4^2 = \left[(-0.0759)^2 + (-0.9541)^2 + (0.2237)^2 + (0.1837)^2\right] = 0.99985 \approx 1$$

Trial 4:

$\beta = 8.4681,\ \alpha_1 = -0.0759,\ \alpha_2 = -0.9541,\ \alpha_3 = 0.2237,\ \alpha_4 = 0.1837$

$$\beta^{IV} = \frac{-8.7192 \times 10^8}{\begin{bmatrix} 4.988 \times 10^7 \times (-0.0759) + 9.977 \times 10^7 \times (-0.9541) + 4.988 \times 10^6 \times (-0.0759) \times (-0.9541) \times 8.4681 \\ -1.886 \times 10^7 \times (0.2237) - 3.1435 \times 10^5 \times (0.1837)^2 \times (8.4681) - 0.47 \times 10^5 \times (0.2237) \times (0.1837)^2 \\ \times (8.4681)^2 - 12.5725 \times 10^6 \times (0.1837) - 1.886 \times 10^6 \times (0.2237) \times (0.1837) \times (8.4681) \end{bmatrix}}$$

$$\therefore \quad \beta^{IV} = 8.4468$$

Here,

$$\therefore \quad \frac{\partial g(x)}{\partial Z_1} = 9.57 \times 10^6$$

$$\therefore \qquad \frac{\partial g(x)}{\partial Z_2} = 96.56 \times 10^6$$

$$\therefore \qquad \frac{\partial g(x)}{\partial Z_3} = -21.90 \times 10^6$$

$$\therefore \qquad \frac{\partial g(x)}{\partial Z_4} = -17.40 \times 10^6$$

$$\Rightarrow \qquad K^2 = (9.57 \times 10^6)^2 + (96.56 \times 10^6)^2 + (-21.90 \times 10^6)^2 + (-17.40 \times 10^6)^2 = 1.019 \times 10^{16}$$

$$\therefore \qquad K = \sqrt{1.019 \times 10^{16}} = 100.98 \times 10^6$$

Now,

$$\alpha_1^{IV} = -\frac{\frac{\partial g(x)}{\partial Z_1}}{K} = -\frac{9.57 \times 10^6}{100.98 \times 10^6} = -0.0947$$

$$\alpha_2^{IV} = -\frac{\frac{\partial g(x)}{\partial Z_2}}{K} = -\frac{96.56 \times 10^6}{100.98 \times 10^6} = -0.9561$$

$$\alpha_3^{IV} = \frac{\frac{\partial g(x)}{\partial Z_3}}{K} = \frac{21.90 \times 10^6}{100.98 \times 10^6} = 0.2169$$

$$\alpha_4^{IV} = \frac{\frac{\partial g(x)}{\partial Z_4}}{K} = \frac{17.40 \times 10^6}{100.98 \times 10^6} = 0.1723$$

Check:

$$\alpha_1^2 + \alpha_2^2 + \alpha_3^2 + \alpha_4^2 = \left[(-0.0947)^2 + (-0.9561)^2 + (0.2169)^2 + (0.1723)^2\right] = 0.99982 \approx 1$$

$\therefore$ Reliability index, $\beta = 8.446$

Note: In the Hasofer-Lind procedure, any error in computations gets self-corrected automatically resulting in increased number of iterations.

7.2 EQUIVALENT NORMAL DISTRIBUTION (RACKWITZ 1976)

This transformation is chosen in such a way that the values of the original density function f_{x_i} and the original distribution function F_{x_i} for the random variable X_i are equal to the corresponding value of equivalent normal density and distribution function at the design point in Fig. 7.6 at the failure surface. It is reasonable in some situations to expect that a given random variable can be considered normally distributed as a good appropriate point X_i^* on the failure surface.

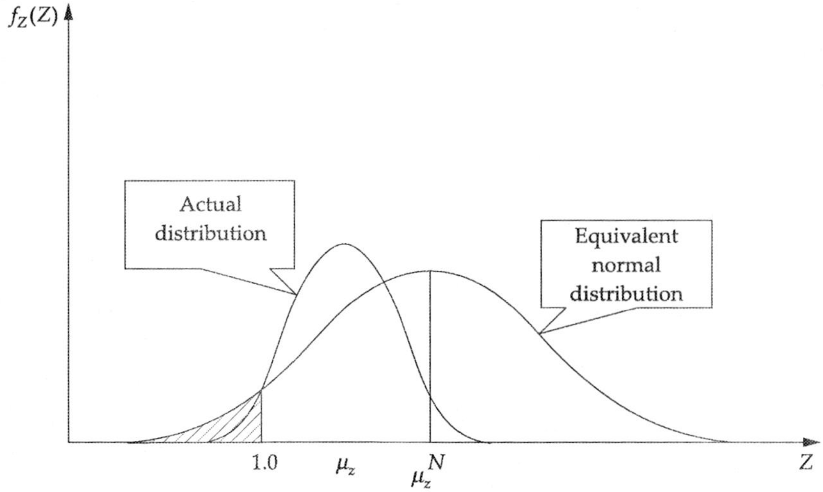

Fig. 7.6 Actual distribution and equivalent normal distribution

The Gaussian probability density function is given by

$$f_X(x) = \frac{1}{\sigma_X \sqrt{2\pi}} \exp\left\{-\frac{1}{2}\left(\frac{x-\mu_X}{\sigma_X}\right)^2\right\} \quad \ldots(7.20)$$

where, μ and σ are the parameters of the normal distribution denoted by $N(\mu, \sigma)$

The cumulative distribution of the normal distribution is given by

$$F_X(x) = \int_{-\infty}^{+X} \frac{1}{\sigma_X \sqrt{2\pi}} \exp\left\{-\frac{1}{2}\left(\frac{x-\mu_X}{\sigma_X}\right)^2\right\} \quad \ldots(7.21)$$

The integral cannot be evaluated in closed form and hence the transformation

$$Z = (x - \mu_x/\sigma_x);$$
$$dx = \sigma_x \, dz$$

$$F_X(x) = \int_{-\infty}^{Z} \frac{1}{\sqrt{2\pi}} \exp\left\{-\frac{1}{2}(Z)^2\right\} dz = \Phi_x\left(\frac{x-\mu_X}{\sigma_X}\right) \quad \ldots(7.22)$$

$$\Phi_X(x) = \frac{1}{\sqrt{2\pi}} \exp\left\{-\frac{1}{2}(Z)^2\right\} \quad \ldots(7.23)$$

The cumulative probability of normal and non-normal variate at the failure point on the surface is given by

$$\Phi\left(\frac{X_i^* - \mu_{X_i}^N}{\sigma_{X_i}^N}\right) = F_{X_i}(x_i^N) \quad \ldots(7.24)$$

Equating the ordinates of the normal and non-normal probability density function at x_i^*

$$f_{X_i}(x_i^*) = \frac{\phi_{X(x)}}{\sigma_{X_i}^N}\left(\frac{X_i^* - \mu_{X_i}^N}{\sigma_{X_i}^N}\right) \qquad \ldots(7.25)$$

Solving Eqs 7.24 and 7.25, the mean and standard deviation of the equivalent normal distribution is obtained for the original density and cumulative distribution function. Solving Eq. 7.24 we get

$$\Phi^{-1}\left\{F_{X_i}(x_i)\right\} = \left(\frac{X_i^* - \mu_{X_i}^N}{\sigma_{X_i}^N}\right) \qquad \ldots(7.26)$$

Substitute LHS of Eq. 7.24, for $\left(\frac{X_i^* - \mu_{X_i}^N}{\sigma_{X_i}^N}\right)$ and we get the equivalent normal standard deviation as in Eq. 7.26.

$$\sigma_{X_i}^N = \frac{\phi_{X(x)}}{f_{X_i}(x_i^*)}\{\Phi^{-1}\{F_{X_i}(x_i^*)\}\} \qquad \ldots(7.27)$$

The equivalent normal mean value of the distribution is given by the expression in Eq. 7.28

$$\mu_{X_i}^N = -\sigma_{X_i}^N\,\Phi^{-1}\{F_{X_i}(x_i)\} + x_i^* \qquad \ldots(7.28)$$

7.3 EQUIVALENT LOG-NORMAL DISTRIBUTION

The Log Normal distribution $N(\lambda, \xi)$ and its parameters are estimated from the known values of the sample mean and standard deviation

$$\mu = \exp\left(\lambda + \frac{1}{2}\xi^2\right) \qquad \ldots(7.29)$$

$$\sigma = \sqrt{\mu^2(e^{\xi^2} - 1)} \qquad \ldots(7.30)$$

$$\xi^2 = \ln\left(1 + \frac{\sigma^2}{\mu^2}\right) \qquad \ldots(7.31)$$

$$\lambda = \ln(\mu) - \frac{1}{2}\xi^2 \qquad \ldots(7.32)$$

The cumulative log-normal distribution evaluated at x^* is given

$$\Phi\left(\frac{\ln x^* - \lambda}{\xi}\right) = F_{LN}(x_i^*) \qquad \ldots(7.33)$$

$$F_{LN}(x_i^*) = \frac{\varphi}{x^*\xi}\left(\frac{\ln x^* - \lambda}{\xi}\right) \qquad \ldots(7.34)$$

The equivalent normal mean and standard deviation values are obtained as

$$\mu_X^N = X^*(1 - \ln X^* + \lambda) \qquad \ldots(7.35)$$

$$\sigma_X^N = X^*\,\xi \qquad \ldots(7.36)$$

Note: 'Φ' denotes cumulative distribution function, gives area under distribution and 'ϕ' represents probability density function, gives ordinate.

7.4 EXPONENTIAL DISTRIBUTION

Determine the equivalent normal distribution using the univariate Rosenblueth transformation in order to determine the equivalent normal distribution for the capacity. Given the mean value of X which is exponentially distributed with mean value of 8 m^3/s.

$$f_X(x) = \frac{1}{8}\exp\left(\frac{-x}{8}\right) = 0.125\exp(-0.125x)$$

$$F_X(x) = 1 - \exp\left(\frac{-x}{8}\right)$$

The mean μ_x^* and the standard deviation σ_x^* of the equivalent normal distribution for the exponential capacity X are found from the assumption that, at failure point x^*

$$\Phi\left(\frac{X_i^* - \mu_{X_i}^N}{\mu_{X_i}^N}\right) = F_{X_i}(x_i^N)$$

where $\Phi(.)$ denoted the cdf of the standard normal variate. Thus, the mean value

$$\mu_{X_i}^N = -\sigma_{X_i}^N \Phi^{-1}\{F_{X_i}(x_i)\} + x_i^*$$

From the previous assumption that at the failure point, by equating the corresponding probability density at the failure point.

$$f_{X_i}(x_i^*) = \frac{\phi\mu_X}{\sigma_{X_i}^N}\left(\frac{X_i^* - \sigma_{X_i}^N}{\sigma_{X_i}^N}\right)$$

By substitution

$$\sigma_{X_i}^N = \frac{\phi_{X(x)}}{f_{X_i}(x_i^*)}\{\Phi^{-1}\{F_{X_i}(x_i^*)\}\}$$

$$\sigma_{X_i}^N = \frac{\phi_{X(x)}}{[0.125\exp(-0.125\times 1)]}\{\Phi^{-1}[1-\exp(-0.125\times 1)]\}$$

$$\sigma_{X_i}^N = \frac{\phi_{X(x)}}{0.110}\{\Phi^{-1}\{-0.1175\}\}$$

$$\sigma_{X_i}^N = \frac{\phi_{X(x)}}{0.110}\{-0.1175\} = 1.79$$

$$\mu_{X_i}^N = -\sigma_{X_i}^N \Phi^{-1}\{F_{X_i}(x_i)\} + x_i^*$$

$$\mu_{X_i}^N = -1.79\,\Phi^{-1}\{[1 - \exp(-0.125\times 1)]\} + 1$$

$$\mu_{X_i}^N = -1.79\,\Phi^{-1}\{0.118\} + 1 = 3.12$$

7.5 GAMMA DISTRIBUTION

Determine the equivalent normal distribution using the univariate Rosenblueth transformation in order to determine the equivalent normal distribution for the capacity. Given, if X is Gamma distributed with mean 8 m^3/s and coefficient of variation is $\frac{1}{\sqrt{2}}$. The parameters of the gamma PDF are found to be by the method of moment

$$r = \left(\frac{1}{V_x}\right)^2 = \left(\frac{1}{\sqrt{2}}\right)^2 = \frac{1}{2}.$$

$$\lambda = \frac{r}{\mu_x} = \frac{\frac{1}{2}}{8} = 0.0625$$

$$\text{PDF } f_X(x) = \left[\frac{\lambda^r}{\Gamma(r)}\right] x^{r-1} \exp(-\lambda x)$$

$$f_X(x) = 0.25^2 x \exp(-0.25x)$$

and for $r = 2$

$$F_X(x) = \int_0^x \frac{\lambda^2}{\Gamma(2)} z^{2-1} \exp(-\lambda z)\, dz$$

$$1 - (1 + \lambda z)\exp^{-\lambda z} = 1 - (1 + 0.25x)\exp^{-0.25x}$$

Using this procedure, one gets for the initial values of $x^* = 1$ m^3/s

$$\sigma_{X_i}^N = \frac{\varphi x_x}{[0.25^2 \times \exp(-0.25 \times 1)]} \{\Phi^{-1}[1 - (1 + 0.25 \times 1)\exp(-0.25)]\} = 1.26$$

$$\mu_{X_i}^N = -1.26\Phi^{-1}\{[1 - (1 + 0.25 \times 1)\exp(-0.25 \times 1)]\} + 1 = 3.44$$

7.6 EQUIVALENT GUMBEL DISTRIBUTION

Determine the equivalent normal distribution using the univariate Rosenblueth transformation in order to determine the equivalent normal distribution for the capacity. Given, if X is Gamma distributed with a mean 5.16 and standard deviation 0.93. The parameters of the gamma PDF are to be found by the method of moments.

$$f_X(x) = \exp(-\exp(-\alpha(x - u))$$

$$f_X(x) = \exp(-\exp(-\alpha(x - u)) - \alpha(x - u))\alpha$$

where $\mu_X = \dfrac{U + 0.57722}{\alpha}$

$$\sigma_X = \frac{\pi}{\sqrt{6}\,\alpha}$$

The parameters of the Gumbel distribution

$$\alpha = \frac{\pi}{\sqrt{6}\,\sigma_X}$$

$$U = \mu_X \alpha - 0.57722$$

The parameters of the Gumbel Extreme value – I distribution is determined as

$$\alpha = \frac{\pi}{\sqrt{6} \times 0.93} = 1.379$$

$$U = 5.16 \times 1.379 - 0.57722 = 6.53$$

$$\sigma_{X_i}^N = \frac{\varphi X_x}{f_{x_i}(5.16)} \{\Phi^{-1}\{f_{x_i}(5.16)\}\}$$

$$\sigma_{X_i}^N = \frac{\varphi X_x}{0.442} \{\Phi^{-1}\{0.570\}\}$$

$$\sigma_{X_i}^N = \frac{\varphi X_x}{0.442} \{0.177\}$$

$$\sigma_{X_i}^N = \frac{1}{0.442} \{0.177\} = 0.889$$

$$\mu_{X_i}^N = -0.889\,\Phi^{-1}\{F_{xi}(5.16)\} + 5.16$$

$$\mu_{X_i}^N = -0.889 \times 0.177 + 5.16 = 5.09$$

7.6.1 Return Period

The probability $p = P(X \geq x_T)$ of occurrence of the event $X \geq x_T$ in any observation may be related to the return period in the following way. For each observation, there are two possible outcomes; either "success" $X \geq x_T$ (probability P) or "failure" $X < x_T$ (probability $1 - P$). Since the observations are independent, the probability of a recurrence interval duration τ is the product of the probabilities of $\tau - 1$ failures followed by one success, that is

$(1 - p)^{\tau-1}\, p$, and the expected value of τ is given by,

$$E(\tau) = \sum_{\tau=1}^{\infty} (1-p)^{\tau-1}\, p \ldots\ldots 1$$

$$= p + 2(1-p)p + 3(1-p)^2 p + 4(1-p)^3 p + \ldots\ldots$$

$$= p[1 + 2(1-p) + 3(1-p)^2 + 4(1-p)^3 + \ldots]$$

The expression within brackets has the form of the power series expansion

$$(1+x)^n = 1 + nx + \left[\frac{n(n-1)}{2}\right] x^2 + \left[\frac{n(n-1)(n-2)}{6}\right] x^3 \text{ with } x = -(1-p) \text{ and } n = 2$$

$$E(\tau) = \frac{p}{[1-(1-p)]^2} = \frac{1}{p}$$

Hence, $E(\tau) = T = \frac{1}{p}$; that is, the probability of occurrence of an event in any observation is the inverse of its return period;

$$P(X \geq x_T) = \frac{1}{T}$$

What is the probability that a T-year return period event will occur at least once in N years? To calculate this, first consider the situation where no T-year event occurs in N years. This would require a sequence of N successive "failures" so that

$$P(X < x_T \text{ each year for } N \text{ years}) = (1 - p)^N$$

The complement of this situation is the case required

$$P(X < x_T \text{ at least once in } N \text{ years}) = 1 - (1 - p)^N$$

Since $p = \frac{1}{T}$

Therefore, $P(X < x_T \text{ at least once in } N \text{ years}) = 1 - \left(1 - \frac{1}{T}\right)^N$

Example 7.6 Estimate the probability that the annual maximum discharge on the X river will exceed 50,000 cfs at least once during the next three years. The probability of occurrence of the event is 0.195.

Solution: $P(Q \geq 50{,}000 \text{ cfs at least once in next 3 years}) = 1 - (1 - 0.195)^3$

$$= 0.478$$

UNSOLVED PROBLEMS

1. Determine the reliability index of a concrete member. The variables are as given below. The performance function is given by $g(x) = A_{st} f_y d - 0.77 A_{st}^2 f_y^2 / b f_{ck}$.

Sl. No.	Variables	Variables Magnitude	Coefficient of variation	Distribution type
1	A_{st}	7390 mm^2	0.05	Normal
2	f_y	415 N/mm^2	0.03	Normal
3	d	1450 mm	0.05	Normal
4	b	300 mm	0.05	Normal
5	f_{ck}	20 N/mm^2	0.05	Normal

2. The performance function with respect to flexure of a beam is given by $Z = (f_y \times Z - 0.125WL^2)$ where f_y is the yield stress, S is the elastic section modulus, W load on the beam and L the span of the beam. Assuming the following data, evaluate the reliability index by the Hasofer-Lind method.

Sl. No.	Variables	Mean value (μ)	Coefficient of variation
1	f_y	380 N/mm^2	0.05
2	Z	100 mm^3	0.05
3	W	0.3 N/mm	0.25
4	L	180 mm	0.05

3. The statically indeterminate beam shown in Fig. 7.7 is loaded by a concentrated force P and the following deflection failure criterion is given by $\Delta_{max} = \frac{1}{192}\frac{PL^3}{EI} \geq \frac{1}{100}L$ where Δ_{max} is the maximum deflection, E the modulus of elasticity and I the relevant moment of inertia. Further, let Δ_{max}, P, L, E and I be the realizations of uncorrelated random variables P, L, E and I with the following mean values and standard deviation. Evaluate the value of the moment of inertia corresponding to the reliability index of value equal to 3.09.

Sl. No.	Variables	Mean value (μ)	Coefficient of variation
1	μ_P	4 kN	0.25
2	μ_L	5 m	0.00
3	μ_E	2 × 10^7 kN/m^2	0.25
4	μ_I	10^{-4}m^4	0.10

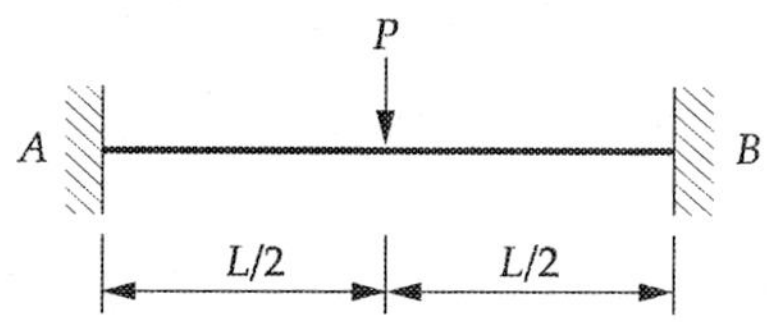

Fig. 7.7 Statically indeterminate beam

4. Determine the mean value of the area of cross section, A of a tension member for a reliability index of 3.09 whose performance function is given by $M = (f_y \times A - P)$, given the following data

Sl. No.	Variables	Mean value (μ)	Standard deviation
1	Yield stress, f_y	272.40 N/mm^2	13.62 N/mm^2
2	Tensile load, P	115.5 kN	10 kN

Assume the standard deviation of the area of cross section as 27.24 N/mm^2.

5. A segment of brick deck is under axial compression, P_a and prestressing force P_f. The condition for safety corresponding to the stress at top is given by

$f_{ck} \geq \frac{P_a}{A} + \frac{P_f}{A} + \frac{P_f \times e \times y}{I} - \frac{M \times y}{I}$. The prestressing force P_f can be expressed as the product of yield stress, f_y and area of tendons A_{st}. Evaluate the safety if the characteristic values and coefficient of variation (CoV) are as follows.

Sl. No.	Variables	Mean value (μ)	CoV
1	f_{ck}	40 N/mm^2	0.05
2	f_y	1570 N/mm^2	0.05
3	Pa	57096000 N	0.20
4	A_{st}	18780.44 mm^2	0.05
5	e	42.9 mm	0.05
6	y	568.4 mm	0.05
7	M	24651 × 10^6 N-mm	0.20
8	A	12259 × 10^3 mm^2	0.05
9	I	6.2282 × 10^{12} mm^4	0.05

8 CHAPTER

Monte Carlo Simulation

8.1 INTRODUCTION

The Monte Carlo Simulation (MCS) is known as the simple random sampling method or statistical trial method, that make realizations based on randomly generated sampling sets for uncertain variables. Application of Monte Carlo method to probabilistic analysis problem is comparatively recent, becoming practical only with the advent of digital computers. Many reliability problems would be impossible to solve without high processing computers. The Monte Carlo method consists of random variables and functions, statistical analysis of trial output and variable reduction techniques.

Monte Carlo analysis is a powerful engineering tool, which enables one to perform a statistical analysis of the uncertainty in structural engineering problems. It is particularly useful for complex problems where numerous random variables are related in an experiment, which is performed by a computer rather than in a structural engineering laboratory.

The computation procedure of MCS is quite simple as given below.

(1) Visualize your problem as a complex natural phenomenon, and try to imitate nature.
(2) Define the limit state in terms of basic variables.
(3) Select and use appropriate probabilistic distribution type for the random variable in terms of probability density function (PDF) or probability mass function (PMF) along with the corresponding parameters.
(4) Generate the values of random variables, according to the joint probability density function or at least according to the marginal distribution with their correlations and sampling set from the distribution.
(5) Conduct simulations by evaluating the problem deterministically for each realization of all the random variables, which is numerical experimentation, using the generated sampling set.
(6) Extract the probabilistic information for N such realizations.
(7) Determine the accuracy and efficiency of simulation.

The following steps are involved, namely, to find the probability of failure by the simulation technique:

Step 1: The sampling set of the corresponding random variables is generated according to the probability density function. The generated set of random variables is used in mathematical modeling $g(x)$ at limit state.

Step 2: Conduct simulations using the generated sampling set, and obtain the response of the structure.

Step 3: The limit state function $g(x)$ is violated, the structure or structural element has failed.

Step 4: The trial is repeated several times to guarantee convergence of the statistical results.

Step 5: If N trials or experiments are conducted using digitally generated data or value and analyzed.

Step 6: The probability of failure $P_f = \frac{N_f}{N}$, where N_f is the number of trials for which $g(x)$ is violated out of N samples.

8.2 GENERATION OF RANDOM VARIABLES

One of the key features in Monte Carlo sampling is the generation of a series of values of one or more random variables with specified probability distribution. The most commonly used generation methods are the following:

(a) The inverse transform method
(b) Composition method
(c) Acceptance and rejection method.

Generation of random numbers to a specific probabilistic distribution is the core or heart of Monte Carlo simulation. All computers generate uniform random numbers in the interval 0 to 1.

8.2.1 Inverse Transform Method (IT)

The next task is to transform the uniform random number, R_{Ui} generated between 0 to 1. The Cumulative Distribution Function (CDF) of the random variable is equated to the generated random numbers, R_{Ui};

$$F_X(x_i) = R_{Ui} \quad \text{...(8.1)}$$

$$x_i = F_X^{-1}(R_{Ui}) \quad \text{...(8.2)}$$

8.2.2 Linear Congruential Generator (LCG)

The most frequently used method for generating random numbers is Linear Congruential Generator. This generator is based on recursive calculations of the residues, m of a linear transformation.

$X_{i+1} = (aX_i + c) \bmod m$, where a, c and m are non-negative integers and mod is the modulus which is defined such that X_{i+1} is the remainder of the division of m into $(ax_i + c)$. If k_i is the integer part of the ratio $(ax_i + c)/m$, that is

$k_i = \text{Int}(ax_i + c/m)$ then the corresponding residue of modulus is

$$X_{i+1} = a \times X_i + c - m \times k_i \qquad \text{...(8.3)}$$

Normalizing the values obtained from Eq. (8.3) by the modulus m, we obtain

$$R_{Ui+1} = \frac{x_{i+1}}{m} \qquad \text{...(8.4)}$$

which constitutes a set of random values on the unit interval (0, 1) with standard uniform distribution (having a mean = 0.5 and standard deviation = 0.289).

Table 8.1 Constants for random number generator

Sl. No.	m (seed number)	a	c
1	7875	421	1663
2	11979	859	2531
3	21870	1291	4621
4	81000	421	17117
5	86436	1093	18257
6	117128	1277	24749
7	121500	4081	25673
8	134456	8121	28411
9	243000	4561	51349
10	259200	7141	54773

A sequence of pseudorandom numbers is cyclic and will repeat itself in at most m steps. To ensure randomness, the period of the cycle should be as long as possible and therefore, in practical applications a large value of m should be assigned in the generation of $U_{i.}$ The selection of values for m, a and c, is the most important step for creating a generator of this sort. Table 8.1 lists some choices for these constants that will yield a large period or cycle. These values have been tested statistically and give satisfactory results. Finally, this choice must be balanced with computational time and required storage space for the array.

Always record the seed number and sample size used at each run and record the execution time. An arbitrary seed value, the generator will produce uniform random numbers between (0, 1). If the seed value is changed, different sets of random numbers are generated. The random numbers are repeated after 10^9 random numbers. The random numbers generated are called pseudorandom numbers.

8.2.3 Generation of Random Variates

The next question that needs to be addressed is to use uniformly distributed pseudo-random numbers to generate values according to a particular probability density. The generation of random variants can be accomplished systematically from the uniform distribution on the interval (0, 1). This is done through the inverse transform method.

Two of the most important continuous distributions are the exponential and normal distributions. The exponential distribution is used to model dynamic loading and applied stress. The normal distribution provides a good representation of many variables and can be used to generate several other common distributions.

A nonstandard uniform distribution on any arbitrary interval (a, b) can be obtained using the inverse transform method. By this method, we solve for the random variable by inverting the cumulative distribution function.

$$F(x) = \begin{Bmatrix} 0 \\ (x-a)/(b-a) \\ 1 \end{Bmatrix} \begin{matrix} x < a \\ a \le x \le b \\ x > b \end{matrix} \qquad \ldots(8.5)$$

and the inverse function in this case is

$$x_i = F_X^{-1}(R_U)$$
$$X = a + (b-a) \times R_U$$

To apply the inverse transform, the distribution function must exist in a form for which the corresponding inverse transform can be found analytically.

Most engineering parameters fall into the normal, log-normal or exponential distribution, as listed in Table 8.2.

Table 8.2 Recursive relations to obtain common random variates

Sl. No.	Distribution	$F_X(x)$	$x = F_X^{-1}(R_U)$
1	Normal	$R_U = F_X(x) = \Phi\left(\frac{x-\mu}{\sigma}\right)$	$x = \mu + \sigma \times R_U$
2	Log-normal	$R_U = F_X(x) = \Phi\left(\frac{\ln x - \lambda}{\xi}\right)$ $\xi^2 = \ln\left[1 + \left(\frac{\sigma^2}{\mu^2}\right)\right]$; $\lambda = \ln \mu - \frac{\xi^2}{2}$	$x = \exp(\lambda + \xi \times R_U)$
3	Exponential	$R_U = 1 - \exp(-(x-c)/a),\ x > 0$ $\mu = c + a;$ $\sigma^2 = a^2$	$x = -a \times \ln(1 - R_U) + c$
4	Uniform	$R_U = F_x(x) = [(x-a)/(b-a)]$	$x = a + (b-a) \times R_U$
5	Gumbel	$R_U = \exp[-\exp[-(x-a)/b]^c$	$x = a - b \ln[-\ln(R_U)]$

Imagine that a random number for a uniform PDF with $a = 0.00$ and $b = 1.00$ has been obtained. Define this random number to be P_i. Locate P_i on the vertical axis of Fig. 8.1.

Now, construct a horizontal line until it intersects the probability distribution function curve. Then construct a vertical line through that intersection point, and the value of x on the bottom horizontal axis is denoted by x_i. That value for x_i is the value of the random number for a uniform PDF over the ranging from a to b.

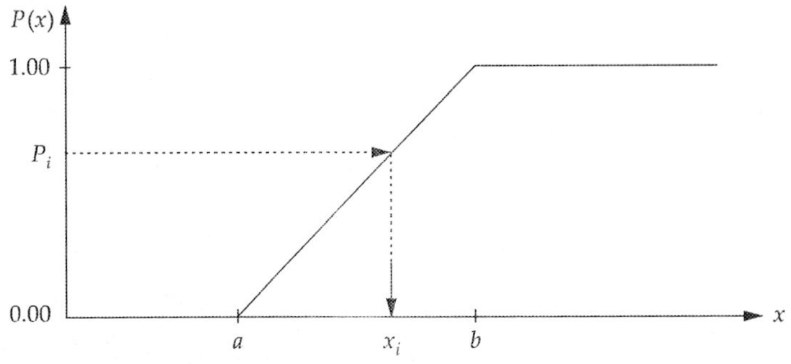

Fig. 8.1 Uniform probability distribution function

This solution can be expressed by the following equation

$$x_i = a + \left[\frac{b-a}{1.00-0.00}\right](P_i - 0.00) \qquad \ldots(8.6)$$

or, alternatively,

$$x_i = a + (b - a)P_i \qquad \ldots(8.7)$$

Consider the equation for the probability distribution function for this uniform PDF, which results from an integration of

$$P(x \le x_i) = \int_{x=a}^{x=x_i} \left(\frac{1}{b-a}\right) dx = \left(\frac{x_i}{b-a}\right) - \left(\frac{a}{b-a}\right)$$

Alternatively,

$$x_i = a + (b - a)P(x \le x_i) \qquad \ldots(8.8)$$

Therefore, the random number can be obtained from Eq. (8.7) or, alternatively, from Eq. (8.8) with $P(x \le x_i) = P_i$.

This latter method indicates that it is first necessary to integrate the PDF and obtain the corresponding probability distribution function. Then successively solve this equation once for each uniform random within a = 0.00 and b = 1.00. The value of the desired random number is obtained after this 0 to 1 random number is set equal to the value of the distribution function.

The fundamental step in Monte Carlo analysis is the development of a set of random numbers. These numbers can be mechanically or electronically generated, but in practice, most random number generation is accomplished by using digital computers. Uniform random numbers have the property that for a selected range of values (e.g., 0.0 to 1.0) the generated random numbers are equally likely to occur anywhere in the range. Table 8.3 lists uniform random numbers for the range 0 to 1.

Table 8.3 Uniform random numbers (0 to 1)

1	**0.86061**	26	**0.15017**	51	**0.42172**	76	**0.48932**
2	**0.92546**	27	**0.74098**	52	**0.95349**	77	**0.54707**
3	**0.41806**	28	**0.58515**	53	**0.16119**	78	**0.64271**
4	**0.28964**	29	**0.70074**	54	**0.58394**	79	**0.66930**
5	**0.14225**	30	**0.09666**	55	**0.95626**	80	**0.27681**
6	**0.44961**	31	**0.97948**	56	**0.20661**	81	**0.90451**
7	**0.24653**	32	**0.65400**	57	**0.24566**	82	**0.79163**
8	**0.21687**	33	**0.67980**	58	**0.94934**	83	**0.42397**
9	**0.56503**	34	**0.46872**	59	**0.16118**	84	**0.68086**
10	**0.40015**	35	**0.12846**	60	**0.01988**	85	**0.82174**
11	**0.83771**	36	**0.12237**	61	**0.27493**	86	**0.94600**
12	**0.73006**	37	**0.17468**	62	**0.03348**	87	**0.26457**
13	**0.56341**	38	**0.21305**	63	**0.38943**	88	**0.31697**
14	**0.82178**	39	**0.82744**	64	**0.36283**	89	**0.05336**
15	**0.32715**	40	**0.20220**	65	**0.41536**	90	**0.82238**
16	**0.68853**	41	**0.98479**	66	**0.30607**	91	**0.97673**
17	**0.74358**	42	**0.53164**	67	**0.14563**	92	**0.72927**
18	**0.24672**	43	**0.58442**	68	**0.44542**	93	**0.68251**
19	**0.90324**	44	**0.11799**	69	**0.53053**	94	**0.23987**
20	**0.79263**	45	**0.29124**	70	**0.58757**	95	**0.02894**
21	**0.44281**	46	**0.73958**	71	**0.17326**	96	**0.87885**
22	**0.70826**	47	**0.51527**	72	**0.10593**	97	**0.80716**
23	**0.22664**	48	**0.63765**	73	**0.72448**	98	**0.14197**
24	**0.62557**	49	**0.52224**	74	**0.44245**	99	**0.74708**
25	**0.48342**	50	**0.46079**	75	**0.37091**	100	**0.80193**

Table 8.4 gives the standard normal random numbers corresponding to the uniform numbers.

Table 8.4 Standard normal random numbers corresponding to the uniform numbers

1	1.08306	26	−1.03571	51	−0.19750	76	−0.02677
2	1.44279	27	0.64637	52	1.67968	77	0.11826
3	−0.20686	28	0.21509	53	−0.98958	78	0.36571
4	−0.55444	29	0.52653	54	0.21198	79	0.43798
5	−1.07027	30	−1.30081	55	1.70884	80	−0.59234
6	−0.12665	31	2.04313	56	−0.81824	81	1.30769
7	−0.68545	32	0.39614	57	−0.68821	82	0.81209
8	−0.78281	33	0.46714	58	1.63849	83	−0.19175
9	0.16373	34	−0.07849	59	−0.98962	84	0.47011
10	−0.25296	35	−1.13370	60	−2.05621	85	0.92202
11	0.98509	36	−1.16322	61	−0.59797	86	1.60725
12	0.61299	37	−0.93583	62	−1.83193	87	−0.62932
13	0.15962	38	−0.79588	63	−0.28081	88	−0.47619
14	0.92217	39	0.94410	64	−0.35090	89	−1.61313
15	−0.44780	40	−0.83379	65	−0.21378	90	0.92447
16	0.49169	41	2.16458	66	−0.50702	91	1.99047
17	0.65442	42	0.07939	67	−1.05536	92	0.61061
18	−0.68485	43	0.21321	68	−0.13724	93	0.47473
19	1.30024	44	−1.18509	69	0.07660	94	−0.70672
20	0.81558	45	−0.54977	70	0.22130	95	−1.89656
21	−0.14385	46	0.64205	71	−0.94136	96	1.16926
22	0.54831	47	0.03829	72	−1.24847	97	0.86748
23	−0.74996	48	0.35218	73	0.59620	98	−1.07151
24	0.32014	49	0.05578	74	−0.14476	99	0.66533
25	−0.04157	50	−0.09844	75	−0.32944	100	0.84854

Table 8.5 Uniform random numbers

52478	22835	33307	73842	67277	32880	76457	94489	82957	04836
80249	16089	01964	21414	72117	91712	11487	67479	13649	94539
94132	15190	08425	70298	02202	80519	23516	86294	32871	89573
56605	86696	37707	90117	17511	27701	35764	88217	70505	75300
58815	01919	22225	38562	45731	91743	99315	70350	78240	22015
69379	89366	50240	49343	31867	81661	41037	59120	44282	66605
75228	79546	65528	48794	73980	87645	22604	49290	08068	54935
14327	93484	49875	12103	77984	97966	08644	07089	18809	33738
90625	98430	03639	76657	26389	99093	51145	59343	22488	67026
06070	44497	21962	48270	68632	68338	39325	35105	42348	14412
33415	72559	19902	40024	74215	93857	04988	24389	22094	89237

Contd.

41999	12790	87990	77646	33177	62684	34119	09212	89973	39638
75908	62356	27342	93069	60284	69329	83998	15037	96165	62149
91323	56853	08468	69550	90860	57946	70370	23114	67185	04633
03428	01736	91578	09165	67708	36704	59481	28243	71395	38607
02333	25192	93932	65485	73266	95972	72606	89242	91968	25721
55696	67106	73369	20689	27707	10432	53118	23692	21450	67362
74838	46105	29798	05504	62588	12700	46093	58754	15780	00361
25833	46204	42441	14284	07858	94467	64358	84445	86230	54172
87260	93170	35494	54207	82683	22976	12257	94522	61364	34228
73595	29104	59346	21213	30923	15747	67104	90389	75901	45606
66224	21746	81973	43832	55932	81707	89193	01511	83257	89931
48078	26348	33935	08981	44947	78208	94370	82235	34382	18908
14168	38881	02968	71715	10814	96338	09439	53864	51951	15691
47813	96995	19524	17227	73490	09448	13156	41802	28217	32658
39404	88593	71327	08978	41241	88350	34760	19507	39102	17168
84131	64236	26803	09167	39695	98995	22498	49489	16808	10807
65097	63684	50298	98391	93703	55438	22718	78013	64409	97879
44552	13101	96263	88862	32977	22191	32112	41046	50771	86355
11997	06462	80215	16900	75972	76712	14861	97496	18986	66671
89716	28633	77208	34231	79158	12531	31612	23543	57480	75667
85258	16576	27023	25722	44809	61284	07636	67054	26665	73238
45790	04380	06893	83032	91230	36690	39612	65695	94966	21734
92386	86028	01737	24812	45158	40744	95550	79951	05457	30445
80321	92435	23677	33356	76405	93136	60668	43458	08562	70311
16964	90116	77618	38200	45273	20442	80655	13676	41471	59063
03060	35414	80332	87759	13961	07849	08970	67354	16026	23225
46517	83209	83758	25428	07686	24628	95824	11554	01428	80580
41481	93999	09645	04406	13666	79199	59323	59115	41436	33185
11580	04688	11925	57414	56554	94938	18151	93058	26924	16181
92862	25355	38189	68819	61797	70112	84563	54657	21490	52086
27419	80915	50829	23146	11641	29047	45806	98176	75455	09782
01450	54579	23503	31250	56057	44450	55982	73182	23666	66578
61200	38309	29934	09351	17290	61419	39377	01770	48134	58599
66047	26430	22415	98215	10413	54380	10492	59665	42368	15138
71899	68860	08150	39941	60556	23386	92449	31012	41277	18925
36567	46306	69777	56251	20007	74448	75234	58915	64903	24311
86135	49654	63467	35906	50560	24921	21109	18652	39797	19964
82155	74998	68901	12964	65056	61967	08628	88194	26741	52840
75099	37473	98759	91653	76447	34010	86452	82362	25185	90842

Example 8.1 Generate a random number and find the random value for a uniform distribution. Given $a = 0.0$ and $b = 272.4053$ N/mm^2.

Solution: Let us choose the 11th value of $R_U = 0.83771$ (uniform random number from Table 8.3)

From Table 8.2

$$x = a + (b - a) R_U$$

$$x = 0.0 + (272.4053 - 0) \times 0.83771 = 228.1966 \text{ N/mm}^2$$

Note: The user can choose any random number, however for this problem we have chosen 11th value from Table.

Example 8.2 Generate a standard normal random number and find the random value. Given $\mu = 272.4053$ N/mm^2, $\sigma = 13.6203$ N/mm^2.

Solution: Let us select the 11th value of $R_U = 0.83771$ (uniform random number from Table 8.3)

$$R_U = F_X(x) = \Phi\left(\frac{x - \mu}{\sigma}\right)$$

$$0.83771 = \Phi\left(\frac{x - \mu}{\sigma}\right)$$

$$x = \mu + \sigma\Phi^{-1}(R_U)$$

$$\Phi^{-1}(0.83771) = 0.98509 \text{ (from Table 8.4)}$$

$$X = 272.4053 + (0.98509 \times 13.6203) = 285.8251 \text{ N/mm}^2$$

Example 8.3 Generate a log-normal random number and find the random value. Given $\mu = 272.4053$ N/mm^2, $\sigma = 13.6203$ N/mm^2.

Solution

$$R_U = F_X(x) = \Phi\left(\frac{\ln x - \lambda}{\xi}\right)$$

Select the 11th value of $R_U = 0.83771$ (uniform random number from Table 8.3)

$$\xi^2 = \ln\left[1 + \left(\frac{\sigma^2}{\mu^2}\right)\right] = \ln[1 + (13.6203^2/272.4053^2)] = 2.496 \times 10^{-3}$$

$$\xi = 0.0499$$

$$\lambda = \ln \mu - \frac{\xi^2}{2} = \ln(272.4053) - (2.496 \times 10^{-3})\ (0.5) = 5.606$$

$$\begin{aligned} x &= \exp(\lambda + \xi.R_u) \\ &= \exp[5.606 + (0.0499)\ (0.83771)] \\ &= \exp(5.6478) \\ &= 283.6667 \text{ N/mm}^2 \end{aligned}$$

Example 8.4 Generate a exponential random number and find the random value. Given $\mu = 272.4053$ N/mm^2, $\sigma = 13.6203$ N/mm^2.

Select the 11th value of $R_U = 0.83771$ (uniform random number from Table 8.3)

$$R_U = 1 - \exp(-(x - c)/a),\ x > 0$$
$$\mu = c + a$$
$$\sigma^2 = a^2$$

The parameters of exponential distribution are, a and c.

$\sigma = a;\ a = 13.6203$

$c = \mu - \sigma;\ c = 272.4053 - 13.6203 = 258.785\ \text{N/mm}^2$

$x = -a \times \ln(1 - R_U) + c = -13.6203 \times \ln(1 - 0.83771) + 258.785 = 283.552\ \text{N/mm}^2$

8.3 NUMERICAL EXPERIMENTATION

Select N random numbers for each basic variable for the limit state chosen. Solve the problem deterministically and generate information on the output response. Compute the sample statistics, histogram, frequency diagram, PDF, CDF and probability of failure (P_f). The accuracy increases with increase in the number N. The extraction of such information could be cumbersome. The number of samples whose value of limit state or performance function is less than zero indicate failure. Therefore, an estimate of

$$P_f = \frac{N_f}{N} = \frac{\text{No. of samples failed}}{\text{Total number of samples}} \qquad \ldots(8.9)$$

8.4 ACCURACY AND EFFICIENCY OF SIMULATION

The variance or coefficient of variation may be estimated by assuming each simulation cycle to constitute a Bernoulli trial, and the number of failures in N trials.

$$\text{CoV}(P_f) = \frac{\sqrt{\dfrac{(1 - P_f)}{N}}}{P_f}$$

Smaller value of $\text{CoV}(P_f)$ is desirable, however $\text{CoV}(P_f)$ approaches zero as N approaches infinity.

To study the error associated with simulation cycles, binomial distribution approximates to normal distribution @ 95% confidence interval of the estimated probability of failure.

$$\text{Error (in \%)} = 200\sqrt{\frac{(1 - P_f)}{P_f}}$$

This equation may be used to determine the sample size N, required in a Monte Carlo simulation for a specified tolerable % error in the estimated P_f.

Example 8.5 A simply supported beam is shown in Fig. 8.2 with a circular cross section and a uniform loading. The deflection at the mid-span of the beam is

$$\Delta = \frac{5Wl^4}{384\,EI}$$

where, Δ = deflection at mid-span of the beam; l = length of beam; E = modulus of elasticity; W = distributed load; I = moment of inertia = $\pi d^4/64$. d = diameter of beam cross section. Assume that l, E and W are deterministic variables and also assume that the diameter is a random variable with a given PDF over the range $d_a \le d \le d_b$. How does one study the uncertainty in displacement?

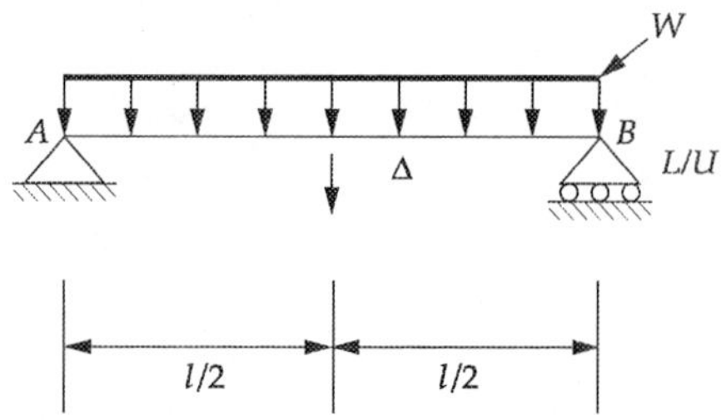

Fig. 8.2 Simply supported beam

Solution: The diameter of the beam is considered a random variable and PDF is considered uniform distribution with the values of $d_a = 5.0$ mm and $d_b = 7.0$ mm. Set the simulation size of random number $N = 10$.

Let us choose the 10 value of R_U(uniform random number from Table (8.3) from 1 to 10)

From Table (8.2)

$$x = a + (b - a)R_U;$$
$$a = d_a = 5.0 \text{ mm}; \; b = d_b = 7.0 \text{ mm}$$

Table 8.6 Computation of deflection of simply supported beam

SI. No.	R_U	$x = 5.0 + (2.0) \times R_U$ Diameter	Inertia (I) $\frac{\pi d^4}{64}$	(Δ Deflection) $\frac{Wl^4}{E} \times 10^{-3}$
1	0.86061	6.7212	100.1757	0.1298
2	0.92546	6.8509	108.1348	0.1202
3	0.41806	5.8361	56.9465	0.2283
4	0.28964	5.5793	47.5645	0.2733
5	0.14225	5.2845	38.2812	0.3396
6	0.44961	5.8992	59.4495	0.2187
7	0.24653	5.4931	44.6917	0.2909
8	0.21687	5.4337	42.7923	0.3038
9	0.56503	6.1301	69.3153	0.1875
10	0.40015	5.8003	55.5612	0.2340
			Total	2.35×10^{-3}

$$\Delta_1 = \left(\frac{5Wl^4}{384E}\right)\left(\frac{1}{I_1}\right) = 0.1298 \times 10^{-4}\left(\frac{Wl^4}{E}\right)$$

$$\text{Mean deflection } \bar{\Delta} = \left(\frac{Wl^4}{E}\right) \times 2.35 \times 10^{-4}$$

By statistical analysis, from Eq. (4.6), we get

Standard deviation $\sigma_\Delta = 6.61 \times 10^{-5}\left(\frac{Wl^4}{E}\right)$

Coefficient of variation = 0.2817

Example 8.6 The cantilever beam is uniformly loaded with 2 kN/m and carries an upward random force with a mean of 20 kN at the end. The coefficient of variation is 15%. Assume that the force has a uniform probability density function. Using the first 20 random numbers in column one of Table 8.3, calculate the 20 corresponding values of the moment 2 m from the end of the beam. Calculate the sample mean and variance of the moment.

Solution: The mean of the force is

$$\overline{P} = \left(\frac{a+b}{2}\right) = 20.0 \text{ kN, and variance is } \sigma_p^2 = \frac{(b-a)^2}{12} = [(20.0)(0.15)]^2 = 9$$

Therefore, the limits on the uniform PDF and its parameters are, a = 14.8037 kN, b = 25.1963 kN.

The moment corresponding to the ith generated value of the moment at the free end of cantilever is $M = 4.0 - (2.0 \times P)$

First the random number is computed for point load P and then moment is computed.

$$P = a + (b - a)R_U$$

$$P = 14.8037 + (25.1963 - 14.8037) \times 0.8606 = 23.7477 \text{ kN}$$

$$M = 4.0 - 2.0 \times P = 4.0 - 2.0 \times 23.7477 = -43.4954 \text{ kN-m}$$

Table 8.7 gives the 20 values of random moment.

Table 8.7 Values of random moments

Sl. No.	Random number (Table 8.3)	Uniform random load P (kN)	Moment (kN-m)	Sl. No.	Random number (Table 8.3)	Uniform random load P (kN)	Moment (kN-m)
1	0.8606	23.7477	−43.4954	11	0.8377	23.5097	−43.0194
2	0.9255	24.4216	−44.8433	12	0.7301	22.3909	−40.7818
3	0.4181	19.1484	−34.2969	13	0.5634	20.6590	−37.3180
4	0.2896	17.8138	−31.6276	14	0.8218	23.3441	−42.6883
5	0.1422	16.2820	−28.5641	15	0.3272	18.2036	−32.4073
6	0.4496	19.4763	−34.9526	16	0.6885	21.9593	−39.9186
7	0.2465	17.3658	−30.7316	17	0.7436	22.5314	−41.0629
8	0.2169	17.0575	−30.1151	18	0.2467	17.3678	−30.7355
9	0.5650	20.6758	−37.3517	19	0.9032	24.1907	−44.3814
10	0.4002	18.9623	−33.9246	20	0.7926	23.0412	−42.0824

The sample mean for these 20 values of moment is $\overline{M} = -37.2$ kN-m

The sample standard deviation is $\sigma_M = 5.15$ (kN-m)

Example 8.7 The maximum ground acceleration, at a distance of R kilometres from an earthquake of Richter magnitude M can be estimated by using the formula a (cm/sec^2) = $1{,}230e^{0.8M}(R + 25)^2$. Use the first 20 numbers, if M is a uniform random variable with a mean of 6 and a coefficient of variation of 10%. Consider R to be a uniform random variable with a mean of 25 km and a coefficient of variation of 10%. Assume M and R to be independent. Perform a Monte Carlo analysis to find the sample mean, sample standard deviation, and sample coefficient of variation of the maximum ground acceleration.

Uniform Random Numbers							
Sl. No.	**Uniform Random Numbers**	**Sl. No.**	**Uniform Random Numbers**	**Sl. No.**	**Uniform Random Numbers**	**Sl. No.**	**Uniform Random Numbers**
1	0.52478	6	0.69379	11	0.33415	16	0.02333
2	0.80249	7	0.75228	12	0.41999	17	0.55696
3	0.94132	8	0.14327	13	0.75908	18	0.74838
4	0.56605	9	0.90625	14	0.91323	19	0.25833
5	0.58815	10	0.06070	15	0.03428	20	0.87260

Consider first the Richter magnitude M, where

$$\overline{M} = \text{mean} = 6 = \left(\frac{a_M + b_M}{2}\right) \text{ and}$$

$$\sigma_M^2 = \text{variance} = [(0.10)(6)]^2 = \frac{(b_M - a_M)^2}{12} = 0.36$$

Therefore, the limits on the Richter magnitudes uniform PDF are

$$a_M = 4.96$$
$$b_M = 7.04$$

It similarly follows for R that the limits on its uniform PDF are

$$\overline{M} = \text{mean} = 25 = \left(\frac{a_R + b_R}{2}\right)$$

$$\sigma_M^2 = \text{variance} = [(0.10)(25)]^2 = \frac{(b_R - a_R)^2}{12} = 6.25$$

On solving the above equations, we get

$$a_R = 20.67 \text{ km}$$
$$b_R = 29.33 \text{ km}$$

Using Eq. (8.6) the first random numbers for M and R are

$$^{(i)}M = 4.96 + \left(\frac{7.04 - 4.96}{1.00 - 0.00}\right)(0.52478 - 0.00000) = 6.05$$

and

$$^{(i)}R = 20.67 + \left(\frac{29.33 - 20.67}{1.00 - 0.00}\right)(0.22835 - 0.00000) = 22.65 \text{ km}$$

Therefore, the first random sample value for the maximum for the maximum ground acceleration is

$$\begin{aligned} ^{(i)}a &= 1230e^{0.80(^{(i)}M)}[^{(i)}R + 25]^{-2} \\ &= 1230e^{0.80(6.05)}[22.65 + 25]^{-2} \\ &= 68.60 \text{ cm/sec}^2 \end{aligned}$$

Table 8.8 lists the first 20 sample values for the maximum ground acceleration. The corresponding sample statistics are

$$\begin{aligned} \text{Sample mean} &= 71.69 \text{ cm/sec}^2 \\ \text{Sample standard deviation} &= 5.69 \text{ cm/sec}^2 \\ \text{Sample coefficient of variation} &= 7.9\% \end{aligned}$$

Table 8.8 Values for the maximum ground acceleration

Sample No.	Richter magnitude	Earthquake distance (km)	Max ground acceleration (cm/sec²)	Sample No.	Richter magnitude	Earthquake distance (km)	Maximum ground acceleration (cm/sec²)
1	6.05	22.65	68.60	11	5.66	26.95	42.02
2	6.63	22.06	111.58	12	5.83	21.78	59.79
3	6.92	21.99	141.02	13	6.54	26.07	88.16
4	6.14	28.18	58.99	14	6.86	25.59	116.07
5	6.18	20.84	82.37	15	5.03	20.82	32.82
6	6.40	28.41	72.31	16	5.01	22.85	29.55
7	6.52	27.56	82.30	17	6.11	26.48	61.99
8	5.26	28.77	28.57	18	6.52	24.66	91.58
9	6.84	29.19	99.99	19	5.50	24.67	40.53
10	5.09	24.52	29.35	20	6.77	28.74	96.16

Note that Monte Carlo simulation is not the linear statistical model, but provides the exact solution for this problem.

Example 8.8 Consider Example 8.7, where the mean and standard deviation of the maximum ground acceleration have been calculated and are:

$$\begin{aligned} \bar{a} &= \text{mean maximum ground acceleration} = 71.69 \text{ cm/sec}^2 \\ \sigma_a &= \text{standard deviation of maximum ground acceleration} = 5.69 \text{ cm/sec}^2 \end{aligned}$$

A structure exists at the site where this maximum ground motion has been estimated. It has been determined by a structural engineer that: No damage will occur to the structure if the maximum ground acceleration is less than a_1.

Moderate damage will occur to the structure if the maximum ground acceleration is between a_1 and a_2.

Major structural damage will occur to the structure if the maximum ground acceleration is between a_2 and a_3.

Collapse of the structure will occur if the maximum ground acceleration is greater than a_3.

(a) Using the 20 Monte Carlo samples generated in Example 8.7, what is the probability that the structure will have (i) no damage, (ii) moderate damage, (iii) major damage, or (iv) will collapse?
Assume $a_1 = 75\ \text{cm/sec}^2$, $a_2 = 100\ \text{cm/sec}^2$, and $a_3 = 125\ \text{cm/sec}^2$.

(b) If the maximum ground acceleration is assumed to have a normal PDF, what is the probability that the structure will have (i) no damage (ii) Moderate damage?

Solution: (a) Monte Carlo analysis enables one to review directly the sample results and to identify the structural damage state. For example, sample number (1) maximum ground acceleration is 68.60 cm/sec^2; therefore, it is less than a_1 and no damage occurs. Sample number (2) maximum ground acceleration is 111.58 cm/sec^2; therefore, it is between a_2 and a_3 and major structural damage occurs. By reviewing the maximum ground acceleration on a sample-by-sample basis, it follows that 11 samples indicate no damage, six moderate damage, two major damage, and one collapse. Therefore, based on the 20 samples in Monte Carlo's analysis, the following probabilities of damage are obtained:

Limit state	No damage	Moderate damage	Major damage	Collapse
	$x < a_1$	$a_1 < x < a_2$	$a_2 < x < a_3$	$x > a_3$
	$a_1 = 75\ \text{cm/sec}^2$	$a_2 = 100\ \text{cm/sec}^2$	$a_3 = 125\ \text{cm/sec}^2$	$a_3 = 125\ \text{cm/sec}^2$
n	11	6	2	1

$$P_1 = \text{probability of no damage} = \frac{11}{20} = 55\%$$

$$P_2 = \text{probability of moderate damage} = \frac{6}{20} = 30\%$$

$$P_3 = \text{probability of major damage} = \frac{2}{20} = 10\%$$

$$P_4 = \text{probability of collapse} = \frac{1}{20} = 5\%$$

Note: For illustrative purposes, the number of samples is only 20. An actual study would use several hundred samples, and therefore a more realistic distribution of samples would result and the calculated probabilities would be more accurate. Also, the structural engineer may wish to be more descriptive in defining damage so that, instead of using only moderate and major damage, one may define more ranges than the four used for this example.

(b) Since the PDF of the maximum ground acceleration has been assumed to be normal; its PDF is completely defined by using the sample mean and variance. That is,

$$f(a) = \frac{1}{5.69\sqrt{2\pi}} \exp\left\{-\frac{1}{2}\left(\frac{a-71.69}{5.69}\right)^2\right\}$$

The probabilities associated with the damages now follow directly:

$$P_1 = Pr[a < a_1] = \int_{-\infty}^{a_1} p(a)da \qquad \text{.......no damage}$$

$$P_2 = Pr[a_1 \le a < a_2] = \int_{a_2}^{a_1} p(a)da \qquad \text{.......moderate damage}$$

$$P_1 = 0.719$$

$$P_2 = 0.281$$

Note: If more than 20 samples were used, the accuracy of the sample mean and sample variance would improve. Also, since a normal PDF has been assumed for the maximum ground acceleration, a linear statistical analysis could be used to calculate the mean and standard deviation and the rest of the solution would follow the same as discussed in solution (b) above.

UNSOLVED PROBLEMS

1. The equation for beam displacement is given as $\Delta = \dfrac{PL^3}{3EI}$ and let P, L, I are deterministic and define a normalized displacement as $\Delta^1 = \dfrac{\Delta}{\dfrac{PL^3}{EI}} = \dfrac{1}{E}$ consider E to be a uniformly distributed random variable with mean and coefficient of variation equal to 24 GPa, and 4.2% respectively.
   (a) Use the first 20 random numbers in column one of Table 8.3 to generate 20 random values of E.
   (b) Use the Monte Carlo method to calculate the sample mean and sample variance of Δ^1.
   (c) Calculate, using a linear statistical analysis, the mean and standard deviation of Δ^1.
   (d) Discuss why your answers to parts (b) and (c) are different.
   (e) Assuming that instead of using just 20 samples in your Monte Carlo analysis you used an infinite number of samples, would your answers to parts (b) and (c) still differ? If so why?
2. A 100 mm wide and 150 mm deep uniform wooden beam is used to support a uniformly distributed load of W kN/m on a simple span 3.0 m. The applied load acts in a plane making an angle θ with the vertical. The maximum moment at the mid-span can be resolved into components acting around y–y and z–z axes.

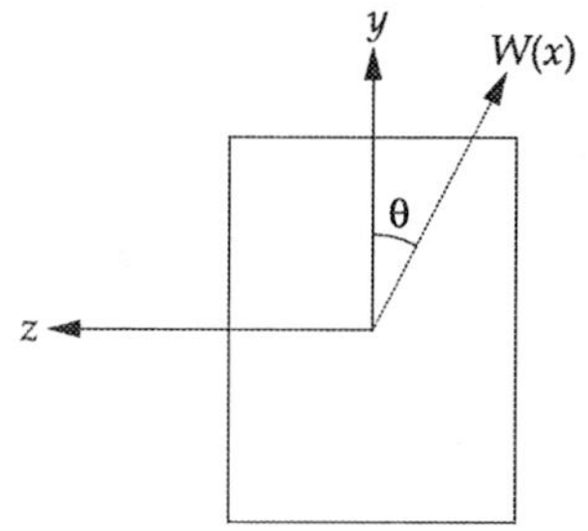

Fig. P.2

The moments are $M_{zz} = \left(\frac{WL^2}{8}\right)\cos\theta$; $M_{yy} = \left(\frac{WL^2}{8}\right)\sin\theta$; the distributed load acts along a line through the centroid of the cross. Consider the geometric properties of the beam to be deterministic.

(a) If the angle θ is deterministic and equal to 30°, calculate, using a linear statistical analysis, the mean and standard deviations of the maximum tensile stress on the cross section. Assume the uniform load to be a random variable with a mean and standard deviation of 150 N/m and 30 N/m respectively.

(b) Now, consider the angle θ to be a random variable, independent of the distributed load W, with a mean of 30° and a standard deviation equal to 5°. Assume that both W, θ have uniform PDF's. Calculate the mean and standard deviations of the maximum tensile stress on the cross section. Use, to generate your random Monte Carlo numbers, the first 20 random numbers in column one of Table 8.3, for W and the first 20 random numbers in column two of Table 8.3 for θ.

3. A series of flexure tests are conducted on reinforced concrete beams. The random variable, denoted by R, is

$$R = \frac{\text{Ultimate test moment}}{\text{Ultimate computed moment}}$$

The ultimate computed moment is calculated by using the code formula,

$$M_u = f_y \times A_{st} \times d - \frac{0.77 \times A_{st}^2 \times f_y^2}{(b \times f_{ck})}$$

M_u = ultimate moment
f_y = yield strength of steel
A_{st} = area of tension steel
d = effective depth
f_{ck} = characteristic strength of concrete

The sample mean and standard deviation of R is 1.15 and 0.08 respectively. Assume that R has normal PDF, generate random numbers, using the first 20 random numbers in column one of Table 8.4, for R.

9

CHAPTER

System Reliability of Structures

9.1 INTRODUCTION

In the earlier chapters the functional safety was assessed in terms of reliability index, and β at component level. It is well known that any structure is an assemblage of various components and the safety of the entire structure will be different from the one at component level. The reliability of a system comprising components depends on the indeterminacy of the system. System reliability analysis depends on whether the system is determinate or indeterminate. Failure of any component in a determinate system leads to failure of the entire system while in an indeterminate system, it continues to perform despite failure of any one of the components. Reliability of a system can be improved in an indeterminate system. The concept of system reliability can be applied to all allied branches in civil engineering—structural engineering, traffic engineering, water supply engineering and so on. It is also adopted in other areas like communication transmission, power transmission, performance of electric/electronic circuits, etc. Irrespective of the application of system reliability, the failure modes of the system will have to be identified, modelled and combined to evaluate the reliability of the system. Assessment of system reliability is a complicated process and depends on factors like performance of the component in a system, redundancy of the system, statistical correlation between the failure modes and component failure and post-failure behaviour of the component. The fundamental concepts required to assess system reliability are discussed in the following sections.

9.2 RELIABILITY SYSTEMS

Redundancy plays an important role in classifying the different types of reliability systems. The failure of some components can lead to failure of the systems or may weaken the system though system failure does not exist. Reliability systems are basically classified into three types. They are (i) series system, (ii) parallel system, and (iii) mixed system.

9.2.1 Series System

This system fails even when one component of the system fails. Performance of this system is assured when only all the components of the system perform satisfactorily. This system is akin to bulbs in series in electrical engineering. A determinate structural system is modelled as a series system. The model for a series system is shown in Fig. 9.1 by a block diagram. The model is also called 'weak link model'.

Fig. 9.1 Block diagram of a series system

Let S_i represent an event of satisfactory performance of a component i, p_{ss} represent the probability of survival of the system. All the components should simultaneously perform satisfactorily for the survival of a series system. The probability of survival, p_{ss} of the system is a joint intersection of probabilities of satisfactory performance of all components in the system.

$$p_{ss} = P(S_1 \cap S_2 \cap S_3 \cap \ldots S_n) \qquad \ldots(9.1)$$

If the events S_1, S_2, S_3,...S_n are statistically independent, and using the probability axioms, we have

$$p_{ss} = P(S_1) \times P(S_2) \times P(S_3) \ldots P(S_n)$$

$$p_{ss} = \prod_{i=1}^{n} P(S_i) \qquad \ldots(9.2)$$

The notation Π is called 'capital pi' refers to product of sequence of numbers. If p_{si} represents the probability of survival of the component i, then

$$p_{ss} = \prod_{i=1}^{n} P(S_i) = \prod_{i=1}^{n} p_{si} = \prod_{i=1}^{n}(1 - p_{fi}) \qquad \ldots(9.3)$$

Here p_{fi} denotes the failure of the component i.

9.2.2 Parallel System

In this type, the system survives even if any one component fails. This is similar to bulbs in parallel normally studied in electrical engineering. The block model for system in parallel is shown in Fig. 9.2.

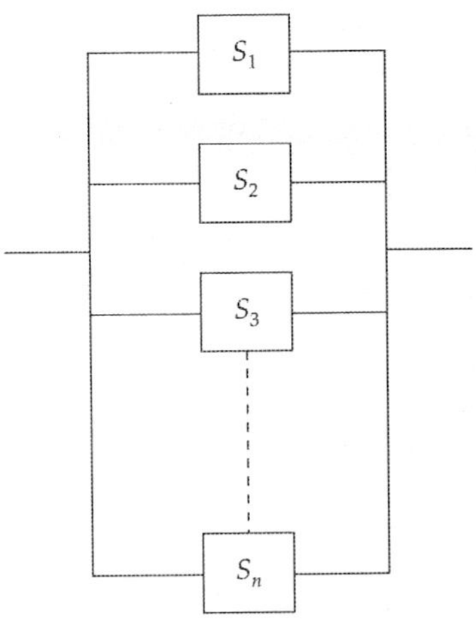

Fig. 9.2 Block diagram for a parallel system

A parallel system will fail when all the components fail. The probability of survival of a parallel system is given by

$$p_{ss} = 1 - p_{fs}$$

Here p_{fs} denotes the probability of failure of the system. As all the components will have to fail, it is a joint intersection event of failure of individual components. Hence, p_{fs} can be expressed as

$$p_{fs} = P\left(S_1^c \cap S_2^c \cap S_3^c \cap \ldots S_n^c\right)$$

$$p_{ss} = 1 - P\left(S_1^c \cap S_2^c \cap S_3^c \cap \ldots S_n^c\right)$$

If events S_i^c are statistically independent, then we have

$$p_{ss} = 1 - P(S_1^c) \times P(S_2^c) \times P(S_3^c) \ldots P(S_n^c)$$

If probability a failure of component is denoted by p_{fi}

$$p_{ss} = 1 - (p_{f1}) \times (p_{f2}) \times (p_{f3}) \ldots (p_{fn})$$

$$p_{ss} = 1 - \prod_{i=1}^{n} p_{fi} \qquad \ldots(9.4)$$

A parallel system is more preferable in any engineering system. In other words, redundancy should exist in engineering structures. The levels of safety in redundant structures are higher than determinate structures and the failure of the entire structure is prolonged. A parallel system exhibits 'ductility' in the structure.

9.2.3 Mixed System

Some systems have a combination of series and parallel systems. Such systems are called as mixed system. Reliability analysis of continuous beam or portal frame under plastic conditions, analysis of redundant structure or a process which has a backup safety component are some cases of mixed system. The block diagram for a mixed system is shown in Fig. 9.3.

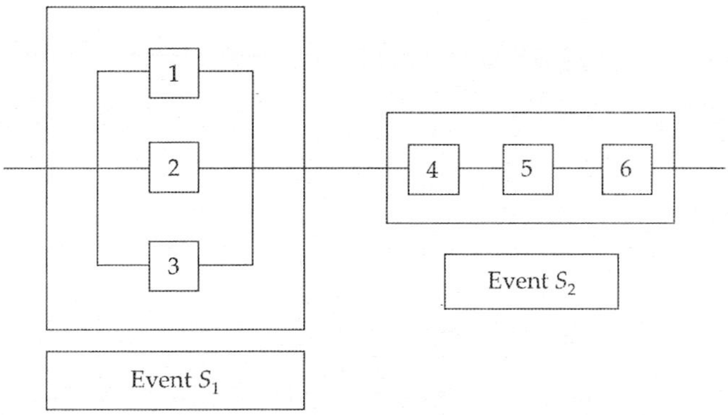

Fig. 9.3 Block diagram of a mixed system

Though only two events: S_1 representing the survival of parallel components 1, 2 and 3, S_2 representing the survival of series components 4, 5 and 6 are shown, a mixed system can have any number of parallel and series paths in any combination. The probability of survival of a mixed system is found out using the basic of series and parallel system explained earlier. For the block diagram shown in Fig. 9.3 the probability of survival is given by

$$\begin{aligned} p_{ss} &= P(S_1 \cap S_2) \\ &= P(S_1) \times P(S_2) \\ &= p_{ss1} \times p_{ss2} \\ &= (1 - p_{fs1}) \times (1 - p_{fs2}) \end{aligned}$$

Example 9.1 Three components 1, 2, 3 are placed in (i) series and (ii) parallel. Evaluate the system reliability. The probability of failure of components 1 and 3 is 0.2 and for the component 2 it is 0.15.

Solution:

Case 1: The block diagram for a series system is shown in Fig. 9.4(a).

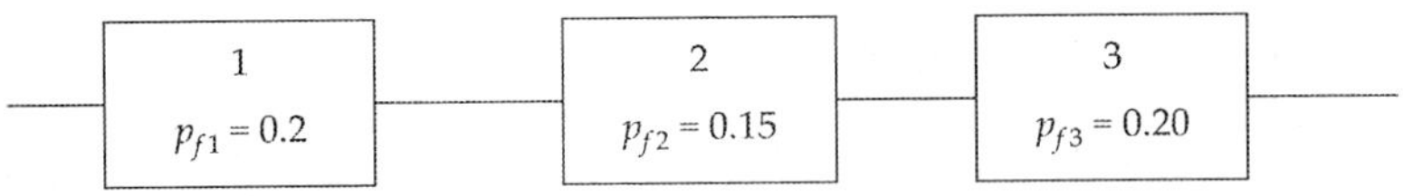

Fig. 9.4(a) Series block diagram for Example 9.1

Using Eq. 9.3 the probability of survival of the system is

$$\begin{aligned} p_{ss} &= \prod_{i=1}^{n} (1 - p_{fi}) \\ &= (1 - p_{f1})(1 - p_{f2})(1 - p_{f3}) \\ &= (1 - 0.2)(1 - 0.15)(1 - 0.2) \\ &= 0.544. \end{aligned}$$

Case 2: If the components are placed in parallel, then the block diagram is shown in Fig. 9.4(b).

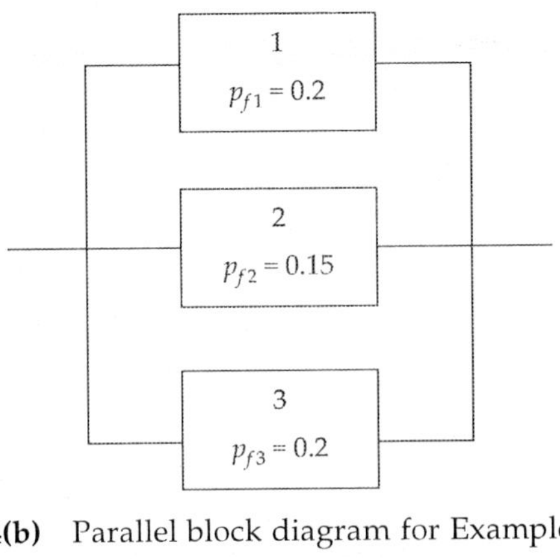

Fig. 9.4(b) Parallel block diagram for Example 9.1

Using Eq. 9.4, the probability of survival is

$$p_{ss} = 1 - \prod_{i=1}^{n} p_{fi}$$
$$= 1 - [(p_{fi})(p_{f2})(p_{f3})]$$
$$= 1 - [(0.2)(0.15)(0.2)]$$
$$= 0.994.$$

It is observed that parallel system has more probability of survival than a series system.

Example 9.2 The block diagram for a mixed system is shown in Fig. 9.5. The probabilities of failure of the components are as follows: Component 1 = 0.2, Component 2 = 0.3, Component 3 = 0.4, Component 4 = 0.5, and Component 5 = 0.6. Evaluate the reliability of the system.

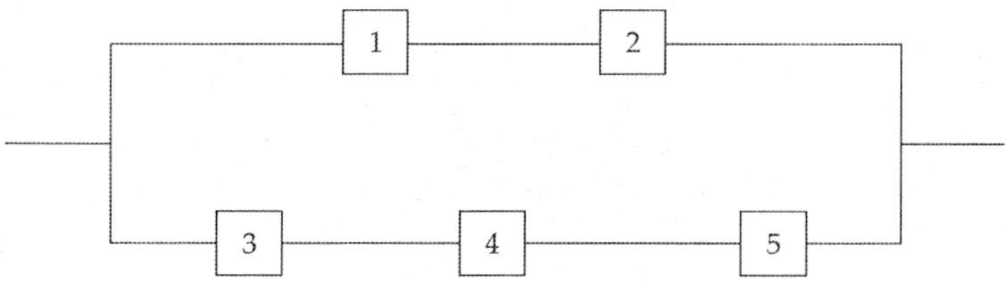

Fig. 9.5 Block diagram for Example 9.2

Solution: Components 1 and 2 are in series. Let the equivalent probability of survival of components 1 and 2 be represented by S_1.

$$p_{ss1} = (1 - 0.2)(1 - 0.3)$$
$$= 0.56$$

The probability of failure of the series system comprising of components 1 and 2 is

$$p_{fs1} = (1 - 0.56) = 0.44$$

Similarly, components 3, 4, and 5 are in series and let the equivalent probability of survival of components 3, 4 and 5 be represented by S_2.

$$p_{ss2} = (1 - 0.4) \times (1 - 0.5) \times (1 - 0.6)$$
$$= 0.12$$

The probability of failure of the series system comprising of components 2, 3 and 4 is

$$p_{fs1} = (1 - 0.12) = 0.88$$

Events S_1 and S_2 are in parallel. The block diagram in Fig. 9.5 is equivalent to parallel system shown in Fig. 9.6.

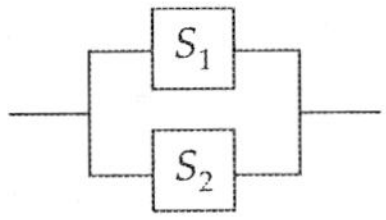

Fig. 9.6 Equivalent parallel system

The probability of survival of the equivalent parallel system is

$$p_{ss} = 1 - [(0.44) \times (0.88)]$$
$$= 0.6128$$

Hence, the reliability of the system is 0.6128.

Example 9.3 A triangular truss having members 1, 2 and 3 of equal length is loaded with a joint load P as shown in Fig. 9.7. If $\mu_p = 30$ kN, $\sigma_p = 3$ kN, $\mu_{R1} = \mu_{R2} = 35$ kN, $\sigma_{R1} = \sigma_{R2} = 3.5$ kN, $\mu_{R3} = 30$ kN, $\sigma_{R3} = 3.0$ kN. Evaluate the reliability of the truss.

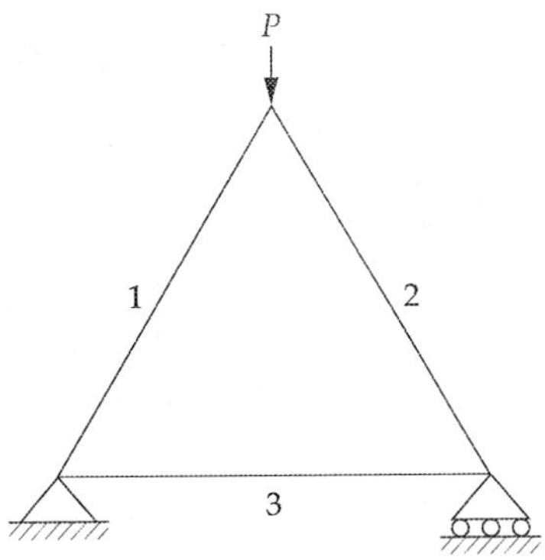

Fig. 9.7 Truss for Example 9.3

Solution: All the sides being equal, interior angle between the members is 60°. Analyzing the truss with load P, the axial forces in the members 1, 2 and 3 are

$$S_1 = S_2 = \frac{P}{\sqrt{3}}$$

$$S_3 = \frac{P}{2\sqrt{3}}$$

Hence,

$$\mu_{s1} = \mu_{s2} = \frac{\mu_P}{\sqrt{3}} = \frac{30}{\sqrt{3}} = 17.3 \text{ kN}$$

$$\sigma_{S1}^2 = \sigma_{S2}^2 = \left(\frac{1}{\sqrt{3}}\right)^2 \sigma_P^2 = \frac{3^2}{3} = 3 \text{ kN}$$

$$\mu_{S3} = \frac{\mu_P}{2\sqrt{3}} = \frac{30}{2\sqrt{3}} = 8.66 \text{ kN}$$

$$\sigma_{S3}^2 = \left(\frac{1}{2\sqrt{3}}\right)^2 \sigma_P^2 = \frac{3^2}{(4)(3)} = \frac{3}{4} = 0.75 \text{ kN}$$

The reliability index, β for members 1 and 3 is given by

$$\beta = \frac{\mu_R - \mu_S}{\sqrt{\sigma_R^2 + \sigma_S^2}}$$

Substituting the following values in the expression for β for members 1 and 2

$$\mu_R = \mu_{R1} = \mu_{R2} = 35 \text{ kN}$$

$$\mu_S = \mu_{S1} = \mu_{S2} = 17.3 \text{ kN}$$

$$\sigma_R = \sigma_{R1} = \sigma_{R2} = 3.5 \text{ kN}$$

$$\sigma_S^2 = \sigma_{S1}^2 = \sigma_{S2}^2 = 3 \text{ kN}$$

$$\beta_{1,2} = \frac{35 - 17.3}{\sqrt{3.5^2 + 3}} = 4.53$$

Probability of failure for members 1 and 2, $p_{f1} = p_{f2} = \Phi(-\beta_{1,2}) = (0.339 \times 10^{-5})$
Similarly, for member 3

$$\mu_R = \mu_{R3} = 30 \text{ kN}$$

$$\mu_S = \mu_{S3} = 8.66 \text{ kN}$$

$$\sigma_R = \sigma_{R3} = 3.0 \text{ kN}$$

$$\sigma_S^2 = \sigma_{S3}^2 = 0.75 \text{ kN}$$

$$\beta_3 = \frac{30 - 8.66}{\sqrt{3.0^2 + 0.75}} = 6.83$$

Probability of failure for member 3, $p_{f3} = \Phi(-\beta_3) = (0.4255 \times 10^{-11})$

The truss being determinate, members 1, 2 and 3 constitute a series system as shown below.

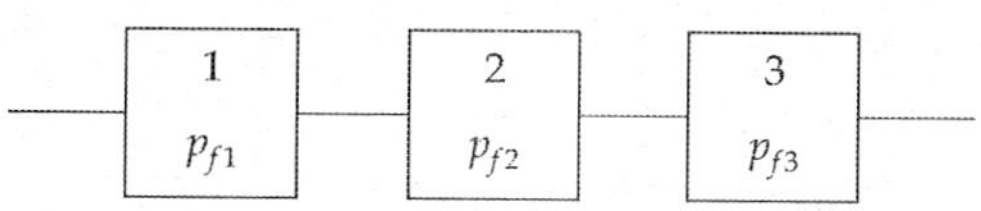

The reliability of the truss is obtained using Eq. 9.3

$$p_{ss} = (1 - p_{f1}) \times (1 - p_{f2}) \times (1 - p_{f3})$$
$$= (1 - (0.339 \times 10^{-5})) \times (1 - (0.339 \times 10^{-5})) \times (1 - (0.4255 \times 10^{-11})$$
$$= 0.9999$$

Example 9.4 A three-span continuous beam as shown in Fig. 9.8 has probability of failure by formation of plastic hinges as follows:

Span 1: Probability of failure for plastic hinges 1 and 2 = 0.2

Span 2: Probability of failure for plastic hinges 3 and 4 = 0.15

Span 3: Probability of failure for plastic hinges 5 and 6 = 0.10.

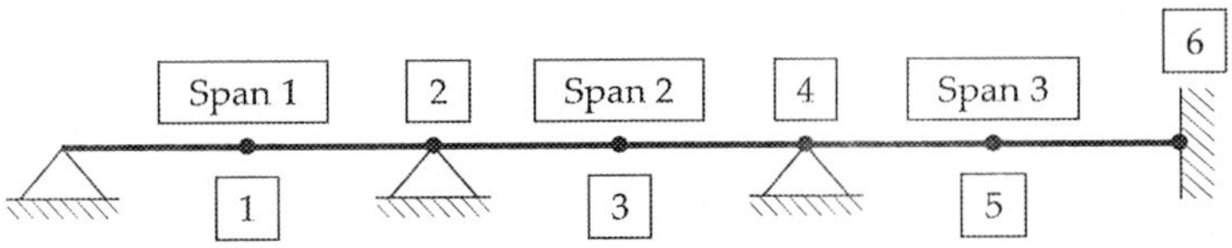

Find the system reliability of the continuous beam.

Solution: Failure of the span 1 is by formation of hinges 1 and 2 and due to redistribution of moments both the hinges will not fail simultaneously. Hence, they form a parallel system. Likewise hinges 2, 3, and 4 in the span 2 form a parallel system. Similarly, in span 3 hinges 4, 5, and 6 form a parallel system.

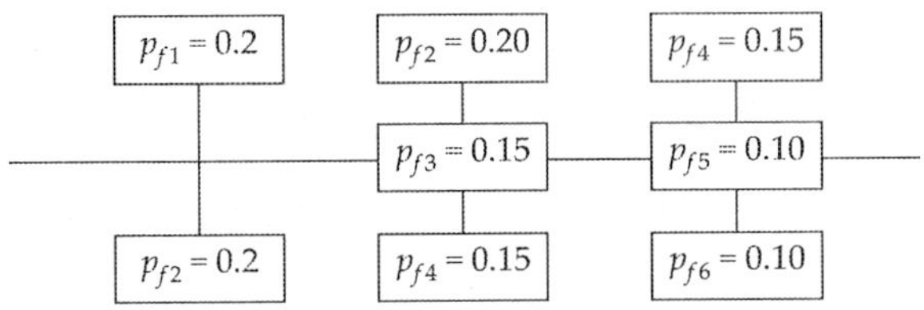

Using Eq. 9.4, the probability of survival of span 1

$$p_{ss1} = 1 - \prod_{i=1}^{n} p_{fi}$$

$$= 1 - [(0.2)(0.2)] = 0.96$$

$$p_{fs1} = 1 - 0.96 = 0.04$$

Similarly, the probabilities of failure for the span 2 and 3 are

$$p_{fs2} = 1 - \{1 - [(0.2)(0.15)(0.15)]\}$$

$$= 0.0045$$

$$p_{fs3} = 1 - \{1 - [(0.15)(0.10)(0.10)]\}$$

$$= 0.0015$$

For the beam to fail any one of the three parallel systems corresponding to spans 1, 2 and 3 can fail. Hence, the three parallel systems of the spans are modelled in series. The resulting system is a mixed or hybrid system as shown below.

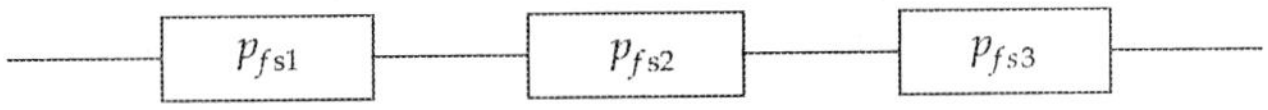

Using Eq. 9.3 the probability of survival of the beam is given by

$$p_{ss} = \prod_{i=1}^{n}(1 - p_{fi})$$

$$\begin{aligned} p_{ss} &= (1 - 0.04)(1 - 0.0045)(1 - 0.0015) \\ &= (0.96)\cdot(0.9955)\cdot(0.9985) \\ &= 0.9542 \end{aligned}$$

9.3 SIMPLE BOUNDS IN RELIABILITY SYSTEMS

A structural system having n failure modes, statistically independent, then the upper bound is given by

$$p_{ss} = 1 - \max p_{fi}$$

The lower bound is given by Eq. 9.3

$$p_{ss} = \prod_{i=1}^{n}(1 - p_{fi})$$

Hence, the reliability of the system is bounded by

$$\text{Lower bound}, p_{ss} \le p_{ss} \le \text{Higher bound}, p_{ss}$$

$$= \prod_{i=1}^{n}(1 - p_{fi}) \le p_{ss} \le (1 - \max p_{fi})$$

The bounds of the probability of failure of system can be written as

$$\max p_{fi} \le p_{fs} \le 1 - \prod_{i=1}^{n}(1 - p_{fi})$$

These bounds are normally wide for practical purposes.
Considering Example 9.4 the bounds of the reliability of the system are given by

$$0.9542 \le p_{ss} \le (1 - 0.04)$$

or

$$0.9542 \le p_{ss} \le 0.96$$

UNSOLVED PROBLEMS

1. Three components with probability of failure 0.1, 0.2 and 0.3 form a system. In what way the components should be assembled for maximum system reliability.
2. Find the reliability of the system shown in Fig. P2.

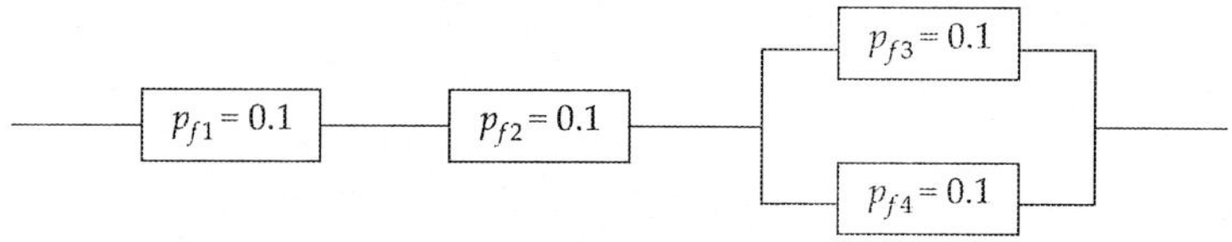

Fig. P2

3. The truss shown in Fig. P3 has members with associated probability of failure. Find the reliability of the truss.

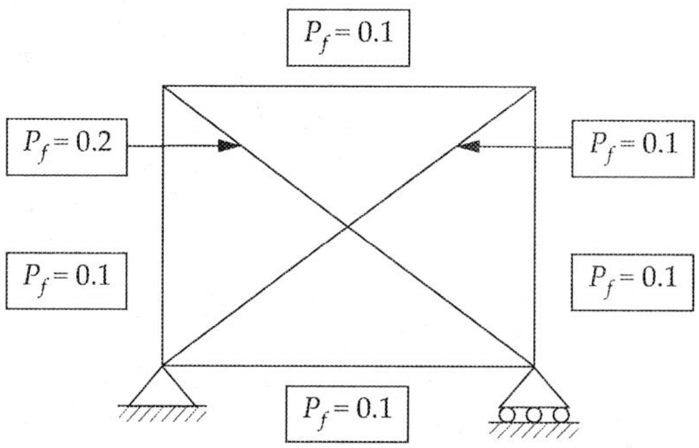

Fig. P3

4. Obtain the reliability of the continuous beam shown in Fig. P4. The probability of failure at all the possible location of plastic hinges is 0.15.

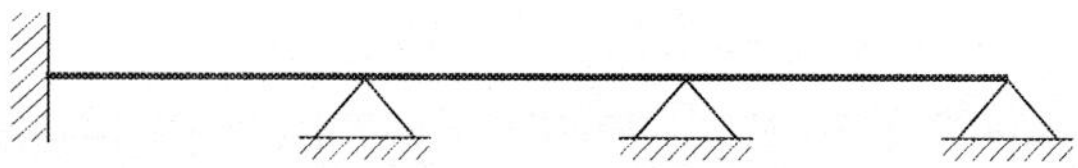

Fig. P4

10
CHAPTER

Load Resistance Factor Design

It is necessary to motivate, emphasize the probabilistic concepts and methods in a meaningful terms to engineers. However, we have theories, we have data, now we need the will.

10.1 INTRODUCTION

The design consists of proportioning the elements of a system so that it satisfies various criteria of performance, safety, serviceability, and durability under various demands. A structure should be designed so that its strength or resistance is greater than the load effects or the applied loads on the structure. There are various sources of uncertainties in geometry, strength and loads. Figure 10.1 shows the structural design with respect to a water filter. Structural engineering is the art of using materials that have properties which can only be estimated to build real structures that can only be approximately analyzed to withstand forces that are not accurately known so that our responsibility with respect to public safety is satisfied.

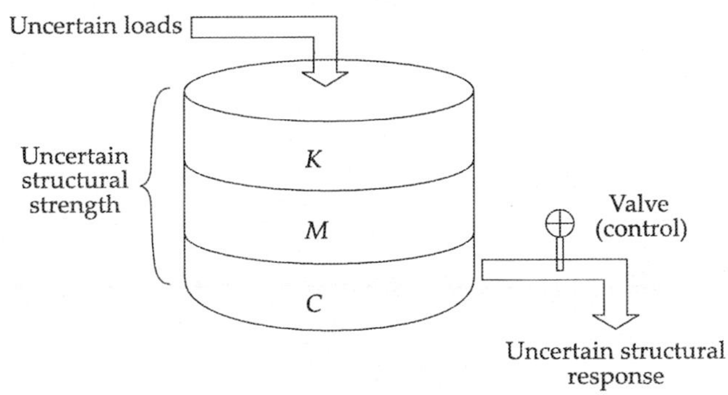

Fig. 10.1 Uncertain loads, strength and response of structures

Prof. Thoft-Christensen & Baker (1970) were responsible for code development and replacement of many simple design rules by scientifically based calculation, derived from experimental and theoretical research. Use of structural reliability theory in determining rational sets of partial safety coefficients, was proposed for the Joint Committee on

Structural Safety (JCSS). Prof. Detlevenson & Madsen (1989) developed the model code for Joint Committee on Structural Safety model code in March 2001 to cover all aspects of structural engineering. Designers want prescriptive rules that enable them to produce designs quickly, safely and efficiently. Code must distinguish between principles and application rules.

10.2 LOAD RESISTANCE FACTOR DESIGN (LRFD)

This is a method for designing structures so that no applicable limit states are violated when the structure is subjected to appropriate load effects and its combinations of factored load. The limit state is a condition in which a structure or a structural component violates one or several limit states. Strength limit states concerns with safety and relate to maximum load carrying capacity (plastic hinge or buckling). Serviceability limit state relates to performance under normal service conditions (excessive deformation or vibration).

Design Philosophy. Three design philosophies are (i) allowable stress design, (ii) plastic design, (iii) LRFD (load resistance factor design).

10.2.1 Allowable Stress Design

The factor of safety in allowable stress design is applied only to the resistance term, and safety is evaluated at the service load.

$$\frac{R_n}{\text{FOS}} \geq \sum_{i=1}^{m} S_{ni} \qquad \text{...(10.1)}$$

10.2.2 Plastic Design: (PD) or Ultimate Strength Design

It is ensured that factored load combinations do not exceed the maximum plastic strength of the structure or component.

$$R_n \geq \gamma \sum_{i=1}^{m} S_{ni} \qquad \text{...(10.2)}$$

10.2.3 Load Resistance Factor Design (LRFD)

It is ensured that the factored load effects do not exceed the factored nominal resistance of the structural members or component. Here there are two safety factors: one applied to the loads, the other to the resistance.

$$\phi R_n \geq \sum_{i=1}^{m} \gamma_i S_{ni} \qquad \text{...(10.3)}$$

It is anticipated that LRFD will soon be universally accepted and will eventually become the standard method of structural design. The new load resistance factor design method was introduced officially by AISC in 1986.

The left side is ϕ multiplied by the resistance which is design strength and the right side represents load combinations; that is, the summation of various loads and loads effects, multiplied by their respective load factors γ_i. The loads are considered from the relevant or governing building codes and the load factors were derived from statistical building load studies.

The design strength is a product of resistance factor ϕ and nominal resistance R_n, the resistance factor is less than unity or equal to one. The greater the scatter in the test data for the given resistance, the lower is its ϕ factor. The load factor is greater than unity γ_i, because greater uncertainty in loads.

The loads, dead loads (DL), live loads (LL), roof loads (LR), snow loads (SL), wind loads (WL), earthquake loads (EQ), represent the loads themselves or the load effects (forces and moments caused by loads). In load combinations, only one load assumes its maximum lifetime value at a time, while others are at their "arbitrary point-in-time" values. Each load combination models the design loading conditions when different load is at its maximum. In determining serviceability limit states (deflections) the unfactored (service) loads are used.

Equations are taken from AISC-LRFD specifications. The required load combinations are adopted as the maximum force obtained by following load combination.

(a) 1.4DL ...(10.4)

(b) 1.2DL + 1.6LL + 0.5 (Lr or SL or R) ...(10.5)

(c) 1.2DL + 1.6 (Lr or S or R) + (0.5LL or 0.8WL) ...(10.6)

(d) 1.2DL + 1.3WL + 0.5LL + 0.5 (Lr or SL or R) ...(10.7)

(e) 1.2DL + 1.5EQ + (0.5LL or 0.2SL) ...(10.8)

(f) 0.9DL – (1.3WL 1.5EQ) ...(10.9)

10.3 DESIGN TENSILE STRENGTH

The structural components, such as truss members, hangars, bridges, towers, bracing systems, cables are subjected to pure tension. Structural steel sections of various shapes are used in the design of tension members. In the past, round bars were used frequently as tension member. When a tensile force is applied through the centroidal axis of a member, the result is uniform tensile stress at the cross section. Tensile forces not acting through the centroid causes bending in addition to tension. The design tensile strength of a structural steel member depends on the appropriate cross-sectional area. Three cross-sectional areas of interest are the gross area (A_g), the net area (A_n), and the effective area (A_e).

Two criteria limit the design tensile strength $\phi_t\, R_n$

(a) *Yielding of gross cross section*

$$\phi_t = 0.9$$

$$R_n = f_y\, A_g \qquad \text{... (10.10)}$$

(b) *Fracture of net cross section*

$$\phi_t = 0.75$$

$$R_n = f_u\, A_e \qquad \text{... (10.11)}$$

ϕ_t = Resistance factor for tension member
R_n = R = Nominal or characteristic strength in N
f_y = Specified minimum yield stress in N/mm^2
f_u = Specified minimum tensile strength in N/mm^2

$$\phi_m R_m \geq \sum_{1}^{n} \gamma_m S_m \quad \text{...(10.12)}$$

$R_m = \mu_R$ = Mean resistance, $S_m = \mu_S$ = Mean load

Bias resistance factor, $$b_R = \frac{R_m}{R_k} = \frac{\mu_R}{R} \quad \text{...(10.13)}$$

Bias load factor, $$b_L = \frac{S_m}{S_k} = \frac{\mu_S}{S} \quad \text{...(10.14)}$$

where ϕ_k, is the characteristic resistance factor

$$R_k = \frac{R_m}{b_R} = \frac{\mu_R}{b_R}; \; S_k = \frac{S_m}{b_L} = \frac{\mu_S}{b_L}$$

R_k, or R, is the characteristic resistance
S_k, or S, is the characteristic load effects or load
γ_k, is the characteristic load factor

$$\phi_m R_m \geq \sum_{1}^{n} \gamma_m S_m$$

The code format. The code format can be obtained by dividing the mean values by the bias factor, for resistance and load effect.

$$\phi_k R_k = \sum_{1}^{n} \gamma_k S_k \quad \text{...(10.15)}$$

The factor of safety. The nominal values of resistance and load cannot be determined with certainty (deterministic)

Factor of safety (FOS) = $\frac{R_k}{S_k} = \frac{R}{S}$ (Ellingwood)

Central safety factor is the ratio of mean resistance to mean load for random or stochastic variable.

Central safety factor (CFS) = $\frac{R_m}{S_m} = \frac{\mu_R}{\mu_S}$

Example 10.1 Analyze by FOSM method, a tie member of roof truss subjected to 300 kN for the limit state of yield of $g(x) = f_y \times A - S$ of the steel member for strength formulation. Develop the LRFD format and codal format for the following data.

Sl. No.	Variables	Characteristic value	CoV	Distribution
1	f_y	250 N/mm^2	0.05	Normal
2	A	1432 mm^2	0.05	Normal
3	$S = P$	300×10^3 N	0.20	Normal

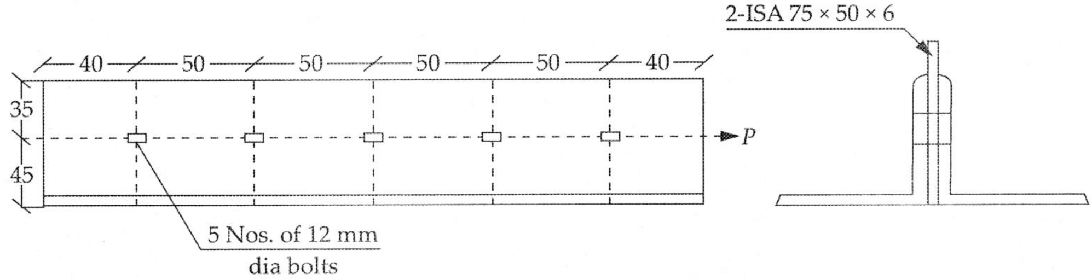

Fig. 10.2 Cross-section of tension member

Solution: **Analysis by FOSM Method**

The strengths are underestimated and loads are overestimated, the mean values are computed from the characteristics value and are tabulated as given below:

Variables	Characteristic value	CoV	Mean value (μ)	Standard deviation (σ)
f_y	250 N/mm^2	0.05	272.40	13.62
A	1432 mm^2	0.05	1560.33	78.01
$S = P$	300×10^3 N	0.20	225.73×10^3	45.14×10^3

The limit state function, $g(x) = f_y A - P$

Here, f_y, A and S are random variables

Here, $R = f_y A$, $S = P$ $\Rightarrow g(x) = R - S$

Mean value,

$$\mu_R = \mu_{f_y} \times \mu_A = 272.4 \times 1560.33 = 4.25 \times 10^5 \text{ N}$$

$$\mu_S = \mu_P = 225.73 \times 10^3 \text{ N}$$

Now, partially differentiating R and S w.r.t variables, we get,

$$\frac{\partial R}{\partial f_y} = A = \mu_A = 1560.33$$

$$\frac{\partial R}{\partial A} = f_y = \mu_{f_y} = 272.40$$

$$\frac{\partial S}{\partial P} = 1.0$$

Standard deviation,

$$\sigma_R^2 = \left(\frac{\partial R}{\partial f_y} \times \sigma_{f_y}\right)^2 + \left(\frac{\partial R}{\partial A} \times \sigma_A\right)^2 = (1560.33 \times 13.62)^2 + (272.4053 \times 78.01)^2$$

$\Rightarrow \quad \sigma_R^2 = 9.03 \times 10^8$

$\therefore \quad \sigma_R = 30 \times 10^3 \text{ N}$

Similarly,

$$\sigma_S^2 = \left(\frac{\partial S}{\partial P} \times \sigma_P\right)^2 = \left((1.0) \times 45.14 \times 10^3\right)^2 = 2.037 \times 10^9$$

$\therefore \quad \sigma_S = 45140 \text{ N}$

Now,

Reliability index (β),

$$\beta = \frac{\mu_R - \mu_S}{\sqrt{\sigma_R^2 + \sigma_S^2}} = \frac{4.25 \times 10^5 - 2.2573 \times 10^5}{\sqrt{(9.03 \times 10^8 + 2.037 \times 10^9)}} = \frac{1.99 \times 10^5}{0.542 \times 10^5} = 3.6716$$

The risk or probability of failure,

$$P_f = \Phi(-\beta) = \Phi(-3.6716)$$

From the standard normal distribution Table (A_3), $\Phi(-3.6716)$, we get,
By using interpolation, we get,

$\therefore \quad P_f = \Phi(-3.6716) = 0.0001210$

$\Rightarrow \quad P_f = 0.0001210$ or [1 in 8260]

Probability of success (P_s),

$\therefore \quad P_S = 1 - P_f = 1 - 0.0001210 = 0.99987$

Resistance factor (ϕ)

$$\phi = [1 - 0.75 \times \beta \times (\text{CoV})_R]$$

$\therefore \quad (\text{CoV})_R = \dfrac{\sigma_R}{\mu_R} = \dfrac{30 \times 10^3}{4.25 \times 10^5} = 0.0705$

$\Rightarrow \quad \phi = [1 - 0.75 \times 3.6716 \times 0.0705] = 0.805$

Resistance factor (γ)

$$\gamma = [1 + 0.75 \times \beta \times (\text{CoV})_S]$$

$\therefore \quad (\text{CoV})_S = \dfrac{\sigma_S}{\mu_S} = \dfrac{45140}{225.73 \times 10^3} = 0.1999 \approx 0.20$

$\Rightarrow \quad \gamma = [1 + 0.75 \times 3.6716 \times 0.20] = 1.55$

Bias resistance factor

Mean resistance, $\mu_R = \mu_{fy} \times \mu_A = 272.4 \times 1560.33 = 4.25 \times 10^5$ N
Characteristic resistance, $R_k = f_y \times A = 250 \times 1432 = 3.58 \times 10^5$ N

$$\Rightarrow \qquad b_R = \frac{\mu_R}{R} = \frac{4.25 \times 10^5}{3.58 \times 10^5} = 1.187$$

Bias load factor

Mean load, $\mu_S = 225.73 \times 10^3$ N
Characteristic load, $S = 300 \times 10^3$ N

$$\Rightarrow \qquad b_L = \frac{\mu_S}{S} = \frac{225.73 \times 10^3}{300 \times 10^3} = 0.752$$

LRFD – Format (mean value)

$$\phi \mu_R \geq \sum_1^n \gamma \mu_S$$

$$0.805\, \mu_R \geq 1.55\, \mu_S$$

LRFD – Code (characteristic or nominal value)

$$0.72\, R_k \geq 2.06\, S_k$$

Comment: The most significant application of probabilistic method is in the calibration of the factors of safety in codes of practice for the limit state of tension in yielding as design by IS: 800–2007. The resistance factor is 0.75 for limit state of fracture as given in AISC-2005 manual.

Example 10.2 A tie member of roof truss subjected to 300 kN for the limit state of yield of $g(x) = f_y A - S$ of the steel member for strength formulation. Determine the area of the section by FOSM method for a reliability index of 3.09.

Sl. No.	Variables	Characteristic value	CoV	Distribution
1	f_y	250 N/mm^2	0.05	Normal
2	A	?	0.05	Normal
3	$S = P$	300×10^3 N	0.20	Normal

Solution:

The limit state of yield of $g(x) = f_y A - S$
Standard deviation = Mean value × CoV

i.e., $$\sigma_A = \mu_A \times (\text{CoV})_A$$

$$\Rightarrow \qquad \mu_A = \frac{\sigma_A}{(\text{CoV})_A} = \frac{\sigma_A}{0.05} = 20\sigma_A$$

Here strengths are underestimated and loads are overestimated
Strengths are f_y and A, load is P

$$\mu_{f_y} = \frac{x_{f_y}}{[1 - 1.645 \times (\text{CoV})_{f_y}]} = \frac{250}{[1 - 1.645 \times 0.05]} = 272.40 \text{ N/mm}^2$$

$$\mu_A = \frac{x_A}{[1 - 1.645 \times (\text{CoV})_A]} = \frac{x_A}{[1 - 1.645 \times 0.05]} = 1.0896\, x_A \text{ mm}^2$$

$$\mu_P = \frac{x_P}{[1 + 1.645 \times (\text{CoV})_P]} = \frac{300 \times 10^3}{[1 + 1.645 \times 0.20]} = 225.73 \times 10^3 \text{ N}$$

Sl. No.	Variables	Characteristic value	CoV	Mean value (μ)	Standard deviation (σ)
1	f_y	250 N/mm^2	0.05	272.40	13.62
2	$S = P$	300×10^3 N	0.20	225.73×10^3	45146
3	A	?	0.05	$20\ \sigma_A$	–

Now,

$$g(x) = R - S$$

Here, $R = f_y A$, $S = P$

Now,
Mean value,

$$\mu_R = \mu_{f_y} \times \mu_Z = 272.40 \times \mu_A = 272.40 \times 20\sigma_A = 5448\sigma_A \text{ N}$$

$$\mu_S = \mu_P = 225.73 \times 10^3 \text{ N}$$

Now, partially differentiating R and S w.r.t variables f_y, A and P, we get,

$$\frac{\partial R}{\partial f_y} = A = \mu_A = 20\sigma_A$$

$$\frac{\partial R}{\partial A} = f_y = \mu_{f_y} = 272.40$$

$$\frac{\partial S}{\partial P} = 1.0$$

Standard deviation,

$$\sigma_R^2 = \left(\frac{\partial R}{\partial f_y} \times \sigma_{fy}\right)^2 + \left(\frac{\partial R}{\partial A} \times \sigma_A\right)^2 = (\mu_A \times 13.62)^2 + (272.40 \times \sigma_A)^2$$

Here, $\mu_A = 20\sigma_A$

$$\Rightarrow \qquad \sigma_R^2 = (20\sigma_A \times 13.62)^2 + (272.40\sigma_A)^2$$

$$\Rightarrow \qquad \sigma_R^2 = 14.80 \times 10^4 \sigma_A^2$$

$$\therefore \qquad \sigma_R = \sqrt{14.80 \times 10^4 \sigma_A^2} = 384.70\sigma_A \text{ N}$$

Similarly,

$$\sigma_S^2 = \left(\frac{\partial S}{\partial P} \times \sigma_P\right)^2 = (1.0 \times 45.14 \times 10^3)^2 = 2.037 \times 10^9$$

$$\therefore \qquad \sigma_S = \sqrt{2.037 \times 10^9} = 45140 \text{ N}$$

Now,

Reliability index (β) is given by

$$\beta = \frac{(\mu_R - \mu_S)}{\sqrt{\sigma_R^2 + \sigma_S^2}}$$

Given $\beta = 3.09$

$$\beta = \frac{(5448\sigma_A - 225.73 \times 10^3)}{\sqrt{(14.80 \times 10^4 \sigma_A^2 + 2.037 \times 10^9)}}$$

$$3.09 = \frac{(5448\sigma_A - 225.73 \times 10^3)}{\sqrt{(14.80 \times 10^4 \sigma_A^2 + 2.037 \times 10^9)}}$$

Now, squaring on both sides we get,

$$\left[\beta \times \left(\sqrt{(14.80 \times 10^4 \sigma_A^2 + 2.037 \times 10^9)}\right)\right]^2 = (5448\sigma_A - 225.73 \times 10^3)^2$$

$$\Rightarrow (3.09)^2 \times (14.80 \times 10^4 \sigma_A^2 + 2.037 \times 10^9) = (5448\sigma_A - 225.73 \times 10^3]^2$$

$$\Rightarrow 141.31 \times 10^4 \sigma_A^2 + 1.944 \times 10^{10} = (5448\sigma_A)^2 + (225.73 \times 10^3)^2 - 2 \times 5448\sigma_A \times 225.73 \times 10^3$$

$$\Rightarrow 2.818 \times 10^7 \sigma_A^2 - 2.459 \times 10^9 \sigma_A + 3.146 \times 10^{10} = 0$$

By solving quadratic equation in σ_A, we get,

$$\therefore \qquad \sigma_{A_{\text{max.}}} = 71.68 \text{ mm}^2$$

$$\therefore \qquad \sigma_{A_{\text{min.}}} = 15.573 \text{ mm}^2$$

We know that,

$$\mu_A = 20\sigma_A$$

$$\Rightarrow \qquad \mu_A = 20\sigma_{A_{\text{max.}}} = 20 \times 71.68 = 1433.6 \text{ mm}^2$$

$$\Rightarrow \qquad \mu_A = 20\sigma_{A_{\text{min.}}} = 20 \times 15.573 = 311.46 \text{ mm}^2$$

Check:

$$\beta = \frac{\mu_R - \mu_S}{\sqrt{\sigma_R^2 + \sigma_S^2}} = \frac{(5448\sigma_A - 225.73 \times 10^3)}{\sqrt{(14.80 \times 10^4 \sigma_A^2 + 2.037 \times 10^9)}}$$

For, $\sigma_{A_{max.}} = 71.68 \text{ mm}^2$

$$\beta = \frac{(5448 \times 71.68 - 225.73 \times 10^3)}{\sqrt{(14.80 \times 10^4 \times (71.68)^2 + 2.037 \times 10^9)}} = \frac{164782.64}{52890.711} = 3.11 \approx 3.09$$

For, $\sigma_{A_{min.}} = 15.573 \text{ mm}^2$

$$\beta = \frac{(5448 \times 15.573 - 225.73 \times 10^3)}{\sqrt{(14.80 \times 10^4 \times (15.573)^2 + 2.037 \times 10^9)}} = \frac{-140888.29}{45529.03} = -3.094$$

Comment: There are two solutions for the area, the one which satisfies the given target reliability is chosen for design. Here the probability density function has no influence on the analysis.

Example 10.3 Analyze by Hasofer-Lind method, a tie member of roof truss subjected to 300 kN for the limit state of yield of $g(x) = f_y A - S$ of the steel member for strength formulation. Analyze for the following data.

Sl. No	Variables	Characteristic value	CoV	Distribution
1	f_y	250 N/mm^2	0.05	Normal
2	A	1432 mm^2	0.05	Normal
3	$S = P$	300×10^3 N	0.20	Normal

Solution:

Hasofer-Lind Method

The limit state function, $g(x) = f_y A - S$ ...(1)

$$R = f_y A,\ S = P$$

The strengths are underestimated and loads are overestimated, the mean values are computed from the characteristic value.

$f_y = 250 \text{ N/mm}^2$, $(\text{CoV})_{fy} = 0.05$, $\mu_{fy} = 272.40 \text{ N/mm}^2$, $\sigma_{fy} = 13.62 \text{ N/mm}^2$

$A = 1432 \text{ mm}^2$, $(\text{CoV})_A = 0.05$, $\mu_A = 1560.33 \text{ mm}^2$, $\sigma_A = 78.01 \text{ mm}^2$

$P = 300 \times 10^3$ N, $\mu_P = 225.73 \times 10^3$ N, $(\text{CoV})_P = 0.20$, $\sigma_P = 45.14 \times 10^3$ N

Normalize the variables

Here, $Z_1 = \dfrac{f_y - \mu_{fy}}{\sigma_{fy}}$; $Z_2 = \dfrac{A - \mu_A}{\sigma_A}$; $Z_3 = \dfrac{S - \mu_S}{\sigma_S}$

$$\Rightarrow \quad f_y = Z_1 \sigma_{fy} + \mu_{fy} \text{ N/mm}^2 \qquad \ldots(2)$$

$$\Rightarrow \quad A = Z_2 \sigma_A + \mu_A \text{ mm}^2 \qquad \ldots(3)$$

$$\Rightarrow \quad P = Z_3 \sigma_P + \mu_P \text{ N} \qquad \ldots(4)$$

Substituting these Eqs (2), (3) and (4) in the Eq. (1)

$$g(x) = (Z_1 \sigma_{fy} + \mu_{fy}) \times (Z_2 \sigma_A + \mu_A) - (Z_3 \sigma_P + \mu_P)$$

$$\Rightarrow g(x) = Z_1 Z_2 \sigma_{fy} \sigma_A + Z_1 \sigma_{fy} \mu_A + Z_2 \sigma_A \mu_{fy} + \mu_{fy} \mu_A - Z_3 \sigma_P - \mu_P$$

$$\Rightarrow g(x) = 13.62 \times 78.01 \times Z_1 Z_2 + 13.62 \times 1560.33 \times Z_1 + 78.01 \times 272.40 \times Z_2 + 272.40 \times 1560.33 - 45.14 \times 10^3 \times Z_3 - 225.73 \times 10^3$$

$$\Rightarrow g(x) = 1062.49 Z_1 Z_2 + 21251.69 Z_1 + 21251.69 Z_2 + 425033.89 - 45.14 \times 10^3 Z_3 - 225.73 \times 10^3$$

$$\Rightarrow g(x) = 21251.69 Z_1 + 21251.69 Z_2 + 1062.49 Z_1 Z_2 - 45.14 \times 10^3 Z_3 + 199303.89 \qquad \ldots(5)$$

At limit state condition, $g(x) = 0$

$$21251.69 Z_1 + 21251.69 Z_2 + 1062.49 Z_1 Z_2 - 45.14 \times 10^3 Z_3 + 199303.89 = 0$$

$$21251.69 Z_1 + 21251.69 Z_2 + 1062.49 Z_1 Z_2 - 45.14 \times 10^3 Z_3 = -199303.89 \qquad \ldots(6)$$

Put $Z_1 = \alpha_1 \beta \quad Z_2 = \alpha_2 \beta \quad Z_3 = \alpha_3 \beta$ in the above Eq. (6)

$$\Rightarrow \quad 21251.69 \alpha_1 \beta + 21251.69 \alpha_2 \beta + 1062.49 \alpha_1 \beta \alpha_2 \beta - 45.14 \times 10^3 \alpha_3 \beta = -199303.89$$

$$\Rightarrow \quad \beta[21251.69 \alpha_1 + 21251.69 \alpha_2 + 1062.49 \alpha_1 \alpha_2 \beta - 45.14 \times 10^3 \alpha_3] = -199303.89$$

$$\therefore \quad \beta = \frac{-199303.89}{[21251.69 \alpha_1 + 21251.69 \alpha_2 + 1062.49 \alpha_1 \alpha_2 \beta - 45.14 \times 10^3 \alpha_3]} \qquad \ldots(7)$$

Differentiating Eq. (5) w.r.t Z_1, Z_2 and Z_3 respectively

$$\frac{\partial g(x)}{\partial Z_1} = 21251.69 + 1062.49 Z_2$$

$$\Rightarrow \quad \frac{\partial g(x)}{\partial Z_1} = 21251.69 + 1062.49 \alpha_2 \beta \qquad \ldots(8)$$

$$\frac{\partial g(x)}{\partial Z_2} = 21251.69 + 1062.49 Z_1$$

$$\Rightarrow \quad \frac{\partial g(x)}{\partial Z_2} = 21251.69 + 1062.49 \alpha_1 \beta \qquad \ldots(9)$$

$$\frac{\partial g(x)}{\partial Z_3} = -45.14 \times 10^3 \qquad \ldots(10)$$

$$\Rightarrow \quad \left[\frac{\partial g(x)}{\partial Z_1}\right]^2 + \left[\frac{\partial g(x)}{\partial Z_2}\right]^2 + \left[\frac{\partial g(x)}{\partial Z_3}\right]^2 = k^2 \qquad \ldots(11)$$

Trial 1:

Put $\beta = 3.09,\ \alpha_1 = \alpha_2 = \frac{1}{\sqrt{3}} = -0.577,\ \alpha_3 = \frac{1}{\sqrt{3}} = 0.577$

Now Eq. (7) becomes

$$\beta^I_{new} = \frac{-199303.89}{\begin{bmatrix} 21251.69 \times (-0.577) + 21251.69 \times (-0.577) + 1062.49 \times (-0.577) \times (-0.577) \times 3.09 \\ -45.14 \times 10^3 \times (0.577) \end{bmatrix}}$$

$\therefore \quad \beta^I_{new} = 4.028$

Here,

$$\Rightarrow \quad \frac{\partial g(x)}{\partial Z_1} = 21251.69 + 1062.49 \times (-0.577) \times 3.09 = 19357.34$$

$$\Rightarrow \quad \frac{\partial g(x)}{\partial Z_2} = 21251.69 + 1062.49 \times (-0.577) \times 3.09 = 19357.34$$

$$\Rightarrow \quad \frac{\partial g(x)}{\partial Z_3} = -45.14 \times 10^3$$

Now, Eq. (11) becomes

$$\Rightarrow \quad K^2 = (19357.34)^2 + (19357.34)^2 + (-45.14 \times 10^3)^2 = 2787.03 \times 10^6$$

$$\therefore \quad K = \sqrt{2787.03 \times 10^6} = 52792.35$$

Now,

$$\alpha^I_{1\,new} = -\frac{\frac{\partial g(x)}{\partial Z_1}}{K} = \frac{19357.34}{52792.35} = -0.3666$$

$$\alpha^I_{2\,new} = -\frac{\frac{\partial g(x)}{\partial Z_2}}{K} = \frac{19357.34}{52792.35} = -0.3666$$

$$\alpha^I_{3\,new} = -\frac{\frac{\partial g(x)}{\partial Z_3}}{K} = \frac{45.14 \times 10^3}{52792.35} = 0.8550$$

Check:

$\alpha_1^2 + \alpha_2^2 + \alpha_3^2 = [(-0.3666)^2 + (-0.3666)^2 + (0.8550)^2] = 0.999816 \approx 1$

Trial 2:

Put $\beta = 4.028,\ \alpha_1 = \alpha_2 = -0.3666,\ \alpha_3 = 0.8550$

Now Eq. (7) becomes

$$\beta^{II} = \frac{-199303.89}{\begin{bmatrix} 21251.69 \times (-0.3666) + 21251.69 \times (-0.3666) + 1062.49 \times (-0.3666) \times (-0.3666) \times 4.028 \\ -45.14 \times 10^3 \times (0.8550) \end{bmatrix}}$$

$\therefore\quad \beta^{II} = 3.719$

Here,

$$\Rightarrow \quad \frac{\partial g(x)}{\partial Z_1} = 21251.69 + 1062.49 \times (-0.3666) \times 4.02 = 19688.42$$

$$\Rightarrow \quad \frac{\partial g(x)}{\partial Z_2} = 21251.69 + 1062.49 \times (-0.3666) \times 4.02 = 19688.42$$

$$\Rightarrow \quad \frac{\partial g(x)}{\partial Z_3} = -45.14 \times 10^3$$

Now, Eq. (11) becomes

$$\Rightarrow \quad K^2 = (19688.42)^2 + (19688.42)^2 + (-45.14 \times 10^3)^2 = 2812.88 \times 10^6$$

$$\therefore \quad K = \sqrt{2812.88 \times 10^6} = 53036.66$$

Now,

$$\alpha_1^{II} = \frac{\frac{\partial g(x)}{\partial Z_1}}{K} = \frac{19688.42}{53036.66} = -0.3712$$

$$\alpha_2^{II} = \frac{\frac{\partial g(x)}{\partial Z_2}}{K} = \frac{19688.42}{53036.66} = -0.3712$$

$$\alpha_3^{II} = \frac{\frac{\partial g(x)}{\partial Z_3}}{K} = \frac{45.14 \times 10^3}{53036.66} = 0.8511$$

Check:

$\alpha_1^2 + \alpha_2^2 + \alpha_3^2 = [(-0.3712)^2 + (-0.3712)^2 + (0.8511)^2] = 0.99995 \approx 1$

Trial 3:

Put $\beta = 3.719,\ \alpha_1 = \alpha_2 = -0.3712,\ \alpha_3 = 0.8511$

Now, Eq. (7) becomes

$$\beta^{III} = \frac{-199303.89}{\left[\begin{array}{r}21251.69\times(-0.3712)+21251.69\times(-0.3712)+1062.49\times(-0.3712)\times(-0.3712)\times 3.719\\ -45.14\times 10^3\times(0.8511)\end{array}\right]}$$

$\therefore\ \beta^{III} = 3.714$

Now,

$$\alpha_1^{III} = -0.370$$

$$\alpha_2^{III} = -0.370$$

$$\alpha_3^{III} = 0.8510$$

Check:

$\alpha_1^2 + \alpha_2^2 + \alpha_3^2 = [(-0.370)^2 + (-0.370)^2 + (0.8510)^2] = 0.99800 \approx 1$

Sl. No.	β	α_1 (f_y)	α_2 (A)	α_3 (S)
1	4.048996	–0.359662	–0.359662	0.860980
2	3.718348	–0.371404	–0.371404	0.850952
3	3.714339	–0.372493	–0.372493	0.849998
4	3.714293	–0.372456	–0.372456	0.850031
5	3.714292	–0.372459	–0.372459	0.850029
6	3.714292	–0.372459	–0.372459	0.850029
7	3.714292	–0.372459	–0.372459	0.850029
8	3.714292	–0.372459	–0.372459	0.850029
9	3.714292	–0.372459	–0.372459	0.850029
10	3.714292	–0.372459	–0.372459	0.850029

Example 10.4 A tie member of roof truss subjected to 300 kN for the limit state of yield of $g(x) = f_y A - S$ of the steel member for strength formulation. Determine the area of the section by Hasofer-Lind method, for a reliability index of 3.09.

Sl. No.	Variables	Characteristic value	CoV	Distribution
1	f_y	250 N/mm^2	0.05	Normal
2	A	?	0.05	Normal
3	$S = P$	300×10^3 N	0.2	Normal

Solution:

Design: Hasofer-Lind Method

Target reliability index, $\beta = 3.09$

The limit state of yield of $g(x) = fy\,A - S$ ...(1)

Here strengths are underestimated and loads are overestimated

Strengths are f_y and A, load is P

$$\mu_{f_y} = \frac{x_{f_y}}{[1 - 1.645 \times (\text{CoV})_{f_y}]} = \frac{250}{[1 - 1.645 \times 0.05]} = 272.40 \text{ N/mm}^2$$

$$\mu_A = \frac{x_A}{[1 - 1.645 \times (\text{CoV})_A]} = \frac{x_A}{[1 - 1.645 \times 0.05]} = 1.0896 x_A \text{ mm}^2$$

$$\mu_P = \frac{x_P}{[1 + 1.645 \times (\text{CoV})_P]} = \frac{300 \times 10^3}{[1 + 1.645 \times 0.20]} = 225.73 \times 10^3 \text{ N}$$

$f_y = 250$ N/mm^2, $(\text{CoV})_{fy} = 0.05$, $\mu_{fy} = 272.40$ N/mm^2, $\sigma_{fy} = 13.62$ N/mm^2

Standard deviation = Mean value × CoV

i.e., $\sigma_A = \mu_A \times (\text{CoV})_A$

$$\Rightarrow \quad \mu_A = \frac{\sigma_A}{(\text{Co V})_A} = \frac{\sigma_A}{0.05} = 20\sigma_A$$

$P = 300 \times 10^3$ N, $\mu_P = 225.73 \times 10^3$ N, $(\text{CoV})_P = 0.20$, $\sigma_P = 45.14 \times 10^3$ N

Normalize the variables

Here, $Z_1 = \dfrac{f_y - \mu_{fy}}{\sigma_{fy}}$; $Z_2 = \dfrac{A - \mu_A}{\sigma_A}$; $Z_3 = \dfrac{S - \mu_S}{\sigma_S}$

$\Rightarrow \quad f_y = Z_1\sigma_{fy} + \mu_{fy}$ N/mm^2 ...(2)

$\Rightarrow \quad A = Z_2\sigma_A + \mu_A$ mm^2 ...(3)

$\Rightarrow \quad P = Z_3\sigma_P + \mu_P$ N ...(4)

Substituting these Eqs (2), (3) and (4) in the Eq. (1)

$$g(x) = (Z_1\sigma_{fy} + \mu_{fy}) \times (Z_2\sigma_A + \mu_A) - (Z_3\sigma_P + \mu_P)$$

$$\Rightarrow g(x) = Z_1 Z_2 \sigma_{fy}\sigma_A + Z_1\sigma_{fy}\mu_A + Z_2\sigma_A\mu_{fy} + \sigma_{fy}\mu_A - Z_3\sigma_P - \mu_P$$

$$\Rightarrow g(x) = Z_1 Z_2 \sigma_{fy}\sigma_A + Z_1\sigma_{fy} \times 20\sigma_A + Z_2\sigma_A\mu_{fy} + \mu_{fy} \times 20\sigma_A - Z_3\sigma_P - \mu_P$$

$$\Rightarrow g(x) = 13.62\sigma_A Z_1 Z_2 + 20 \times 13.62\sigma_A Z_1 + 272.40\sigma_A Z_2 + 20 \times 272.40\sigma_A - 45.14 \times 10^3 Z_3 - 225.73 \times 10^3$$

$$\Rightarrow g(x) = 13.62\sigma_A Z_1 Z_2 + 272.40\sigma_A Z_1 + 272.40\sigma_A Z_2 + 5448\sigma_A - 45.14 \times 10^3 Z_3 - 225.73 \times 10^3$$

$$\Rightarrow g(x) = 272.40\sigma_A Z_1 + 272.40\sigma_A Z_2 + 13.62\sigma_A Z_1 Z_2 + 5448\sigma_A - 45.14 \times 10^3 Z_3 - 225.73 \times 10^3$$

...(5)

At limit state condition, $g(x) = 0$

$$\Rightarrow \quad 272.40\sigma_A Z_1 + 272.40\sigma_A Z_2 + 13.62\sigma_A Z_1 Z_2 + 5448\sigma_A - 45.14\times 10^3 Z_3 - 225.73\times 10^3 = 0$$

$$\sigma_A[272.40Z_1 + 272.40Z_2 + 13.62Z_1Z_2 + 5448] = 45.14\times 10^3 Z_3 + 225.73\times 10^3$$

Put $Z_1 = \alpha_1\beta \quad Z_2 = \alpha_2\beta \quad Z_3 = \alpha_3\beta$

$$\Rightarrow \quad \sigma_A[272.40\alpha_1\beta + 272.40\alpha_2\beta + 13.62\alpha_1\alpha_2\beta^2 + 5448] = 45.14\times 10^3\alpha_3\beta + 225.73\times 10^3$$

$$\therefore \quad \sigma_A = \frac{45.14\times 10^3\alpha_3\beta + 225.73\times 10^3}{[272.40\alpha_1\beta + 272.40\alpha_2\beta + 13.62\alpha_1\alpha_2\beta^2 + 5448]} \qquad \text{...(6)}$$

Differentiating Eq. (5) w.r.t Z_1, Z_2 and Z_3 respectively

$$\frac{\partial g(x)}{\partial Z_1} = 272.40\sigma_A + 13.62\sigma_A Z_2$$

$$\Rightarrow \quad \frac{\partial g(x)}{\partial Z_2} = 272.40\sigma_A + 13.62\sigma_A\sigma_2\beta \qquad \text{...(7)}$$

$$\frac{\partial g(x)}{\partial Z_2} = 272.40\sigma_A + 13.62\sigma_A Z_1$$

$$\Rightarrow \quad \frac{\partial g(x)}{\partial Z_2} = 272.40\sigma_A + 13.62\sigma_A\sigma_1\beta \qquad \text{...(8)}$$

$$\frac{\partial g(x)}{\partial Z_3} = -45.14\times 10^3 \qquad \text{...(9)}$$

$$\Rightarrow \quad \left[\frac{\partial g(x)}{\partial Z_1}\right]^2 + \left[\frac{\partial g(x)}{\partial Z_2}\right]^2 + \left[\frac{\partial g(x)}{\partial Z_3}\right]^2 = k^2 \qquad \text{...(10)}$$

Trial 1:

Put $\beta = 3.09,\ \alpha_1 = \alpha_2 = -\frac{1}{\sqrt{3}},\ \alpha_3 = \frac{1}{\sqrt{3}} = 0.577$

Now Eq. (6) becomes

$$\therefore \sigma_A = \frac{45.14\times 10^3\times 0.577\times 3.09 + 225.73\times 10^3}{[272.40\times(-0.577)\times 3.09 + 272.40\times(-0.577)\times 3.09 + 13.62\times(-0.577)^2\times 3.09^2 + 5448]}$$

$\therefore \sigma_A = 67.746\ \text{mm}^2$

$\Rightarrow \mu_A = 20\sigma_A = 20\times 67.746 = 1354.93\ \text{mm}^2$

Now, Eqs (7), (8) and (9) becomes,

$$\Rightarrow \quad \frac{\partial g(x)}{\partial Z_1} = 272.40 \times 67.746 + 13.62 \times 67.746 \times (-0.577) \times 3.09 = 16809.04$$

$$\Rightarrow \quad \frac{\partial g(x)}{\partial Z_2} = 272.40 \times 67.746 + 13.62 \times 67.746 \times (-0.577) \times 3.09 = 16809.04$$

$$\frac{\partial g(x)}{\partial Z_3} = -45.14 \times 10^3$$

Now, Eq. (10) becomes

$$\Rightarrow \quad K^2 = (16809.04)^2 + (16809.04)^2 + (-45.14 \times 10^3)^2 = 2602.70 \times 10^6$$

$$\therefore \quad K = \sqrt{2602.70 \times 10^6} = 51016.73$$

Now,

$$\alpha^I_{1\,\text{new}} = \frac{\frac{\partial g(x)}{\partial Z_1}}{K} = \frac{16809.04}{51016.73} = -0.3294$$

$$\alpha^I_{2\,\text{new}} = \frac{\frac{\partial g(x)}{\partial Z_2}}{K} = \frac{16809.04}{51016.73} = -0.3294$$

$$\alpha^I_{3\,\text{new}} = \frac{\frac{\partial g(x)}{\partial Z_3}}{K} = \frac{45.14 \times 10^3}{51016.73} = 0.8848$$

Check:

$$\alpha_1^2 + \alpha_2^2 + \alpha_3^2 = [(-0.3294)^2 + (-0.3294)^2 + (0.8848)^2] = 0.999879 \approx 1$$

Trial 2:

Put $\beta = 3.09,\ \alpha_1 = \alpha_2 = -0.3294,\ \alpha_3 = 0.8848$

Now, Eq. (6) becomes

$$\therefore\ \sigma_A = \frac{45.14 \times 10^3 \times 0.8848 \times 3.09 + 225.73 \times 10^3}{[272.40 \times (-0.3294) \times 3.09 + 272.40 \times (-0.3294) \times 3.09 + 13.62 \times (-0.3294)^2 \times 3.09^2 + 5448]}$$

$$\therefore\ \sigma_A = 71.145 \text{ mm}^2$$

$$\Rightarrow \mu_A = 20\sigma_A = 20 \times 71.145 = 1422.91 \text{ mm}^2$$

Now, Eqs (7), (8) and (9) becomes,

$$\Rightarrow \quad \frac{\partial g(x)}{\partial Z_1} = 272.40 \times 71.145 + 13.62 \times 71.145 \times (-0.3294) \times 3.09 = 18393.58$$

$$\Rightarrow \quad \frac{\partial g(x)}{\partial Z_2} = 272.40 \times 71.145 + 13.62 \times 71.145 \times (-0.3294) \times 3.09 = 18393.58$$

$$\frac{\partial g(x)}{\partial Z_3} = -45.14 \times 10^3$$

Now, Eq. (10) becomes

$$\Rightarrow \quad K^2 = (18393.58)^2 + (18393.58)^2 + (-45.14 \times 10^3)^2 = 2714.26 \times 10^6$$

$$\therefore \quad K = \sqrt{2714.26 \times 10^6} = 52098.63$$

Now,

$$\alpha_1^{II} = \frac{\frac{\partial g(x)}{\partial Z_1}}{K} = \frac{18393.58}{52098.63} = -0.3530$$

$$\alpha_2^{II} = \frac{\frac{\partial g(x)}{\partial Z_2}}{K} = \frac{18393.58}{52098.63} = -0.3530$$

$$\alpha_3^{II} = \frac{\frac{\partial g(x)}{\partial Z_3}}{K} = \frac{45.14 \times 10^3}{52098.63} = 0.8664$$

Check:

$$\alpha_1^2 + \alpha_2^2 + \alpha_3^2 = [(-0.3530)^2 + (-0.3530)^2 + (0.8664)^2] = 0.99986 \approx 1$$

Note: $f_y = (0.3530)^2 \times 100 = 12.46\%$

$A = (0.3530)^2 \times 100 = 12.46\%$

$S = (0.8664)^2 \times 100 = 75.06\%$

Comment: The sensitivity coefficient $\alpha\_$ factors reflects how sensitive the reliability is to the basic variables. The term importance factors are defined as the square of $\alpha\_$ factors. The basic variables f_y, A, S, and their contribution to the performance function are quantified by importance factors. The computed importance factors are 12.46%, 12.46% and 75.06% respectively.

Sl. No.	β	Area, A	α_1 (f_y)	α_2 (A)	α_3 (S)
1	3.09	1354.93	−0.3294	−0.3294	0.8848
2		1422.91	−0.3530	−0.3530	0.8664
3		1423.37	−0.3521	−0.3521	0.8671
4		1423.37	−0.3521	−0.3521	0.8671
5		1423.37	−0.3521	−0.3521	0.8671
6		1423.37	−0.3521	−0.3521	0.8671
7		1423.37	−0.3521	−0.3521	0.8671
8		1423.37	−0.3521	−0.3521	0.8671
9		1423.37	−0.3521	−0.3521	0.8671
10		1423.37	−0.3521	−0.3521	0.8671

Example 10.5 Analyze by FOSM method, for the given data. The top chord of a bridge truss has an effective length of 4 m and the cross section as shown in Fig. 10.3. The limit state of buckling $g(x) = P_{cr} - S$; where $P_{cr} = \dfrac{\pi^2 E A r^2}{L^2}$, $S = P$. S is the load, treat E, A, r, S and L as random variables. Compute the reliability index, resistance factor, load factor, bias resistance factor and bias load factor.

Sl. No.	Variables	Characteristic value	CoV	Distribution
1	E	2×10^5 N/mm^2	0.05	Normal
2	A	9880 mm^2	0.05	Normal
3	r	76.29 mm	0.05	Normal
4	L	4000 mm	0.05	Normal
5	P	1100×10^3 N	0.20	Normal

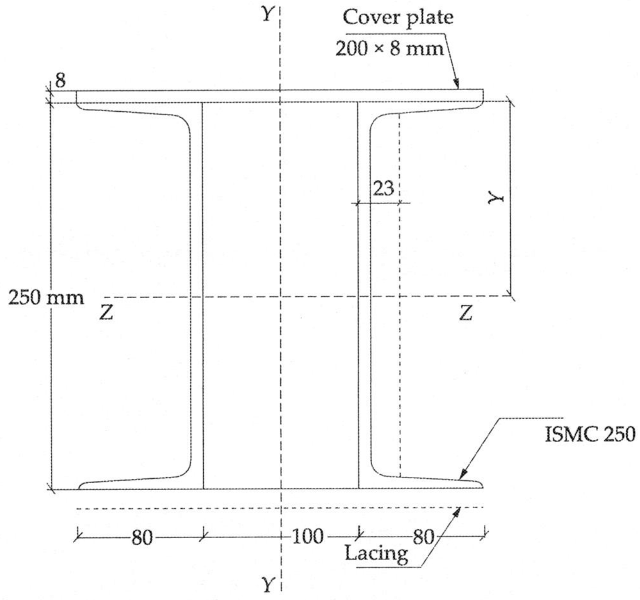

Fig. 10.3 Cross section of compression member

Solution:

Analysis—FOSM Method

The strengths are underestimated and loads are overestimated, the mean values are computed from the characteristics value and are tabulated below:

Sl. No.	Variables	Characteristic value	CoV	Mean value (μ)	Standard deviation (σ)
1	E	2×10^5 N/mm^2	0.05	2.179×10^5	10896
2	A	9880 mm^2	0.05	10765.46	538.27
3	r	76.29 mm	0.05	83.127	4.156
4	L	4000 mm	0.05	4358.48	217.92
5	P	1100×10^3 N	0.20	827.69×10^3	165.53×10^3

The limit state function, $g(x) = P_{cr} - S$

$$\Rightarrow \qquad g(x) = R - S$$

Here, $R = P_{cr} = \dfrac{\pi^2 E A r^2}{L^2}$, $S = P$

Now, the mean and standard derivations are

Mean value,

$$\mu_R = \mu_{P_{cr}} = \frac{\pi^2 \mu_E \mu_A \mu_{r^2}}{\mu_{L^2}} = \frac{\pi^2 \times 2.179 \times 10^5 \times 10765.46 \times (83.127)^2}{(4358.48)^2} = 8.421 \times 10^6 \text{ N}$$

$$\mu_S = \mu_P = 827.69 \times 10^3 \text{ N}$$

Now, partially differentiating R and S w.r.t variables, we get,

Here, E, A, r, S and L are random variables.

$$\frac{\partial R}{\partial E} = \frac{\pi^2 \mu_A \mu_{r^2}}{\mu_{L^2}} = \frac{\pi^2 \times 10765.46 \times (83.127)^2}{(4358.48)^2} = 38.649$$

$$\frac{\partial R}{\partial A} = \frac{\pi^2 \mu_E \mu_{r^2}}{\mu_{L^2}} = \frac{\pi^2 \times 2.179 \times 10^5 \times (83.127)^2}{(4358.48)^2} = 782.30$$

$$\frac{\partial R}{\partial r} = \frac{2\pi^2 \mu_E \mu_A \mu_r}{\mu_{L^2}} = \frac{2\pi^2 \times 2.179 \times 10^5 \times 10765.46 \times 83.127}{(4358.48)^2} = 2.02 \times 10^5$$

$$\frac{\partial R}{\partial L} = -\frac{2\pi^2 \mu_E \mu_A \mu_{r^2}}{\mu_{L^3}} = -\frac{2\pi^2 \times 2.179 \times 10^5 \times 10765.46 \times (83.127)^2}{(4358.48)^3} = -3864.54$$

$$\frac{\partial S}{\partial P} = -1$$

Standard deviation

$$\sigma_R^2 = \left(\frac{\partial R}{\partial E} \times \sigma_E\right)^2 + \left(\frac{\partial R}{\partial A} \times \sigma_A\right)^2 + \left(\frac{\partial R}{\partial r} \times \sigma_r\right)^2 + \left(\frac{\partial R}{\partial L} \times \sigma_L\right)^2$$

$$\sigma_R^2 = (38.649 \times 10896)^2 + (782.30 \times 538.27)^2 + (2.02 \times 10^5 \times 4.156)^2 + (-3864.54 \times 217.92)^2$$

$$\Rightarrow \sigma_R^2 = 1.7683 \times 10^{12}$$

$$\therefore \sigma_R = \sqrt{1.7683 \times 10^{12}} = 1.329 \times 10^6 \text{ N}$$

Similarly,

$$\sigma_S^2 = \left(\frac{\partial S}{\partial P} \times \sigma_P\right)^2 = (1)^2 \times (165.53 \times 10^3)^2 = 2.74 \times 10^{10}$$

$$\therefore \quad \sigma_S = \sqrt{2.74 \times 10^{10}} = 165.53 \times 10^3 \text{ N}$$

(a) *Reliability index (β),*

$$\beta = \frac{\mu_R - \mu_S}{\sqrt{\sigma_R^2 + \sigma_S^2}} = \frac{(8.421 \times 10^6 - 827.69 \times 10^3)}{\sqrt{(1.7683 \times 10^{12} + 2.74 \times 10^{10})}} = 5.66$$

(b) *Resistance factor (ϕ),*

$$\phi = [1 - 0.75 \times \beta \times (\text{CoV})_R]$$

$$\therefore \quad (\text{CoV})_R = \frac{\sigma_R}{\mu_R} = \frac{1.329 \times 10^6}{8.421 \times 10^6} = 0.1578$$

$$\Rightarrow \quad \phi = [1 - 0.75 \times 5.66 \times 0.1578] = 0.33$$

(c) *Load factor (γ),*

$$\gamma = [1 + 0.75 \times \beta \times (\text{CoV})_S]$$

$$\therefore \quad (\text{CoV})_S = \frac{\sigma_S}{\mu_S} = \frac{165.53 \times 10^3}{827.69 \times 10^3} = 0.199 \approx 0.20$$

$$\Rightarrow \quad \gamma = [1 + 0.75 \times 5.66 \times 0.20] = 1.844$$

(d) *Bias resistance factor,*
Mean resistance,

$$\mu_R = \mu_{P_{cr}} = \frac{\pi^2 \mu_E \mu_A \mu_{r^2}}{\mu_{L^2}} = \frac{\pi^2 \times 2.179 \times 10^5 \times 10765.46 \times (83.127)^2}{(4358.48)^2} = 8.421 \times 10^6 \text{ N}$$

Characteristic resistance,

$$R_k = P_{cr} = \frac{\pi^2 EAr^2}{L^2} = \frac{\pi^2 \times 2 \times 10^5 \times 9880 \times (76.29)^2}{(4000)^2} = 7.094 \times 10^6 \text{ N}$$

$$\therefore \quad b_R = \frac{\mu_R}{R_k} = \frac{8.421 \times 10^6}{7.094 \times 10^6} = 1.187$$

(e) *Bias load factor,*
Mean load, $\mu_S = 827.69 \times 10^3$
Characteristic load, $S_k = 1100 \times 10^3$

$$\therefore \quad b_L = \frac{\mu_S}{S_k} = \frac{827.69 \times 10^3}{1100 \times 10^3} = 0.7524$$

Example 10.6 Design by FOSM method, for the given data. The top chord of a bridge truss has an effective length of 4 m and the cross section as shown in Fig. 10.3. The limit state of buckling $g(x) = P_{cr} - S$; where $P_{cr} = \dfrac{\pi^2 EAr^2}{L^2}$, $S = P$. S is the load, treat E, A, r, S and L as random variables. Determine the area of the column, required for a reliability index of 3.09.

Sl. No.	Variables	Characteristic value	CoV	Distribution
1	E	2×10^5 N/mm^2	0.05	Normal
2	A	?	0.05	Normal
3	r	76.29 mm	0.05	Normal
4	L	4000 mm	0.05	Normal
5	P	1100×10^3 N	0.2	Normal

Solution:

Design:

The limit state of yield of $g(x) = P_{cr} - S$

Target Reliability Index, $\beta = 3.09$

Standard deviation = Mean × $(\text{CoV})_A$

$$\Rightarrow \quad \mu_A = \frac{\sigma_A}{(\text{CoV})_A} = \frac{\sigma_A}{0.05} = 20\sigma_A$$

Now, the mean and standard deviations are:

Mean value,

$$\mu_R = \mu_{P_{cr}} = \frac{\pi^2 \mu_E \mu_A \mu_{r^2}}{\mu_{L^2}} = \frac{\pi^2 \times 2.179 \times 10^5 \times 20\sigma_A \times (83.127)^2}{(4358.48)^2} = 0.156 \times 10^5 \sigma_A \text{ N}$$

$$\mu_S = \mu_P = 827.69 \times 10^3 \text{ N}$$

Now, partially differentiating R and S w.r.t variables, we get,

Here, E, A, r, S and L are random variables

$$\frac{\partial R}{\partial E} = \frac{\pi^2 \mu_A \mu_{r^2}}{\mu_{L^2}} = \frac{\pi^2 \times 20\sigma_A \times (83.127)^2}{(4358.48)^2} = 0.0718\sigma_A$$

$$\frac{\partial R}{\partial A} = \frac{\pi^2 \mu_E \mu_{r^2}}{\mu_{L^2}} = \frac{\pi^2 \times 2.179 \times 10^5 \times (83.127)^2}{(4358.48)^2} = 782.30$$

$$\frac{\partial R}{\partial r} = \frac{2\pi^2 \mu_E \mu_A \mu_r}{\mu_{L^2}} = \frac{2\pi^2 \times 2.179 \times 10^5 \times 20\sigma_A \times 83.127}{(4358.48)^2} = 376.43\sigma_A$$

$$\frac{\partial R}{\partial L} = -\frac{2\pi^2 \mu_E \mu_A \mu_{r^2}}{\mu_{L^3}} = -\frac{2\pi^2 \times 2.179 \times 10^5 \times 20\sigma_A \times (83.127)^2}{(4358.48)^3} = -7.18\sigma_A$$

$$\frac{\partial S}{\partial P} = 1$$

Standard deviation,

$$\sigma_R^2 = \left(\frac{\partial R}{\partial E} \times \sigma_E\right)^2 + \left(\frac{\partial R}{\partial A} \times \sigma_A\right)^2 + \left(\frac{\partial R}{\partial r} \times \sigma_r\right)^2 + \left(\frac{\partial R}{\partial L} \times \sigma_L\right)^2$$

$$\sigma_R^2 = (0.0718\sigma_A \times 10896)^2 + (782.30 \times \sigma_A)^2 + (376.43\sigma_A \times 4.156)^2 + (-7.18\sigma_A \times 217.92)^2$$

$$\Rightarrow \sigma_R^2 = 6.12 \times 10^6 \sigma_A^2$$

Similarly,

$$\sigma_s^2 = \left(\frac{\partial S}{\partial P} \times \sigma_P\right)^2 = (1)^2 \times (165.53 \times 10^3)^2 = 2.74 \times 10^{10}$$

$$\therefore \quad \sigma_S = \sqrt{2.74 \times 10^{10}} = 165.53 \times 10^3 \text{ N}$$

Now,

Reliability index (β),

Given $\beta = 3.09$

$$\beta = \frac{\mu_R - \mu_S}{\sqrt{\sigma_R^2 + \sigma_S^2}}$$

$$3.09 = \frac{(0.156 \times 10^5 \sigma_A - 827.69 \times 10^3)}{\sqrt{(6.12 \times 10^6 \sigma_A^2 + 2.74 \times 10^{10})}}$$

Now, squaring on both sides, we get,

$$\left[3.09\times\sqrt{(6.12\times10^{6}\sigma_A^2+2.74\times10^{10})}\right]^2=(0.156\times10^{5}\sigma_A-827.69\times10^{3})^2$$

$$\Rightarrow \quad (3.09)^2\times(6.12\times10^{6}\sigma_A^2+2.74\times10^{10})=(0.156\times10^{5}\sigma_A-827.69\times10^{3})^2$$

$$\Rightarrow \quad 58.43\times10^{6}\sigma_A^2+26.12\times10^{10}=(0.156\times10^{5}\sigma_A)^2+(827.69\times10^{3})^2-2\times0.156\times10^{5}\sigma_A\times827.69\times10^{3}$$

$$\Rightarrow \quad 1.845\times10^{8}\sigma_A^2-2.58\times10^{10}\sigma_A+4.234\times10^{11}=0$$

By solving quadratic equation in σ_A, we get,

$$\therefore \quad \sigma_{A_{max.}}=120.84\text{ mm}^2$$

$$\therefore \quad \sigma_{A_{min.}}=18.98\text{ mm}^2$$

We know that,

$$\mu_A=20\mu_A$$

$$\Rightarrow \quad \mu_A=20\sigma_{A_{max.}}=20\times120.84=2416.8\text{ mm}^2$$

$$\Rightarrow \quad \mu_A=20\sigma_{A_{min.}}=20\times18.98=379.6\text{ mm}^2$$

Check:

$$\mu_R=0.156\times10^{5}\times120.84=1.885\times10^{6}\text{ N}$$

$$\Rightarrow \quad \sigma_R^2=6.12\times10^{6}\times(120.84)^2=8.936\times10^{10}$$

$$\beta=\frac{\mu_R-\mu_S}{\sqrt{\sigma_R^2+\sigma_S^2}}=\frac{1.885\times10^{6}-827.6\times10^{3}}{\sqrt{8.936\times10^{10}+2.74\times10^{10}}}=3.094\approx3.09$$

Example 10.7 Analyze by Hasofer-Lind method, for the given data. The top chord of a bridge truss has an effective length of 4 m and the cross-section as shown in Fig. (10.3). The limit state of buckling $g(x) = P_{cr} - S$; where $P_{cr}=\dfrac{\pi^2 EAr^2}{L^2}$, $S = P$. S is the load, treat E, A, r, S and L as random variables. Compute the reliability index, resistance factor, load factor, bias resistance factor and bias load factor.

Sl. No.	Variables	Characteristic value	CoV	Distribution
1	E	2×10^5 N/mm^2	0.05	Normal
2	A	9880 mm^2	0.05	Normal
3	r	76.29 mm	0.05	Normal
4	L	4000 mm	0.05	Normal
5	P	1100×10^3 N	0.20	Normal

Solution:

Analysis: Hasofer-Lind Method

The limit state function, $g(x) = P_{cr} - S$

$E = 2 \times 10^5$ N/mm^2, $(\text{CoV})_E = 0.05$, $\mu_E = 2.1792 \times 10^5$ N/mm^2, $\sigma_E = 10896$ N/mm^2

$A = 9880$ mm^2, $(\text{CoV})_A = 0.05$, $\mu_A = 10765.46$ mm^2, $\sigma_A = 538.27$ mm^2

$r = 76.29$ mm, $(\text{CoV})_r = 0.05$, $\mu_r = 83.127$ mm, $\sigma_r = 4.156$ mm

$L = 4000$ mm, $(\text{CoV})_L = 0.05$, $\mu_L = 4358.48$ mm, $\sigma_L = 217.92$ mm

$S = 1100 \times 10^3$ N, $\mu_S = 827.69 \times 10^3$ N, $(\text{CoV})_S = 0.20$, $\sigma_P = 165.53 \times 10^3$ N

Normalize the variables

$$Z_1 = \frac{E - \mu_E}{\sigma_E}; \qquad E = Z_1\sigma_E + \mu_E$$

$$Z_2 = \frac{A - \mu_A}{\sigma_A}; \qquad A = Z_2\sigma_A + \mu_A$$

$$Z_3 = \frac{r - \mu_r}{\sigma_r}; \qquad r = Z_3\sigma_r + \mu_r$$

$$Z_4 = \frac{L - \mu_L}{\sigma_L}; \qquad L = Z_4\sigma_L + \mu_L$$

$$Z_5 = \frac{P - \mu_P}{\sigma_P}; \qquad P = Z_5\sigma_P + \mu_P$$

Substituting these values in Eq. (10.1)

$$g(x) = P_{cr} - S$$

$$R = P_{cr} = \frac{\pi^2 E A r^2}{L^2}; \; S = P$$

$$g(x) = \frac{\pi^2 E A r^2}{L^2} - P \qquad (1)$$

At limit state, $g(x) = 0$

$$0 = \frac{\pi^2 E A r^2}{L^2} - P$$

$$\pi^2 E A r^2 - P L^2 = 0$$

$$\pi^2[(Z_1\sigma_E + \mu_E)(Z_2\sigma_A + \mu_A)(Z_3\sigma_r + \mu_r)^2] - (Z_5\sigma_P + \mu_P)(Z_4\sigma_L + \mu_L)^2 \; \pi^2[(Z_1Z_2\sigma_E\sigma_A + Z_1\mu_A\sigma_E + Z_2\mu_E\sigma_A + \mu_E\mu_A)(Z_3^2\sigma_r^2 + \mu_r^2 + 2Z_3\mu_r\sigma_r)] - (Z_5\sigma_P + \mu_P)(Z_4^2\sigma_L^2 + \mu_L^2 + 2Z_4\mu_L\sigma_L) = 0 \quad (2)$$

$$\pi^2[(Z_1Z_2Z_3^2\sigma_E\sigma_A\sigma_r^2 + Z_1Z_3^2\mu_A\sigma_E\sigma_r^2 + Z_2Z_3^2\mu_E\sigma_A\sigma_r^2 + Z_3^2\mu_E\mu_A\sigma_r^2 + Z_1Z_2\mu_r^2\sigma_E\sigma_A + Z_1\mu_A\mu_r^2\sigma_E + Z_2\mu_E\mu_r^2\sigma_A + \mu_E\mu_A\mu_r^2 + 2Z_1Z_2Z_3\mu_r\sigma_E\sigma_A\sigma_r + 2Z_1Z_3\mu_A\mu_r\sigma_E\sigma_r + 2Z_2Z_3\mu_E\mu_r\sigma_A\sigma_r + 2Z_3\mu_E\mu_A\mu_r\sigma_r)] - (Z_4^2Z_5\sigma_P\sigma_L^2 + Z_5\mu_L^2\sigma_P + 2Z_4Z_5\mu_L\sigma_L\sigma_P) + (Z_4^2\mu_P\sigma_L^2 + \mu_P\mu_L^2 + 2Z_4\mu_P\mu_L\sigma_L) = 0$$

Put, $Z_1 = \alpha_1\beta$; $Z_2 = \alpha_2\beta$; $Z_3 = \alpha_3\beta$; $Z_4 = \alpha_4\beta$; $Z_5 = \alpha_5\beta$;

$$\pi^2[(\alpha_1\beta\alpha_2\beta\alpha_3^2\beta^2\sigma_E\sigma_A\sigma_r^2 + \alpha_1\beta\alpha_3^2\beta^2\mu_A\sigma_E\sigma_r^2 + \alpha_2\beta\alpha_3^2\beta^2\mu_E\sigma_A\sigma_r^2 + \alpha_3^2\beta^2\mu_E\mu_A\sigma_r^2 + \alpha_1\beta\alpha_2\beta\mu_r^2\sigma_E\sigma_A + \alpha_1\beta\mu_A\mu_r^2\sigma_E + \alpha_2\beta\mu_E\mu_r^2\sigma_A + \mu_E\mu_A\mu_r^2 + 2\alpha_1\beta\alpha_2\beta\alpha_3\beta\mu_r\sigma_E\sigma_A\sigma_r + 2\alpha_1\beta\alpha_3\beta\mu_A\mu_r\sigma_E\sigma_r + 2\alpha_2\beta\alpha_3\beta\mu_E\mu_r\sigma_A\sigma_r + 2\alpha_3\beta\mu_E\mu_A\mu_r\sigma_r)] - [(\alpha_4^2\beta^2\alpha_5\beta\sigma_P\sigma_L^2 + \alpha_5\beta\mu_L^2\sigma_P + 2\alpha_4\beta\alpha_5\beta\mu_L\sigma_L\sigma_P)(\alpha_4^2\beta^2\mu_P\sigma_L^2 + \mu_P\mu_L^2 + 2\alpha_4\beta\mu_P\mu_L\sigma_L)] = 0$$

$$\beta\{\pi^2[(\alpha_1\alpha_2\beta\alpha_3^2\beta^2\sigma_E\sigma_A\sigma_r^2 + \alpha_1\alpha_3^2\beta^2\mu_A\sigma_E\sigma_r^2 + \alpha_2\alpha_3^2\beta^2\mu_E\sigma_A\sigma_r^2 + \alpha_3^2\beta^2\mu_E\mu_A\sigma_r^2 + \alpha_1\alpha_2\beta\mu_r^2\sigma_E\sigma_A + \alpha_1\mu_A\mu_r^2\sigma_E + \alpha_2\mu_E\mu_r^2\sigma_A + 2\alpha_1\beta\alpha_2\alpha_3\beta\mu_r\sigma_E\sigma_A\sigma_r + 2\alpha_1\alpha_3\beta\mu_A\mu_r\sigma_E\sigma_r + 2\alpha_2\alpha_3\beta\mu_E\mu_r\sigma_A\sigma_r + 2\alpha_3\beta\mu_E\mu_A\mu_r\sigma_r)] - (\alpha_4^2\beta^2\alpha_5\sigma_P\sigma_L^2 + \alpha_5\mu_L^2\sigma_P + 2\alpha_4\alpha_5\beta\mu_L\sigma_L\sigma_P)(\alpha_4^2\beta^2\mu_P\sigma_L^2 + 2\alpha_4\mu_P\mu_L\sigma_L)\} = (\mu_P\mu_L^2 - \pi^2\mu_E\mu_A\mu_r^2)$$

$$\beta = \frac{\mu_P\mu_L^2 - (3.14)^2 \times \mu_E\mu_A\mu_r^2}{\begin{aligned}&9.869[\alpha_1\alpha_2\alpha_3^2\beta^3\sigma_E\sigma_A\sigma_r^2 + \alpha_1\alpha_3^2\beta^2\mu_A\sigma_E\sigma_r^2 + \alpha_2\alpha_3^2\beta^2\mu_E\sigma_A\sigma_r^2 + \alpha_3^2\beta\mu_E\mu_A\sigma_r^2\\ &+ \alpha_1\alpha_2\beta\mu_r^2\sigma_E\sigma_A + \alpha_1\mu_A\mu_r^2\sigma_E + \alpha_2\mu_E\mu_r^2\sigma_A + 2\alpha_1\alpha_2\alpha_3\beta^2\mu_r\sigma_E\sigma_A\sigma_r + 2\alpha_1\alpha_3\beta\mu_A\mu_r\sigma_E\sigma_r\\ &+ 2\alpha_2\alpha_3\beta\mu_E\mu_r\sigma_A\sigma_r + 2\alpha_3\mu_E\mu_A\mu_r\sigma_r] - (\alpha_4^2\alpha_5\beta^2\sigma_P\sigma_L^2 - \alpha_5\mu_L^2\sigma_P - 2\alpha_4\alpha_5\beta\mu_L\sigma_P\sigma_L\\ &- \alpha_4^2\beta\mu_P\sigma_L^2 - 2\alpha_4\mu_P\mu_L\sigma_L)\end{aligned}}$$

$$\beta = \frac{827.69 \times 10^3 \times (4358.48)^2 - (3.14)^2 \times 2.179 \times 10^5 \times 10765.46 \times (83.127)^2}{\begin{aligned}&9.869[\alpha_1\alpha_2\alpha_3^2\beta^3 \times 10896 \times 538.27 \times (4.156)^2 + \alpha_1\alpha_3^2\beta^2 \times 10765.46 \times 10896 \times (4.156)^2\\ &+ \alpha_2\alpha_3^2\beta^2 \times 2.179 \times 10^5 \times 538.27 \times (4.156)^2 + \alpha_3^2\beta \times 2.179 \times 10^5 \times 10765.46 \times (4.156)^2 + \alpha_1\alpha_2\beta\\ &\times ((83.127)^2 \times 10896 \times 538.27 + \alpha_1 \times 10765.46 \times (83.127)^2 \times 10896 + \alpha_2 \times 2.179 \times 10^5\\ &\times (83.127)^2 \times 538.27 + 2\alpha_1\alpha_2\alpha_3\beta^2 \times 83.127 \times 10896 \times 538.27 \times 4.156 + 2\alpha_1\alpha_3\beta \times 10765.46\\ &\times 83.127 \times 10896 \times 4.156 + 2\alpha_2\alpha_3\beta \times 2.179 \times 10^5 \times 83.127 \times 538.27 \times 4.156 + 2\alpha_3 \times 2.179 \times 10^5\\ &\times 10765.46 \times 83.127 \times 4.156] - \alpha_4^2\alpha_5\beta^2 \times 165.53 \times 10^3 \times (217.92)^2 - \alpha_5(4358.48)^2 \times 165.53 \times 10^3\\ &- 2\alpha_4\alpha_5\beta \times 4358.48 \times 165.53 \times 10^3 \times 217.92 - \alpha_4^2\beta \times 827.69 \times 10^3 \times (217.92)^2 - 2\alpha_4 \times 827.69\\ &\times 10^3 \times 4358.48 \times 217.92\end{aligned}}$$

$$\beta = \frac{1.572 \times 10^{13} - 1.599 \times 10^{14}}{\begin{array}{r}[9.99 \times 10^{8}\alpha_1\alpha_2\alpha_3^2\beta^3 + 1.99 \times 10^{10}\alpha_1\alpha_3^2\beta^2 + 1.99 \times 10^{10}\alpha_2\alpha_3^2\beta^2 + 3.99 \times 10^{11}\alpha_3^2\beta \\ + 3.99 \times 10^{11}\alpha_1\alpha_2\beta + 7.99 \times 10^{12}\alpha_1 + 7.99 \times 10^{12}\alpha_2 + 3.99 \times 10^{10}\alpha_1\alpha_2\alpha_3\beta^2 + 7.99 \times 10^{11} \\ \alpha_1\alpha_3\beta + 7.99 \times 10^{11}\alpha_2\alpha_3\beta + 1.599 \times 10^{13}\alpha_3 - 7.86 \times 10^{9}\alpha_4^2\alpha_5\beta^2 - 3.14 \times 10^{12}\alpha_5 \\ - 3.14 \times 10^{11}\alpha_4\alpha_5\beta - 3.93 \times 10^{10}\alpha_4^2\beta - 1.57 \times 10^{12}\alpha_4]\end{array}}$$

$$\beta = \frac{-1.4418 \times 10^{14}}{\begin{array}{r}[7.99 \times 10^{12}\alpha_1 + 7.99 \times 10^{12}\alpha_2 + 1.599 \times 10^{13}\alpha_3 - 1.57 \times 10^{12}\alpha_4 - 3.14 \times 10^{12}\alpha_5 + 3.99 \times 10^{11}\alpha_1\alpha_2\beta \\ + 7.99 \times 10^{11}\alpha_2\alpha_3\beta + 7.99 \times 10^{11}\alpha_1\alpha_3\beta - 3.14 \times 10^{11}\alpha_4\alpha_5\beta + 3.99 \times 10^{10}\alpha_1\alpha_2\alpha_3\beta^2 + 9.99 \\ \times 10^{8}\alpha_1\alpha_2\alpha_3{}^2\beta^3 + 1.99 \times 10^{10}\alpha_1\alpha_3{}^2\beta^2 + 1.99 \times 10^{10}\alpha_2\alpha_3{}^2\beta^2 - 7.86 \times 10^{9}\alpha_4{}^2\alpha_5\beta^2 + 3.99 \\ \times 10^{11}\alpha_3{}^2\beta - 3.93 \times 10^{10}\alpha_4{}^2\beta]\end{array}} \tag{3}$$

from equation

$$\alpha_1 = \frac{\partial g(x)}{\partial Z_1} = -\frac{\begin{array}{r}[\,7.99 \times 10^{12} + 3.99 \times 10^{11}\alpha_2\beta + 7.99 \times 10^{11}\alpha_3\beta + 3.99 \times 10^{10}\alpha_2\alpha_3\beta^2 \\ + 9.99 \times 10^{8}\alpha_2\alpha_3^2\beta^3 + 1.99 \times 10^{10}\alpha_3^2\beta^2]\end{array}}{K} \tag{4}$$

$$\alpha_2 = \frac{\partial g(x)}{\partial Z_2} = -\frac{\begin{array}{c}[7.99 \times 10^{12} + 3.99 \times 10^{11}\alpha_1\beta + 7.99 \times 10^{11}\alpha_3\beta + 3.99 \times 10^{10}\alpha_1\alpha_3\beta^2 \\ + 9.99 \times 10^{8}\alpha_1\alpha_3^2\beta^3 + 1.99 \times 10^{10}\alpha_3^2\beta^2]\end{array}}{K} \tag{5}$$

$$\alpha_3 = \frac{\partial g(x)}{\partial Z_3} = -\frac{\begin{array}{l}[1.599 \times 10^{13} + 7.99 \times 10^{11}\alpha_2\beta + 7.99 \times 10^{11}\alpha_1\beta + 3.99 \times 10^{10}\alpha_1\alpha_2\beta^2 \\ + 19.98 \times 10^{8}\alpha_1\alpha_2\alpha_3\beta^3 + 3.98 \times 10^{10}\alpha_1\alpha_3\beta^2 + 3.98 \times 10^{10}\alpha_2\alpha_3\beta^2 \\ + 7.98 \times 10^{11}\alpha_3\beta]\end{array}}{K} \tag{6}$$

$$\alpha_4 = \frac{\partial g(x)}{\partial Z_4} = \frac{[1.57 \times 10^{12} + 3.14 \times 10^{11}\alpha_5\beta + 15.72 \times 10^{9}\alpha_4\alpha_5\beta^2 + 7.86 \times 10^{10}\alpha_4\beta}{K} \tag{7}$$

$$\alpha_5 = \frac{\partial g(x)}{\partial Z_5} = \frac{[3.14 \times 10^{12} + 3.14 \times 10^{11}\alpha_4\beta + 7.86 \times 10^{9}\alpha_4^2\beta^2}{K} \tag{8}$$

Trial I, $\beta = 3.09$

$\alpha_1 = \alpha_2 = \alpha_3 = \alpha_4 = \alpha_5 = 1/\sqrt{5} = 0.447$

$\alpha_1 = -0.447,\ \alpha_2 = -0.447,\ \alpha_3 = -0.447,\ \alpha_4 = 0.447,\ \alpha_5 = 0.447$

from Eq. (3)

$$\beta = \frac{-1.4418\times10^{14}}{\begin{aligned}&[7.99\times10^{12}\times(-0.447)+7.99\times10^{12}\times(-0.447)+1.599\times10^{13}\times(-0.447)-1.57\times10^{12}\\ &\times(0.447)-3.14\times10^{12}\times(0.447)+3.99\times10^{11}\times(-0.447)(-0.447)\times3.09+7.99\times10^{11}\\ &\times(-0.447)(-0.447)\times3.09+7.99\times10^{11}\times(-0.447)(-0.447)\times3.09-3.14\times10^{11}\\ &\times(0.447)(0.447)\times3.09+3.99\times10^{10}\times(-0.447)(-0.447)(-0.447)\times3.09^2+9.99\times10^{8}\\ &\times(-0.447)(-0.447)(-0.447)^2\times3.09^3+1.99\times10^{10}(-0.447)(-0.447)^2\times3.09^2+1.99\times10^{10}\\ &\times(-0.447)(-0.447)^2\times3.09^2-7.86\times10^{9}(0.447)^2(0.447)\times3.09^2+3.99\times10^{11}\\ &\times(-0.447)^2\times3.09-3.93\times10^{10}\times(0.447)^2\times3.09]\end{aligned}}$$

$\beta = 9.483$

Now equation becomes

$$\alpha_1 = \frac{\partial g(x)}{\partial Z_1} = -\frac{\begin{aligned}&[7.99\times10^{12}+3.99\times10^{11}\times(-0.447)\times3.09+7.99\times10^{11}\times(-0.447)\times3.09+3.99\times10^{10}\\ &\times(-0.447)(-0.447)\times3.09^2+9.99\times10^{8}\times(-0.447)(-0.447)^2\times3.09^3+1.99\times10^{10}\\ &\times(-0.447)^2\times3.09^2]\end{aligned}}{K}$$

$$\alpha_2 = \frac{\partial g(x)}{\partial Z_2} = -\frac{\begin{aligned}&[7.99\times10^{12}+3.99\times10^{11}\times(-0.447)\times3.09+7.99\times10^{11}\times(-0.447)\times3.09+3.99\times10^{10}\\ &\times(-0.447)(-0.447)\times3.09^2+9.99\times10^{8}\times(-0.447)(-0.447)^2\\ &\times3.09^3+1.99\times10^{10}\times(-0.447)^2\times3.09^2]\end{aligned}}{K}$$

$$\alpha_3 = \frac{\partial g(x)}{\partial Z_3} = -\frac{\begin{aligned}&[1.599\times10^{13}+7.99\times10^{11}\times(-0.447)\times3.09+7.99\times10^{11}\times(-0.447)\times3.09+3.99\times10^{10}\\ &\times(-0.447)(-0.447)\times3.09^2+19.98\times10^{8}\times(-0.447)(-0.447)(-0.447)\\ &\times3.09^3+3.98\times10^{10}\times(-0.447)(-0.447)\times3.09^2+3.98\times10^{10}\times(-0.447)(-0.447)\\ &\times3.09^2+7.98\times10^{11}(-0.447)\times3.09]\end{aligned}}{K}$$

$$\alpha_4 = \frac{\partial g(x)}{\partial Z_4} = \frac{\begin{aligned}&[1.57\times10^{12}+3.14\times10^{11}\times0.447\times3.09+15.72\times10^{9}\times0.447\times0.447\times3.09^2\\ &+7.86\times10^{10}\times0.447\times3.09]\end{aligned}}{K}$$

$$\alpha_5 = \frac{\partial g(x)}{\partial Z_5} = \frac{[3.14\times10^{12}+3.14\times10^{11}\times0.447\times3.09+7.86\times10^{9}\times0.447^2\times3.09^2}{K}$$

$$\alpha_1 = -\frac{[6.446\times10^{12}]}{K}$$

$$\alpha_2 = \frac{-6.446 \times 10^{12}}{K}$$

$$\alpha_3 = \frac{1.29 \times 10^{13}}{K}$$

$$\alpha_4 = \frac{[2.142 \times 10^{12}]}{K}$$

$$\alpha_5 = \frac{[3.558 \times 10^{12}]}{K}$$

$$\alpha_1^2 + \alpha_2^2 + \alpha_3^2 + \alpha_4^2 + \alpha_5^2 = \begin{bmatrix} [-6.446 \times 10^{12}]^2 + [-6.522 \times 10^{12}]^2 + [1.023 \times 10^{13}]^2 \\ + [2.142 \times 10^{12}]^2 + [3.558 \times 10^{12}]^2 \end{bmatrix} = K^2$$

$$K^2 = 2.66 \times 10^{26}$$

Now,

$$\alpha_1 = -(6.446 \times 10^{12}/K = 1.633 \times 10^{13}) = -0.394$$
$$\alpha_2 = -(6.446 \times 10^{12}/K = 1.633 \times 10^{13}) = -0.394$$
$$\alpha_3 = (1.023 \times 10^{13}/K = 1.633 \times 10^{13}) = 0.789$$
$$\alpha_4 = (2.142 \times 10^{12}/K = 1.633 \times 10^{13}) = 0.131$$
$$\alpha_5 = (3.558 \times 10^{12}/K = 1.633 \times 10^{13}) = 0.217$$

Sl. No.	B	α_1 (*E*)	α_2 (*A*)	α_3 (*r*)	α_4 (*L*)	α_5 (*S*)
1	9.4830	–0.394	–0.394	–0.789	0.131	0.217
2	12.774684	–0.408240	–0.408240	–0.816481	0.002098	0.005882
3	14.552594	–0.294178	–0.294178	–0.908988	0.011617	0.022921
4	15.605550	–0.205933	–0.205933	–0.956027	0.016167	0.030568
5	16.288105	–0.147853	–0.147853	–0.976999	0.019911	0.036812
6	16.767520	–0.114479	–0.114479	–0.985623	0.023330	0.042340
7	17.123992	–0.095042	–0.095042	–0.989420	0.026685	0.047649
8	17.398337	–0.082477	–0.082477	–0.991310	0.030062	0.052870
9	17.614389	–0.073571	–0.073571	–0.992321	0.033423	0.057935
10	17.787486	–0.066907	–0.066907	0.992855	0.036706	0.062762
11	17.928063	–0.061762	–0.061762	–0.993102	0.039862	0.067301
12	18.043486	–0.057710	–0.057710	–0.993170	0.042857	0.071518

Example 10.8 The top chord of a bridge truss has an effective length of 4 m and the cross section as shown in Fig. 10.3. The limit state of buckling $g(x) = P_{cr} - S$; where $P_{cr} = \dfrac{\pi^2 E A r^2}{L^2}$, $S = P.S$ is the load, treat E, A, r, S and L as random variables. Determine the area of the column by Hasofer-Lind method, for a reliability index of 3.09.

Sl. No.	Variables	Characteristic value	CoV	Distribution
1	E	2×10^5 N/mm^2	0.05	Normal
2	A	?	0.05	Normal
3	r	76.29 mm	0.05	Normal
4	L	4000 mm	0.05	Normal
5	P	1100×10^3 N	0.2	Normal

Solution:

Design: Hasofer-Lind Method

Target reliability index, $\beta = 3.09$

Given, $(\text{CoV})_A = 0.050$

Standard deviation = Mean × $(\text{CoV})_A$

$$\mu_A = \frac{A}{1 - 1.645 \times (\text{CoV})_A} = \frac{A}{1 - 1.645 \times (0.05)} = 1.089A$$

Normalize the variables

$$Z_1 = \frac{E - \mu_E}{\sigma_E}; \qquad E = Z_1\sigma_E + \mu_E$$

$$Z_2 = \frac{A - \mu_A}{\sigma_A}; \qquad A = Z_2\sigma_A + \mu_A$$

$$Z_3 = \frac{r - \mu_r}{\sigma_r}; \qquad r = Z_3\sigma_r + \mu_r$$

$$Z_4 = \frac{L - \mu_L}{\sigma_L}; \qquad L = Z_4\sigma_L + \mu_L$$

$$Z_5 = \frac{P - \mu_P}{\sigma_P}; \qquad P = Z_5\sigma_P + \mu_P$$

$$g(x) = P_{cr} - S$$

$$R = P_{cr} = \frac{\pi^2 E A r^2}{L^2}; \; S = P$$

$$g(x) = \frac{\pi^2 EAr^2}{L^2} - P$$

At limit state, $g(x) = 0$

$$0 = \frac{\pi^2 EAr^2}{L^2} - P$$

$$\pi^2 EAr^2 - PL^2 = 0$$

$$\pi^2[(Z_1\sigma_E + \mu_E)(Z_2\sigma_A + \mu_A)(Z_3\sigma_r + \mu_r)^2] - (Z_5\sigma_P + \mu_P)(Z_4\sigma_L + \mu_L)^2$$

$$\pi^2[(Z_1Z_2\sigma_E\sigma_A + Z_1\mu_A\sigma_E + Z_2\mu_E\sigma_A + \mu_E\mu_A)(Z_3^2\sigma_r^2 + \mu_r^2 + 2Z_3\mu_r\sigma_r)] - (Z_5\sigma_P + \mu_P)(Z_4^2\sigma_L^2 + \mu_L^2 + 2Z_4\mu_L\sigma_L) = 0$$

$$\pi^2[(Z_1Z_2Z_3^2\sigma_E\sigma_A\sigma_r^2 + Z_1Z_3^2\mu_A\sigma_E\sigma_r^2 + Z_2Z_3^2\mu_E\sigma_A\sigma_r^2 + Z_3^2\mu_E\mu_A\sigma_r^2 + Z_1Z_2\mu_r^2\sigma_E\sigma_A + Z_1\mu_A\mu_r^2\sigma_E + Z_2\mu_E\mu_r^2\sigma_A + \mu_E\mu_A\mu_r^2 + 2Z_1Z_2Z_3\mu_r\sigma_E\sigma_A\sigma_r + 2Z_1Z_3\mu_A\mu_r\sigma_E\sigma_r + 2Z_2Z_3\mu_E\mu_r\sigma_A\sigma_r + 2Z_3\mu_E\mu_A\mu_r\sigma_r)] - (Z_4^2Z_5\sigma_P\sigma_L^2 + Z_5\mu_L^2\sigma_P + 2Z_4Z_5\mu_L\sigma_L\sigma_P)(Z_4^2\mu_P\sigma_L^2 + \mu_P\mu_L^2 + 2Z_4\mu_P\mu_L\sigma_L) = 0$$

Here, $\sigma_A = A \times (\text{CoV})_A = 0.05A;\qquad \mu_A = 1.089A$

$$\pi^2[(Z_1Z_2Z_3^2\sigma_E \times 0.05A\,\sigma_r^2 + Z_1Z_3^2 \times 1.089A\sigma_E\sigma_r^2 + Z_2Z_3^2\mu_E \times 0.05A\,\sigma_r^2 + Z_3^2\mu_E \times 1.089A\,\sigma_r^2 + Z_1Z_2\mu_r^2\sigma_E \times 0.05A + Z_1 \times 1.089A\,\mu_r^2\sigma_E + Z_2\mu_E\mu_r^2 \times 0.05A + \mu_E \times 1.089A\,\mu_r^2 + 2Z_1Z_2Z_3\mu_r\sigma_E \times 0.05A\sigma_r + 2Z_1Z_3 \times 1.089A\,\mu_r\sigma_E\sigma_r + 2Z_2Z_3\mu_E\mu_r \times 0.05A\sigma_r + Z_3\mu_E \times 1.089A\,\mu_r\sigma_r)] - (Z_4^2Z_5\sigma_P\sigma_L^2 + Z_5\mu_L^2\sigma_P + 2Z_4Z_5\mu_L\sigma_L\sigma_P)(Z_4^2\mu_P\sigma_L^2 + \mu_P\mu_L^2 + 2Z_4\mu_P\mu_L\sigma_L) = 0$$

$$A[(0.493 \times Z_1Z_2Z_3^2\sigma_E\sigma_r^2 + 10.75 \times Z_1Z_3^2\sigma_E\sigma_r^2 + 0.493 \times Z_2Z_3^2\sigma_r^2 + 10.75 \times Z_3^2\sigma_r^2 + 0.493 \times Z_1Z_2\mu_r^2\sigma_E + 10.75 \times Z_1\mu_r^2\sigma_E + 0.493 \times Z_2\mu_E\mu_r^2 + 10.75 \times \mu\mu_r^2 + 0.986 \times Z_1Z_2Z_3\mu_r\sigma_E\sigma_r + 21.50 \times Z_1Z_3\mu_r\sigma_E\sigma_r + 0.986 \times Z_2Z_3\mu_E\mu_r\sigma_r + 10.75 \times Z_3\mu_E\mu_r\sigma_r)] - (Z_4^2Z_5\sigma_P\sigma_L^2 + Z_5\mu_L^2\sigma_P + 2Z_4Z_5\mu_L\sigma_L\sigma_P)(Z_4^2\mu_P\sigma_L^2 + \mu_P\mu_L^2 + 2Z_4\mu_P\mu_L\sigma_L) = 0$$

Put, $Z_1 = \alpha_1\beta;\qquad Z_2 = \alpha_2\beta;\qquad Z_3 = \alpha_3\beta;\qquad Z_4 = \alpha_4\beta;\qquad Z_5 = \alpha_5\beta;$

$$A = \frac{[\alpha_4^2\beta^3\alpha_5\sigma_P\sigma_L^2 + \alpha_5\beta\mu_L^2\sigma_P + 2\alpha_4\alpha_5\beta^2\mu_L\sigma_P\sigma_L + \alpha_4^2\beta^2\mu_P\sigma_L^2 + \mu_P\mu_L^2 + 2\alpha_4\beta\mu_P\mu_L\sigma_L]}{\begin{gathered}[0.493 \times \alpha_1\alpha_2\alpha_3^2\beta^4\sigma_E\sigma_r^2 + 10.75 \times \alpha_1\alpha_3^2\beta^3\sigma_E\sigma_r^2 + 0.493 \times \alpha_2\alpha_3^2\beta^3\mu_E\sigma_r^2 \\ + 10.75 \times \alpha_3^2\beta^2\mu_E\sigma_r^2 + 0.493 \times \alpha_1\alpha_2\beta^2\mu_r^2\sigma_E + 10.75 \times \alpha_1\beta\mu_r^2\sigma_E + 0.493 \\ \times \alpha_2\mu_E\beta\mu_r^2 + 10.75 \times \mu_E\mu_r^2 + 0.986 \times \alpha_1\alpha_2\alpha_3\beta^3\mu_r\sigma_E\sigma_r + 21.50 \times \alpha_1\alpha_3\beta^2\mu_r\sigma_E\sigma_r \\ + 0.986 \times \alpha_2\alpha_3\beta^2\mu_E\mu_r\sigma_r + 10.75 \times \alpha_3\beta\mu_E\mu_r\sigma_r]\end{gathered}}$$

$$\alpha_1 = \frac{\partial g(x)}{\partial Z_1} = -\frac{\begin{matrix}[0.493 \times Z_2 Z_3^2 \sigma_E A \sigma_r^2 + 10.75 \times Z_3^2 A \sigma_E \sigma_r^2 + 0.493 \times Z_2 \mu_r^2 \sigma_E A \\ + 10.75 \times A \mu_r^2 \sigma_E + 0.986 \times Z_2 Z_3 A \mu_r \sigma_E \sigma_r + 21.50 \times Z_3 A \mu_r \sigma_E \sigma_r]\end{matrix}}{K}$$

$$\alpha_1 = \frac{\partial g(x)}{\partial Z_1} = -\frac{\begin{matrix}[0.493 \times \alpha_2 \alpha_3^2 \beta^3 \sigma_E A \sigma_r^2 + 10.75 \times \alpha_3^2 \beta^2 A \sigma_E \sigma_r^2 + 0.493 \times \alpha_2 \beta \mu_r^2 \sigma_E A \\ + 10.75 \times A \mu_r^2 \sigma_E + 0.986 \times \alpha_2 \alpha_3 \beta^2 A \mu_r \sigma_E \sigma_r + 21.50 \times \alpha_3 \beta A \mu_r \sigma_E \sigma_r]\end{matrix}}{K}$$

$$\alpha_2 = \frac{\partial g(x)}{\partial Z_2} = -\frac{\begin{matrix}[0.493 \times Z_2 Z_3^2 A \sigma_E \sigma_r^2 + 0.493 \times Z_3^2 A \mu_E \sigma_r^2 + 0.493 \times Z_1 A \mu_r^2 \sigma_E \\ + 0.493 \times A \mu_E \mu_r^2 + 0.986 \times Z_1 Z_3 A \mu_r \sigma_E \sigma_r + 0.986 \times Z_3 A \mu_E \mu_r \sigma_r]\end{matrix}}{K}$$

$$\alpha_2 = \frac{\partial g(x)}{\partial Z_2} = -\frac{\begin{matrix}[0.493 \times \alpha_1 \alpha_3^2 \beta^3 A \sigma_E \sigma_r^2 + 0.493 \times \alpha_3^2 \beta^2 A \mu_E \sigma_r^2 + 0.493 \times \alpha_1 \beta A \mu_r^2 \sigma_E \\ + 0.493 \times A \mu_E \mu_r^2 + 0.986 \times \alpha_1 \alpha_3 \beta^2 A \mu_r \sigma_E \sigma_r + 0.986 \times \alpha_3 \beta A \mu_E \mu_r \sigma_r]\end{matrix}}{K}$$

$$\alpha_3 = \frac{\partial g(x)}{\partial Z_3} = -\frac{\begin{matrix}[0.986 \times Z_1 Z_2 Z_3 A \sigma_E \sigma_r^2 + 21.50 \times Z_1 Z_3 A \sigma_E \sigma_r^2 + 0.986 \times Z_2 Z_3 A \mu_E \sigma_r^2 \\ + 21.50 \times Z_3 A \mu_E \sigma_r^2 + 0.986 \times Z_1 Z_2 A \mu_r \sigma_E \sigma_r + 21.50 \times Z_1 A \mu_r \sigma_E \sigma_r \\ + 0.986 \times Z_2 A \mu_E \mu_r \sigma_r + 10.75 \times A \mu_E \mu_r \sigma_r]\end{matrix}}{K}$$

$$\alpha_3 = \frac{\partial g(x)}{\partial Z_3} = -\frac{\begin{matrix}[0.986 \times \alpha_1 \alpha_2 \alpha_3 \beta^3 A \sigma_E \sigma_r^2 + 21.50 \times \alpha_1 \alpha_3 \beta^2 A \sigma_E \sigma_r^2 + 0.986 \times \alpha_2 \alpha_3 \beta^2 \\ A \mu_E \sigma_r^2 + 21.50 \times \alpha_3 \beta A \mu_E \sigma_r^2 + 0.986 \times \alpha_1 \alpha_2 \beta^2 A \mu_r \sigma_E \sigma_r + 21.50 \times \alpha_1 \beta \\ A \mu_r \sigma_E \sigma_r + 0.986 \times \alpha_2 \beta A \mu_E \mu_r \sigma_r + 10.75 \times A \mu_E \mu_r \sigma_r]\end{matrix}}{K}$$

$$\alpha_4 = \frac{\partial g(x)}{\partial Z_4} = \frac{[2 Z_4 Z_5 \sigma_P \sigma_L^2 + 2 Z_5 \mu_L \sigma_P \sigma_L + 2 Z_4 \mu_P \sigma_L^2 + 2 \mu_P \mu_L \sigma_L]}{K}$$

$$\alpha_4 = \frac{\partial g(x)}{\partial Z_4} = \frac{[2 \alpha_4 \alpha_5 \beta^2 \sigma_P \sigma_L^2 + 2 \alpha_5 \beta \mu_L \sigma_P \sigma_L + 2 \alpha_4 \beta \mu_P \sigma_L^2 + 2 \mu_P \mu_L \sigma_L]}{K}$$

$$\alpha_5 = \frac{\partial g(x)}{\partial Z_5} = \frac{[Z_4^2 \sigma_P \sigma_L^2 + \mu_L^2 \sigma_P + 2 Z_4 \mu_L \sigma_P \sigma_L]}{K}$$

$$\alpha_5 = \frac{\partial g(x)}{\partial Z_5} = \frac{[\alpha_4^2 \beta^2 \sigma_P \sigma_L^2 + \mu_L^2 \sigma_P + 2 \alpha_4 \beta \mu_L \sigma_P \sigma_L]}{K}$$

Trial I, β = 3.09

$$\alpha_1 = \alpha_2 = \alpha_3 = \alpha_4 = \alpha_5 = 1/\sqrt{5} = 0.447$$

$$\alpha_1 = -0.447,\ \alpha_2 = -0.447,\ \alpha_3 = -0.447,\ \alpha_4 = 0.447,\ \alpha_5 = 0.447$$

$$A = \frac{\begin{gathered}[(0.447)^2 \times 3.09^3 \times 0.447 \times 165.53 \times 10^3 \times (217.92)^2 + 0.447 \times 3.09 \\ \times (4358.48)^2 \times 165.53 \times 10^3 + 2 \times 0.447 \times 0.447 \times 4358.48 \times 165.53 \times \\ 10^3 \times 217.92 + 0.447^2 \times 3.09^2 \times 827.69 \times 10^3 \times (217.92)^2 + 827.69 \times 10^3 \\ \times (4358.48)^2\, 2 \times 0.447 \times 3.09 \times 827.69 \times 10^3 \times 4358.48 \times 217.92]\end{gathered}}{\begin{gathered}[0.493 \times (-0.447)(-0.447)(-0.447)^2 \times 3.09^4 \times 10896 \times (4.156)^2 + 10.75 \times (-0.447)(-0.447)^2 \\ \times (3.09)^3 \times 10896 \times (4.156)^2 + 0.493 \times (-0.447)(-0.447)^2 \times (3.09)^3 \times 2.1792 \times 10^5 (4.156)^2 \\ + 10.75 \times (-0.447)^2 \times (3.09)^2 \times 2.1792 \times 10^5 (4.156)^2 + 0.493 \times (-0.447)(-0.447) \times (3.09)^2 \\ \times (83.127)^2 \times 10896 + 10.75 \times (-0.447) \times 3.09 \times (83.127)^2 \times 10896 + 0.493 \times (-0.447) \times 3.0 \\ \times 2.1792 \times 10^5 \times (83.127)^2 + 10.75 \times 2.1792 \times 10^5 \times (83.127)^2 + 0.986 \times (-0.447)(-0.447)(-0.447) \\ \times 3.09^3 \times 83.127 \times 10896 \times 4.156 + 21.50 \times (-0.447)(-0.447) \times 3.09^2 \times 83.127 \times 10896 \times 4.156 \\ + 0.986 \times (-0.447)(-0.447) \times 3.09^2 \times 2.1792 \times 10^5 \times 83.127 \times 4.156 + 10.75 \times (-0.447) \times 3.09 \\ \times 2.1792 \times 10^5 \times 83.127 \times 4.156]\end{gathered}}$$

A = 1717.74 mm^2

$$\alpha_1 = -\frac{\begin{gathered}[0.493 \times (-0.447)(-0.447)^2 \times 3.09^3 \times 10896 \times 1717.74 \times (4.156)^2 + 10.75 \\ \times (-0.447)^2 \times 3.09^2 \times 1717.74 \times 10896 \times (4.156)^2 + 0.493 \times (-0.447) \times 3.09 \\ \times (83.127)^2 \times 10896 \times 1717.74 + 10.75 \times 1717.74 \times (83.127)^2 \times 10896 + 0.986 \\ \times (-0.447)(-0.447) \times 3.09^2 \times 1717.74 \times 83.127 \times 10896 \times 4.156 + 21.50 \\ \times (-0.447) \times 3.09 \times 1717.74 \times 83.127 \times 10896 \times 4.156]\end{gathered}}{K}$$

$$\alpha_1 = -\frac{[1.1549 \times 10^{12}]}{K}$$

$$\alpha_2 = -\frac{\begin{gathered}[0.493 \times (-0.447)(-0.447)^2 \times 3.09^3 \times 1717.74 \times 10896 \times (4.156)^2 + 0.493 \\ \times (-0.447)^2 \times 3.09^2 \times 1717.74 \times 2.1792 \times 10^5 \times (4.156)^2 + 0.493 \times (-0.447) \\ \times 3.09 \times 1717.74 \times (83.127)^2 \times 10896 + 0.493 \times 1717.74 \times 2.1792 \times 10^5 \\ \times (83.127)^2 + 0.986 \times (-0.447)(-0.447) \times 3.09^2 \times 1717.74 \times 83.127 \times 10896 \\ \times 4.156 + 0.986 \times (-0.447) \times 3.09 \times 1717.74 \times 2.1792 \times 10^5 \times 83.127 \times 4.156]\end{gathered}}{K}$$

$$\alpha_2 = -\frac{[1.0528 \times 10^{12}]}{K}$$

$$\alpha_3 = -\frac{\begin{array}{r}[0.986 \times (-0.447)(-0.447)(-0.447) \times 3.09^3 \times 1717.74 \times 10896 \times (4.156)^2 \\ +21.50 \times (-0.447)(-0.447) \times 3.09^2 \times 1717.74 \times 10896 \times (4.156)^2 + 0.986 \\ \times (-0.447)(-0.447) \times 3.09^2 \times 1717.74 \times 2.1792 \times 10^5 \times (4.156)^2 + 21.50 \\ \times (-0.447) \times 3.09 \times 1717.74 \times 2.1792 \times 10^5 \times (4.156)^2 + 0.986 \times (-0.447) \\ (-0.447) \times 3.09^2 \times 1717.74 \times (83.127) \times 10896 \times 4.156 + 21.50 \times (-0.477) \\ \times 3.09 \times 1717.74 \times 83.127 \times 10896 \times 4.156 + 0.986 \times (-0.447) \times 3.09 \\ \times 1717.74 \times 2.1792 \times 10^5 \times 83.127 \times 4.156 + 10.75 \times 1717.74 \times 2.1792 \times 10^5 \\ \times 83.127 \times 4.156]\end{array}}{K}$$

$$\alpha_3 = -\frac{[8.87 \times 10^{11}]}{K}$$

$$\alpha_4 = \frac{\begin{array}{r}[2 \times 0.447 \times 0.447 \times 3.09^2 \times 165.53 \times 10^3 \times 217.92^2 + 2 \times 0.447 \times 3.09 \times 4358.48 \\ 165.53 \times 10^3 \times 217.92 + 2 \times 0.447 \times 3.09 \times 827.69 \times 10^3 \times (217.92)^2 + 2 \times 827.69 \\ \times 10^3 \times 4358.48 \times 217.92]\end{array}}{K}$$

$$\alpha_4 = \frac{[2.185 \times 10^{12}]}{K}$$

$$\alpha_5 = \frac{\partial g(x)}{\partial Z_5} = \frac{\begin{array}{r}[(0.447)^2 \times 3.09^2 \times 165.53 \times 10^3 \times (217.92)^2 + (4358.48)^2 \times 165.53 \times 10^3 \\ + 2 \times 0.447 \times 3.09 \times 4358.48 \times 165.53 \times 10^3 \times 217.92]\end{array}}{K}$$

$$\alpha_5 = \frac{[3.625 \times 10^{12}]}{K}$$

$$\alpha_1^2 + \alpha_2^2 + \alpha_3^2 + \alpha_4^2 + \alpha_5^2 = \left[\begin{array}{r}\left[-1.1549 \times 10^{12}\right]^2 + \left[-1.0528 \times 10^{12}\right]^2 + \left[8.87 \times 10^{11}\right]^2 \\ + \left[2.185 \times 10^{12}\right]^2 + \left[3.625 \times 10^{12}\right]^2\end{array}\right] = K^2$$

$$K = 4.598 \times 10^{12}$$

Now,

$$\alpha_1 = -(1.1549 \times 10^{12}/4.598 \times 10^{12}) = -0.2551$$

$$\alpha_2 = -(1.0528 \times 10^{12}/4.598 \times 10^{12}) = -0.2289$$

$$\alpha_3 = (8.87 \times 10^{11}/4.598 \times 10^{12}) = -0.1928$$

$$\alpha_4 = (2.185 \times 10^{12}/4.598 \times 10^{12}) = 0.4753$$

$$\alpha_5 = (3.625 \times 10^{12}/4.598 \times 10^{12}) = 0.7882$$

Trial II, $\beta = 3.09$

$\alpha_1 = -0.2551,\ \alpha_2 = -0.2289,\ \alpha_3 = -0.1928,\ \alpha_4 = 0.4753,\ \alpha_5 = 0.7882$

$A = 1838.53 \text{ mm}^2$

$\mu_A = 1.089A = 1.089 \times 1838.53 = 2002.159 \text{ mm}^2$

Sl. No.	β	(A)	α_1 (E)	α_2 (A)	α_3 (r)	α_4 (L)	α_5 (S)
1		1254.658437	–0.001212	–0.001212	–0.002423	0.983977	0.178271
2		1421.236741	–0.001221	–0.001221	–0.002443	0.974997	0.222196
3		1421.635372	–0.001220	–0.001220	–0.002440	0.975184	0.221377
4		1421.633434	–0.001220	–0.001220	–0.002440	0.975180	0.221392
5		1421.633470	–0.001220	–0.001220	–0.002440	0.975180	0.221392
6	3.09	1421.633469	–0.001220	–0.001220	–0.002440	0.975180	0.221392
7		1421.633469	–0.001220	–0.001220	–0.002440	0.975180	0.221392
8		1421.633469	–0.001220	–0.001220	–0.002440	0.975180	0.221392
9		1421.633469	–0.001220	–0.001220	–0.002440	0.975180	0.221392
10		1421.633469	–0.001220	–0.001220	–0.002440	0.975180	0.221392
11		1421.633469	–0.001220	–0.001220	–0.002440	0.975180	0.221392

Example 10.9 Analyze by Taylor series method, for the given data. The top chord of a bridge truss has an effective length of 4 m and the cross section as shown in Fig. 10.3. The limit state of buckling $g(x) = P_{cr} - S$; where $P_{cr} = \dfrac{\pi^2 E A r^2}{L^2}$, $S = P$. 'S' is the load, treat E, A, r, S and L as random variables. Compute the reliability index by (a) first order linear expansion (b) Second order expansion.

SI. No.	Variables	Characteristic value	CoV	Distribution
1	E	2×10^5 N/mm^2	0.05	Normal
2	A	9880 mm^2	0.05	Normal
3	r	76.29 mm	0.05	Normal
4	L	4000 mm	0.05	Normal
5	P	1100×10^3 N	0.20	Normal

Solution:

SI. No.	Variables	Characteristic value	CoV	Mean value (μ)	Standard deviation (σ)
1	E	2×10^5 N/mm^2	0.05	2.179×10^5 N/mm^2	10896 N/mm^2
2	A	9880 mm^2	0.05	10765.46 mm^2	538.27 mm^2
3	r	76.29 mm	0.05	83.127 mm	4.156 mm
4	L	4000 mm	0.05	4358.48 mm	217.92 mm
5	P	1100×10^3 N	0.20	827.69×10^3 N	165.53×10^3 N

$$g(\mu_x) = \frac{\pi^2 \mu_E \mu_A \mu_r^2}{\mu_L^2} - \mu_P$$

$$g(\mu_x) = \frac{\pi^2 \times 217924.27 \times 10765.46 \times 83.1272^2}{4358.4854^2} - 827689.9925$$

$$g(\mu_x) = 7586507.21$$

$$\frac{\partial g(x)}{\partial E} = \frac{\pi^2 \mu_A \mu_r^2}{\mu_L^2} = 38.681$$

$$\frac{\partial^2 g(x)}{\partial E^2} = 0$$

$$\frac{\partial^2 g(x)}{\partial E \partial A} = \frac{\pi^2 \mu_r^2}{\mu_L^2} = 0.0036$$

$$\frac{\partial^2 g(x)}{\partial E \partial r} = \frac{2 \times \pi^2 \mu_A \mu_r}{\mu_L^2} = 0.9289A$$

$$\frac{\partial^2 g(x)}{\partial E \partial L} = \frac{-2 \times \pi^2 \mu_A \mu_r^2}{\mu_L^3} = -0.01772$$

$$\frac{\partial^2 g(x)}{\partial E \partial P} = 0$$

$$\frac{\partial g(x)}{\partial A} = \frac{\pi^2 \mu_E \mu_r^2}{\mu_L^2} = 783.015$$

$$\frac{\partial^2 g(x)}{\partial A \partial E} = \frac{\pi^2 \mu_r^2}{\mu_L^2} = 0.0036$$

$$\frac{\partial^2 g(x)}{\partial A^2} = 0$$

$$\frac{\partial^2 g(x)}{\partial A \partial r} = \frac{2\times\pi^2 \mu_E \mu_r}{\mu_L^2} = \frac{2\times\pi^2\times 217.9242\times 10^3\times 83.127}{4358.4854^2} = 18.8047$$

$$\frac{\partial^2 g(x)}{\partial A \partial L} = \frac{-2\times\pi^2 \mu_E \mu_r^2}{\mu_L^3} = -0.3586$$

$$\frac{\partial^2 g(x)}{\partial A \partial P} = 0$$

$$\frac{\partial g(x)}{\partial r} = \frac{2\times\pi^2 \mu_E \mu_A \mu_r}{\mu_L^2} = 202810.043$$

$$\frac{\partial^2 g(x)}{\partial r \partial E} = \frac{2\times\pi^2 \mu_A \mu_r}{\mu_L^2} = 0.9289$$

$$\frac{\partial^2 g(x)}{\partial r \partial A} = \frac{2\times\pi^2 \mu_E \mu_r}{\mu_L^2} = 18.8047$$

$$\frac{\partial^2 g(x)}{\partial r^2} = \frac{2\times\pi^2 \mu_E \mu_A}{\mu_L^2} = 2435.3212$$

$$\frac{\partial^2 g(x)}{\partial r \partial L} = \frac{-4\times\pi^2 \mu_E \mu_A \mu_r}{\mu_L^3} = -92.8953$$

$$\frac{\partial^2 g(x)}{\partial r \partial P} = 0$$

$$\frac{\partial g(x)}{\partial L} = \frac{-2\times\pi^2 \mu_E \mu_A \mu_r^2}{\mu_L^3} = -3868.095$$

$$\frac{\partial^2 g(x)}{\partial L \partial E} = \frac{-2\times\pi^2 \mu_A \mu_r^2}{\mu_L^3} = -0.01772$$

$$\frac{\partial^2 g(x)}{\partial L \partial A} = \frac{-2\times\pi^2 \mu_E \mu_r^2}{\mu_L^3} = -0.3587$$

$$\frac{\partial^2 g(x)}{\partial L \partial r} = \frac{-4\times\pi^2 \mu_E \mu_A \mu_r}{\mu_L^3} = -932.8953$$

$$\frac{\partial^2 g(x)}{\partial L^2} = \frac{+6\times\pi^2 \mu_E \mu_A \mu_r^2}{\mu_L^4} = 2.6576$$

$$\frac{\partial^2 g(x)}{\partial L \partial P} = 0$$

$$\frac{\partial g(x)}{\partial P} = -1$$

$$\frac{\partial^2 g(x)}{\partial P \partial E} = \frac{\partial^2 g(x)}{\partial P \partial A} = \frac{\partial^2 g(x)}{\partial P \partial r} = \frac{\partial^2 g(x)}{\partial P \partial L} = \frac{\partial^2 g(x)}{\partial P^2} = 0$$

Hessian Matrix

$$H = \begin{bmatrix} 0 & 0.0036 & 0.9289 & -0.01772 & 0 \\ 0.0036 & 0 & 18.8047 & -0.3586 & 0 \\ 0.9289 & 18.8047 & 2435.3212 & -92.8953 & 0 \\ -0.01772 & -0.3587 & -92.8953 & 2.6576 & 0 \\ 0 & 0 & 0 & 0 & 0 \end{bmatrix}$$

$$\nabla g^T\{x - \mu_x\} = \begin{Bmatrix} 0.0039A \\ 781.5908 \\ 20.4895A \\ -0.3908A \\ -1 \end{Bmatrix} \times \begin{Bmatrix} -17924.27 \\ -0.0896A \\ -6.8372 \\ -358.4854 \\ +272310.1 \end{Bmatrix}$$

$$\nabla g^T\{x - \mu_x\} = \langle 0.0039A \;\; 781.5908 \;\; 20.4895A \;\; -0.3908A \;\; -1 \rangle \times \begin{Bmatrix} -17924.27 \\ -0.0896A \\ -6.8372 \\ -358.4854 \\ +272310.1 \end{Bmatrix}$$

$$\begin{aligned} \nabla g^T\{x - \mu_x\} = {} & (0.0039A \times -17924.27) + (781.5908 \times -0.0896A) + \\ & (20.4895A \times -6.8372) + (-0.3908A \times -358.4854) + \\ & (-1 \times 272310.1) \end{aligned}$$

$$\nabla g^T\{x - \mu_x\} = -1656445.5$$

$$\sigma_{g(\mu_x)} = \sqrt{\left(\frac{\partial g(\mu_x)}{\partial \mu_E}\right)^2 \sigma_E^2 + \left(\frac{\partial g(\mu_x)}{\partial \mu_A}\right)^2 \sigma_A^2 + \left(\frac{\partial g(\mu_x)}{\partial \mu_r}\right)^2 \sigma_r^2 + \left(\frac{\partial g(\mu_x)}{\partial \mu_L}\right)^2 \sigma_L^2 + \left(\frac{\partial g(\mu_x)}{\partial \mu_S}\right)^2 \sigma_S^2}$$

$$\sigma_{g(\mu_x)} = \sqrt{\begin{aligned} & (38.61 \times 10896.21)^2 + (781.59 \times 538.27)^2 + (202441.465 \times 4.16)^2 + \\ & \qquad (-3861.0648 \times 217.9643)^2 + (-1 \times 165.5379 \times 10^3)^2 \end{aligned}}$$

$$\sigma_{g(\mu_x)} = 1340660.577$$

$$g(x) = g(\mu_x) + \nabla g^T\{x - \mu_x\}$$

$$g(x) = 7586507.421 - 1656445.5 = 5930061.9$$

$$\beta = \frac{g(x)}{\sigma_{g(\mu_x)}} = \frac{5930061.9}{1340660.577} = 4.425$$

Second Order Analysis

$$\{x - x_i\}^T[H]\{x - x_i\} = \langle -17924.27 \;\; -885.46 \;\; -6.84 \;\; -358.49 \;\; 272310.01\rangle \times$$

$$\begin{bmatrix} 0 & 0.0036 & 0.9289 & -0.01772 & 0 \\ 0.0036 & 0 & 18.8047 & -0.3586 & 0 \\ 0.9289 & 18.8047 & 2435.3212 & -92.8953 & 0 \\ -0.01772 & -0.3587 & -932.8953 & 2.6576 & 0 \\ 0 & 0 & 0 & 0 & 0 \end{bmatrix} \times \begin{Bmatrix} -17924.27 \\ -885.46 \\ -6.84 \\ -358.49 \\ +272310.01 \end{Bmatrix}$$

$$\{x - x_i\}^T[H]\{x - x_i\} = \langle -3.1757 \;\; -64.29 \;\; -16650.81 \;\; 317.5725 \;\; 0\rangle \times \begin{Bmatrix} -17924.27 \\ -885.46 \\ -6.84 \\ -358.49 \\ +272310.01 \end{Bmatrix}$$

$$\{x - \mu_x\}^T[H]\{x - \mu_x\} = 113845.1428$$

$$g(x) = g(\mu_x) + \nabla g^T\{x - \mu_x\} + \frac{1}{2}\{x - \mu_x\}^T[H]\{x - \mu_x\}$$

$$g(x) = 7586507.421 + (-1656445.5) + \frac{1}{2}(113845.1428) = 5986984.51$$

$$\beta = \frac{g(x)}{\sigma_{g(x_m)}} = \frac{5986984.51}{1340660.58} = 4.4657$$

Example 10.10 Design by Taylor series method, for the given data. The top chord of a bridge truss has an effective length of 4 m and the cross section as shown in Fig. 10.3. The limit state of buckling $g(x) = P_{cr} - S$; where $P_{cr} = \dfrac{\pi^2 E A r^2}{L^2}$, $S = P$. 'S' is the load, treat E, A, r, S and L as random variables. Determine the area of the column, required for a reliability index of 3.09, for (a) consider first order terms and (b) second order terms.

Sl. No.	Variables	Characteristic value	CoV	Distribution
1	E	2×10^5 N/mm^2	0.05	Normal
2	A	?	0.05	Normal
3	r	76.29 mm	0.05	Normal
4	L	4000 mm	0.05	Normal
5	P	1100×10^3 N	0.20	Normal

Solution:

$$\mu_A = \frac{A}{1-1.645(0.05)} = 1.08962A$$

SI. No.	Variables	Characteristic value	CoV	Mean value (μ)	Standard deviation (σ)
1	E	2×10^5 N/mm^2	0.05	2.179×10^5 N/mm^2	10896 N/mm^2
2	**A**	**A**	**0.05**	**1.08962A**	**0.0545A**
3	r	76.29 mm	0.05	83.127 mm	4.156 mm
4	L	4000 mm	0.05	4358.48 mm	217.92 mm
5	P	1100×10^3 N	0.2	827.69×10^3 N	165.53×10^3 N

$$g(\mu_x) = \frac{\pi^2 \mu_E \mu_A \mu_r^2}{\mu_L^2} - \mu_P$$

$$g(\mu_x) = \frac{\pi^2 \times 217924.27 \times 1.08962A \times 83.1272^2}{4358.4854^2} - 827689.9925$$

$$g(\mu_x) = 851.6214A - 827689.9925$$

$$\frac{\partial g(x)}{\partial E} = \frac{\pi^2 \mu_A \mu_r^2}{\mu_L^2} = 0.0039A$$

$$\frac{\partial^2 g(x)}{\partial E^2} = 0$$

$$\frac{\partial^2 g(x)}{\partial E \partial A} = \frac{\pi^2 \mu_r^2}{\mu_L^2} = 0.0036$$

$$\frac{\partial^2 g(x)}{\partial E \partial r} = \frac{2 \times \pi^2 \mu_A \mu_r}{\mu_L^2} = 9.4022 \times 10^{-5} A$$

$$\frac{\partial^2 g(x)}{\partial E \partial L} = \frac{-2\times\pi^2\mu_A\mu_r^2}{\mu_L^3} = -1.7933\times 10^{-6}A$$

$$\frac{\partial^2 g(x)}{\partial E \partial P} = 0$$

$$\frac{\partial g(x)}{\partial A} = \frac{\pi^2\mu_E\mu_r^2}{\mu_L^2} = 781.5908$$

$$\frac{\partial^2 g(x)}{\partial A \partial E} = \frac{\pi^2\mu_r^2}{\mu_L^2} = 0.0036$$

$$\frac{\partial^2 g(x)}{\partial A^2} = 0$$

$$\frac{\partial^2 g(x)}{\partial A \partial r} = \frac{2\times\pi^2\mu_E\mu_r}{\mu_L^2} = \frac{2\times\pi^2\times 217.9242\times 10^3\times 83.127}{4358.4854^2} = 18.8047$$

$$\frac{\partial^2 g(x)}{\partial A \partial L} = \frac{-2\times\pi^2\mu_E\mu_r^2}{\mu_L^3} = -0.3586$$

$$\frac{\partial^2 g(x)}{\partial A \partial P} = 0$$

$$\frac{\partial g(x)}{\partial r} = \frac{2\times\pi^2\mu_E\mu_A\mu_r}{\mu_L^2} = 20.4895A$$

$$\frac{\partial^2 g(x)}{\partial r \partial E} = \frac{2\times\pi^2\mu_A\mu_r}{\mu_L^2} = 9.4022\times 10^{-5}A$$

$$\frac{\partial^2 g(x)}{\partial r \partial A} = \frac{2\times\pi^2\mu_E\mu_r}{\mu_L^2} = 18.8047$$

$$\frac{\partial^2 g(x)}{\partial r^2} = \frac{2\times\pi^2\mu_E\mu_A}{\mu_L^2} = 0.2464A$$

$$\frac{\partial^2 g(x)}{\partial r \partial L} = \frac{-4\times\pi^2\mu_E\mu_A\mu_r}{\mu_L^3} = -0.0094A$$

$$\frac{\partial^2 g(x)}{\partial r \partial P} = 0$$

$$\frac{\partial g(x)}{\partial L} = \frac{-2\times\pi^2\mu_E\mu_A\mu_r^2}{\mu_L^3} = -0.3908A$$

$$\frac{\partial^2 g(x)}{\partial L \partial E} = \frac{-2\times\pi^2\mu_A\mu_r^2}{\mu_L^3} = -1.7932\times10^{-6}A$$

$$\frac{\partial^2 g(x)}{\partial L \partial A} = \frac{-2\times\pi^2\mu_E\mu_r^2}{\mu_L^3} = -0.3586$$

$$\frac{\partial^2 g(x)}{\partial L \partial r} = \frac{-4\times\pi^2\mu_E\mu_A\mu_r}{\mu_L^3} = -0.0094A$$

$$\frac{\partial^2 g(x)}{\partial L^2} = \frac{+6\times\pi^2\mu_E\mu_A\mu_r^2}{\mu_L^4} = 2.6898\times10^{-4}A$$

$$\frac{\partial^2 g(x)}{\partial L \partial P} = 0$$

$$\frac{\partial g(x)}{\partial P} = -1$$

$$\frac{\partial^2 g(x)}{\partial P \partial E} = \frac{\partial^2 g(x)}{\partial P \partial A} = \frac{\partial^2 g(x)}{\partial P \partial r} = \frac{\partial^2 g(x)}{\partial P \partial L} = \frac{\partial^2 g(x)}{\partial P^2} = 0$$

$$\nabla g^T\{x-\mu_x\} = \begin{Bmatrix} 0.0039A \\ 781.5908 \\ 20.4895A \\ -0.3908A \\ -1 \end{Bmatrix}^T \times \begin{Bmatrix} -17924.27 \\ -0.0896A \\ -6.8372 \\ -358.4854 \\ +272310.1 \end{Bmatrix}$$

$$\nabla g^T\{x-\mu_x\} = \langle\, 0.0039A \;\; 781.5908 \;\; 20.4895A \;\; -0.3908A \;\; -1 \rangle \times \begin{Bmatrix} -17924.27 \\ -0.0896A \\ -6.8372 \\ -358.4854 \\ +272310.1 \end{Bmatrix}$$

$$\nabla g^T\{x-\mu_x\} = (0.0039A\times -17924.27) + (781.5908\times -0.0896A) + (20.4895A\times -6.8372) + (-0.3908A\times -358.4854) + (-1\times 272310.1)$$

$$\nabla g^T\{x-\mu_x\} = -139.9299A - 272310.1$$

$$\sigma_{g(\mu_x)} = \sqrt{\left(\frac{\partial g(\mu_x)}{\partial \mu_E}\right)^2\sigma_E^2 + \left(\frac{\partial g(\mu_x)}{\partial \mu_A}\right)^2\sigma_A^2 + \left(\frac{\partial g(\mu_x)}{\partial \mu_r}\right)^2\sigma_r^2 + \left(\frac{\partial g(\mu_x)}{\partial \mu_L}\right)^2\sigma_L^2 + \left(\frac{\partial g(\mu_x)}{\partial \mu_S}\right)^2\sigma_S^2}$$

$$\sigma_{g(\mu_x)} = \sqrt{\begin{aligned}&(0.0039A\times 10896.2)^2 + (781.5908\times 0.0545A)^2 + (20.4835A\times 4.15636)^2 + \\ &\qquad (-0.3908A\times 217.9643)^2 + (-1\times 165.5379\times 10^3)^2\end{aligned}}$$

$$\sigma_{g(\mu_x)} = \sqrt{18124.3026A^2 + 2.7402\times 10^{10}}$$

$$g(x) = g(\mu_x) + \nabla g^T\{x-\mu_x\}$$

$$g(x) = 851.6214A - 827689.9925 - 139.9299A - 272310.1$$

$$g(x) = 711.6915A - 1100000.0925$$

$$\beta = \frac{g(x)}{\sigma_{g(\mu_x)}} = \frac{711.6915A - 1100000.0925}{\sqrt{18124.3026A^2 + 2.7402\times 10^{10}}} = 4.424$$

On solving we get $A = 9493\ \text{mm}^2$

Second Order Analysis

$\{x-\mu_x\}^T[H]\{x-\mu_x\} =$

$$\begin{Bmatrix}-17924.27\\ -0.0896A\\ -6.8372\\ -358.4854\\ +272310.1\end{Bmatrix}^T \times \begin{Bmatrix} 0 & 0.0036 & 9.4022\times 10^{-5}A & -1.7933\times 10^{-6}A & 0\\ 0.0036 & 0 & 18.8047 & -0.3586 & 0\\ 9.4022\times 10^{-5}A & 18.8047 & 0.2464A & -0.0094A & 0\\ -1.7932\times 10^{-6}A & -0.3586 & -0.0094A & 2.6898\times 10^{-4}A & 0\\ 0 & 0 & 0 & 0 & 0\end{Bmatrix} \times \begin{Bmatrix}-17924.27\\ -0.0896A\\ -6.8372\\ -358.4854\\ +272310.1\end{Bmatrix}$$

$$\{x-\mu_x\}^T[H]\{x-\mu_x\} = \langle -3.2257\times 10^{-4}A \quad -64.546 \quad -1.6851A \quad 0.0321A \quad 0\rangle \times \begin{Bmatrix}-17924.27\\ -0.0896A\\ -6.8372\\ -358.4854\\ +272310.1\end{Bmatrix}$$

$$\{x-\mu_x\}^T[H]\{x-\mu_x\} = 11.5791A$$

$$g(x) = g(\mu_x) + \nabla g^T\{x-\mu_x\} + \frac{1}{2}\{x-\mu_x\}^T[H]\{x-\mu_x\}$$

$$g(x) = 711.6915A - 1100000.0925 + \frac{1}{2}[11.5719A]$$

$$g(x) = 717.4775A - 1100000.0925$$

$$\beta = \frac{g(x)}{\sigma_{g(\mu_x)}} = \frac{717.4775A - 1100000.0925}{\sqrt{18124.3026A^2 - 2.7402 \times 10^{10}}} = 4.4657$$

On solving we get $A = 10699$ mm^2

Example 10.11 The moment capacity of a rectangular reinforced concrete beam is given by the performance function $g(x) = R - S$. $R = A_{st} f_y d - \frac{0.77 A_{st}^2 f_y^2}{b f_{ck}}$; $S = M_{ex}$

Determine the reliability index, resistance factor, load factor, bias resistance factor and bias load factor, by FOSM method.

Sl. No.	Variables	Characteristic value	CoV	Distribution
1	A_{st}	1432 mm^2	0.05	Normal
2	f_y	250 N/mm^2	0.05	Normal
3	d	600 mm	0.05	Normal
4	b	300 mm	0.05	Normal
5	f_{ck}	20 N/mm^2	0.05	Normal
6	M_{ex}	200 × 10^3 N-m	0.20	Normal

Solution:

Analysis—FOSM Method

The strengths are underestimated and loads are overestimated, the mean values are computed from the characteristic values and are tabulated below:

Sl. No.	Variables	Characteristic value	CoV	Mean value (μ)	Standard deviation (σ)
1	A_{st}	1432 mm^2	0.05	1560.34	78.02
2	f_y	250 N/mm^2	0.05	272.41	13.62
3	d	600 mm	0.05	653.77	32.69
4	b	300 mm	0.05	326.88	16.35
5	f_{ck}	20 N/mm^2	0.05	21.79	1.09
6	M_{ex}	200 × 10^6 N-mm	0.20	1.504 × 10^8	3 × 10^7

$M_{ex} = 200 \times 10^3$ N-m

$M_{ex} = 200 \times 10^6$ N-mm

The limit state of failure of $g(x) = R - S$

where, $R = A_{st} f_y d - \frac{0.77 A_{st}^2 f_y^2}{b f_{ck}}$, $S = M_{ex}$

Now, the mean and standard deviations are:

Mean value,

$$\mu_R = \mu_{A_{st}}\mu_{f_y}\mu_d - \frac{0.77\mu_{A_{st}^2}\mu_{f_y^2}}{\mu_b\mu_{f_{ck}}}$$

$$\Rightarrow \quad \mu_R = 1560.34 \times 272.41 \times 653.77 - \frac{0.77 \times (1560.34)^2 \times (272.41)^2}{326.88 \times 21.79} = 2.583 \times 10^8 \text{ N-mm}$$

$$\mu_S = \mu_{M_{ex}} = 1.504 \times 10^8 \text{ N-mm}$$

Now, partially differentiating *R* and *S* w.r.t variables, we get,

$$\frac{\partial R}{\partial A_{st}} = \mu_{f_y}\mu_d - \frac{2 \times 0.77\mu_{A_{st}}\mu_{f_y^2}}{\mu_b\mu_{f_{ck}}}$$

$$\Rightarrow \quad \frac{\partial R}{\partial A_{st}} = 272.41 \times 653.77 - \frac{2 \times 0.77 \times 1560.34 \times (272.41)^2}{326.88 \times 21.79} = 1.53 \times 10^5$$

$$\frac{\partial R}{\partial f_y} = \mu_{A_{st}}\mu_d - \frac{2 \times 0.77\mu_{A_{st}^2}\mu_{f_y}}{\mu_b\mu_{f_{ck}}}$$

$$\Rightarrow \quad \frac{\partial R}{\partial f_y} = 1560.34 \times 653.77 - \frac{2 \times 0.77 \times (1560.34)^2 \times 272.41}{326.88 \times 21.79} = 8.76 \times 10^5$$

$$\frac{\partial R}{\partial d} = \mu_{A_{st}}\mu_{f_y} = 1560.34 \times 272.41 = 4.25 \times 10^5$$

$$\frac{\partial R}{\partial b} = \frac{0.77\mu_{A_{st}^2}\mu_{f_y^2}}{\mu_{b^2}\mu_{f_{ck}}} = \frac{0.77 \times (1560.34)^2 \times (272.41)^2}{(326.88)^2 \times 21.79} = 5.97 \times 10^4$$

$$\frac{\partial R}{\partial f_{ck}} = \frac{0.77\mu_{A_{st}^2}\mu_{f_y^2}}{\mu_b\mu_{f_{ck}^2}} = \frac{0.77 \times (1560.34)^2 \times (272.41)^2}{326.88 \times (21.79)^2} = 8.96 \times 10^5$$

$$\frac{\partial S}{\partial \mu_{M_{ex}}} = 1$$

Standard deviation,

$$\sigma_R^2 = \left(\frac{\partial R}{\partial A_{st}} \times \sigma_{A_{st}}\right)^2 + \left(\frac{\partial R}{\partial f_y} \times \sigma_{fY}\right)^2 + \left(\frac{\partial R}{\partial d} \times \sigma_d\right)^2 + \left(\frac{\partial R}{\partial b} \times \sigma_b\right)^2 + \left(\frac{\partial R}{\partial f_{ck}} \times \sigma_{fCK}\right)^2$$

$$\sigma_R^2 = (1.53 \times 10^5 \times 78.02)^2 + (8.76 \times 10^5 \times 13.62)^2 + (4.25 \times 10^5 \times 32.69)^2 + (5.97 \times 10^4 \times 16.35)^2 + (8.96 \times 10^5 \times 1.09)^2$$

$$\Rightarrow \quad \sigma_R^2 = 4.80 \times 10^{14}$$

$$\therefore \quad \sigma_R = \sqrt{4.80 \times 10^{14}} = 2.19 \times 10^7 \text{ N-mm}$$

Similarly,

$$\sigma_S^2 = \left(\frac{\partial S}{\partial M_{ex}} \times \sigma_{M_{ex}} \right)^2 = (1 \times 3 \times 10^7)^2 = 9 \times 10^{14}$$

$$\therefore \quad \sigma_S = \sqrt{9 \times 10^{14}} = 3 \times 10^7 \text{ N-mm}$$

(a) *Reliability index (β),*

$$\beta = \frac{\mu_R - \mu_S}{\sqrt{\sigma_R^2 + \sigma_S^2}} = \frac{(2.583 \times 10^8 - 1.504 \times 10^8)}{\sqrt{(4.80 \times 10^{14} + 9 \times 10^{14})}} = 2.897$$

(b) *Resistance factor (ϕ),*

$$\phi = [1 - 0.75 \times \beta \times (\text{CoV})_R]$$

$$\therefore \quad (\text{CoV})_R = \frac{\sigma_R}{\mu_R} = \frac{2.19 \times 10^7}{2.583 \times 10^8} = 0.0848$$

$$\Rightarrow \quad \phi = [1 - 0.75 \times 2.897 \times 0.0848] = 0.815$$

(c) *Load factor (γ),*

$$\gamma = [1 + 0.75 \times \beta \times (\text{CoV})_S]$$

$$\therefore \quad (\text{CoV})_S = \frac{\sigma_S}{\mu_S} = \frac{3 \times 10^7}{1.504 \times 10^8} = 0.199 \approx 0.2$$

$$\Rightarrow \quad \gamma = [1 + 0.75 \times 2.897 \times 0.2] = 1.434$$

(d) *Bias resistance factor,*

Mean resistance,

$$\mu_R = \mu_{A_{st}} \mu_{f_y} \mu_d - \frac{0.77 \mu_{A_{st}^2} \mu_{f_y^2}}{\mu_b \mu_{f_{ck}}}$$

$$\Rightarrow \mu_R = 1560.34 \times 272.41 \times 653.77 - \frac{0.77 \times (1560.34)^2 \times (272.41)^2}{326.88 \times 21.79} = 2.583 \times 10^8 \text{ N-mm}$$

Characteristic resistance,

$$R = A_{st} f_y d - \frac{0.77 A_{st}^2 f_y^2}{b f_{ck}}$$

$$\Rightarrow \qquad R_k = 1432 \times 250 \times 600 - \frac{0.77 \times (1432)^2 \times (250)^2}{300 \times 20} = 1.983 \times 10^8 \text{ N-mm}$$

$$\therefore \qquad b_R = \frac{\mu_R}{R_k} = \frac{2.583 \times 10^8}{1.983 \times 10^8} = 1.302$$

(e) *Bias load factor,*

Mean load, $\mu_S = 1.504 \times 10^8$

Characteristic load, $S_k = 2 \times 10^8$

$$\therefore \qquad b_L = \frac{\mu_S}{S_k} = \frac{1.504 \times 10^8}{2 \times 10^8} = 0.752$$

Example 10.12 Design by FOSM method, the moment capacity of reinforced concrete member given by the performance function $g(x) = R - S$; where $R = A_{st} f_y d - \dfrac{0.77 A_{st}^2 f_y^2}{b f_{ck}}$; $S = M_{ex}$ the member for strength formulation. Determine the area of the section for a reliability index of 3.09.

Solution:

Design:

The limit state of yield of $g(x) = R - S$

Target reliability index, $\beta = 3.09$

Standard deviation = Mean × $(\text{CoV})_{Ast}$

$$\Rightarrow \qquad \mu_A = \frac{\sigma_A}{(\text{CoV})_{A_{st}}} = \frac{\sigma_A}{0.05} = 20\sigma_{A_{st}}$$

Now, the mean and standard deviations are:

Mean value,

$$\mu_R = \mu_{A_{st}} \mu_{f_y} \mu_d - \frac{0.77 \mu_{A_{st}^2} \mu_{f_y^2}}{\mu_b \mu_{f_{ck}}}$$

$$\mu_R = 20\sigma_{A_{st}} \times 272.41 \times 653.77 - \frac{0.77 \times (20\sigma_{A_{st}})^2 \times (272.41)^2}{326.88 \times 21.79}$$

$$\Rightarrow \qquad \mu_R = 3.56 \times 10^6 \sigma_{A_{st}} \times 3208.86 \times \sigma_{A_{st}}^2 \text{ N-mm}$$

$$\mu_S = \mu_{M_{ex}} = 1.504 \times 10^8 \text{ N-mm}$$

Now, partially differentiating R and S w.r.t variables, we get,

$$\frac{\partial R}{\partial A_{st}} = \mu_{f_y}\mu_d - \frac{2\times 0.77\mu_{A_{st}}\mu_{f_y^2}}{\mu_b\mu_{f_{ck}}}$$

$$\Rightarrow \quad \frac{\partial R}{\partial A_{st}} = 272.41\times 653.77 - \frac{2\times 0.77\times 20\sigma_{A_{st}}\times(272.41)^2}{326.88\times 21.79} = 1.78\times 10^5 - 320.88\sigma_{A_{st}}$$

$$\frac{\partial R}{\partial f_y} = \mu_{A_{st}}\mu_d - \frac{2\times 0.77\mu_{A_{st}^2}\mu_{f_y}}{\mu_b\mu_{f_{ck}}}$$

$$\Rightarrow \quad \frac{\partial R}{\partial f_y} = 20\sigma_{A_{st}}\times 653.77 - \frac{2\times 0.77\times(20\sigma_{A_{st}})^2\times 272.41}{326.88\times 21.79} = 13075.40\sigma_{A_{st}} - 23.55\sigma_{A_{st}}^2$$

$$\frac{\partial R}{\partial d} = \mu_{A_{st}}\mu_{f_y} = 20\sigma_{A_{st}}\times 272.41 = 5448.2\sigma_{A_{st}}$$

$$\frac{\partial R}{\partial b} = \frac{0.77\mu_{A_{st}^2}\mu_{f_y^2}}{\mu_{b^2}\mu_{f_{ck}}} = \frac{0.77\times(20\sigma_{A_{st}})^2\times(272.41)^2}{(326.88)^2\times 21.79} = 9.816\sigma_{A_{st}}^2$$

$$\frac{\partial R}{\partial f_{ck}} = \frac{0.77\mu_{A_{st}^2}\mu_{f_y^2}}{\mu_b\mu_{f_{ck}^2}} = \frac{0.77\times(20\sigma_{A_{st}})^2\times(272.41)^2}{326.88\times(21.79)^2} = 147.26\sigma_{A_{st}}^2$$

$$\frac{\partial S}{\partial \mu_{M_{ex}}} = 1$$

Standard déviation,

$$\sigma_R^2 = \left(\frac{\partial R}{\partial A_{st}}\times\sigma_{A_{st}}\right)^2 + \left(\frac{\partial R}{\partial f_y}\times\sigma_{f_y}\right)^2 + \left(\frac{\partial R}{\partial d}\times\sigma_d\right)^2 + \left(\frac{\partial R}{\partial b}\times\sigma_b\right)^2 + \left(\frac{\partial R}{\partial f_{ck}}\times\sigma_{f_{CK}}\right)^2$$

$$\sigma_R^2 = \left((1.78\times 10^5 - 320.88\sigma_{A_{st}})\times\sigma_{A_{st}}\right)^2 + \left((13075.40\sigma_{A_{st}} - 23.55\sigma_{A_{st}}^2)\times 13.62\right)^2$$
$$+\left(5448.2\sigma_{A_{st}}\times 32.69\right)^2 + \left(9.816\sigma_{A_{st}}^2\times 16.35\right)^2 + \left(147.26\sigma_{A_{st}}^2\times 1.09\right)^2$$

$$\Rightarrow \quad \sigma_R^2 = 2.573\times 10^5\sigma_{A_{st}}^4 - 2.28\times 10^8\sigma_{A_{st}}^3 + 9.511\times 10^{10}\sigma_{A_{st}}^2$$

Similarly,

$$\sigma_S^2 = \left(\frac{\partial S}{\partial M_{ex}}\times\sigma_{M_{ex}}\right)^2 = (1\times 3\times 10^7)^2 = 9\times 10^{14}$$

$$\therefore \quad \sigma_S = \sqrt{9\times 10^{14}} = 3\times 10^7 \text{ N-mm}$$

Now,
Reliability index (β),
Here $\beta = 3.09$

$$\beta = \frac{\mu_R - \mu_S}{\sqrt{\sigma_R^2 + \sigma_S^2}}$$

$$3.09 = \frac{[(3.56\times10^6\sigma_{A_{st}} \times 3208.86\times\sigma_{A_{st}}^2) - 1.504\times10^8]}{\sqrt{(2.573\times10^5\sigma_{A_{st}}^4 - 2.28\times10^8\sigma_{A_{st}}^3 + 9.511\times10^{10}\sigma_{A_{st}}^2 + 9\times10^{14})}}$$

Now, squaring on both sides, we get,

$$(3.09)^2 = \frac{[(3.56\times10^6\sigma_{A_{st}} \times 3208.86\times\sigma_{A_{st}}^2) - 1.504\times10^8]}{(2.573\times10^5\sigma_{A_{st}}^4 - 2.28\times10^8\sigma_{A_{st}}^3 + 9.511\times10^{10}\sigma_{A_{st}}^2 + 9\times10^{14})}$$

On simplifying we get,

$$7.834\times10^6\sigma_{A_{st}}^4 - 2.066\times10^{10}\sigma_{A_{st}}^3 + 1.27\times10^{13}\sigma_{A_{st}}^2 - 1.07\times10^{15}\sigma_{A_{st}} + 1.401\times10^{16} = 0$$

Solving the quartic equation we get,

$$\sigma_{Ast_1} = -3.1613\times10^3\ \text{mm}^2$$
$$\sigma_{Ast_2} = 0.4279\times10^3\ \text{mm}^2$$
$$\sigma_{Ast_3} = 0.0822\times10^3\ \text{mm}^2$$
$$\sigma_{Ast_4} = 0.0161\times10^3\ \text{mm}^2$$

$$\therefore \quad \mu_A = \sigma_{Ast_3} = 20\times0.0822\times10^3 = 1644\ \text{mm}^2$$

Example 10.13 A simply supported beam is to carry a load of 20 N/mm, and the effective span of the beam is 6500 mm, the section is ISMB 350 @ 0.524 N/mm. Compute the reliability index, resistance factor, load factor, bias resistance factor and bias load factor. Assume all the variables are random. The performance function for the limit state of strength is given as $g(x) = f_y Z - \dfrac{(W_g + W_L)L^2}{8}$.

Sl. No.	Variables	Characteristic value	CoV	Distribution
1	W_g	0.524 N/mm	0.10	Normal
2	W_L	20 N/mm	0.15	Normal
3	L	6500 mm	0.05	Normal
4	f_y	250 N/mm^2	0.05	Normal
5	Z	1363.6 × 10^4 mm^3	0.05	Normal

Solution: **Analysis—FOSM Method**

The strengths are underestimated and loads are overestimated, the mean values are computed from the characteristic values and are tabulated below:

Sl. No.	Variables	Characteristic value	CoV	Mean value (μ)	Standard deviation (σ)
1	W_g	0.524 N/mm	0.10	0.45	4.5×10^{-2}
2	W_L	20 N/mm	0.15	16.04	2.40
3	L	6500 mm	0.05	7082.53	354.12
4	f_y	250 N/mm^2	0.05	272.40	13.62
5	Z	1363.6×10^4 mm^3	0.05	1485.80×10^4	74.29×10^4

The limit state function, $g(x) = f_y Z - \dfrac{(W_g + W_L)L^2}{8}$

$$R = f_y Z, \quad S = \frac{(W_g + W_L)L^2}{8}$$

Now, the mean and standard deviations are:

Mean value,

$$\mu_R = \mu_{f_y} \times \mu_Z = 272.40 \times 1485.80 \times 10^4 = 4.04 \times 10^9 \text{ N-mm}$$

$$\mu_S = \frac{(\mu_{W_g} \times \mu_{W_L}) \times \mu_{L^2}}{8} = \frac{(0.45 + 16.04) \times (7082.53)^2}{8} = 103.40 \times 10^6 \text{ N-mm}$$

Now, partially differentiating R and S w.r.t variables, we get,

$$\frac{\partial R}{\partial f_y} = \mu_Z = 1485.80 \times 10^4$$

$$\frac{\partial R}{\partial Z} = \mu_{f_y} = 272.40$$

$$\frac{\partial S}{\partial W_g} = \frac{\mu_{L^2}}{8} = \frac{(7082.53)^2}{8} = 6.27 \times 10^6$$

$$\frac{\partial S}{\partial W_L} = \frac{\mu_{L^2}}{8} = \frac{(7082.53)^2}{8} = 6.27 \times 10^6$$

$$\frac{\partial S}{\partial L} = \frac{2 \times \mu_L(\mu_{W_g} + \mu_{W_L})}{8} = \frac{2 \times 7082.53(0.45 + 16.04)}{8} = 29197.72$$

Standard deviation,

$$\sigma_R^2 = \left(\frac{\partial R}{\partial f_y} \times \sigma_{f_y}\right)^2 + \left(\frac{\partial R}{\partial Z} \times \sigma_Z\right)^2$$

$$= (272.40 \times 74.29 \times 10^4)^2 + (1485.80 \times 10^4 \times 13.62)^2$$

$$= 8.19 \times 10^{16}$$

$$\therefore \quad \sigma_R = \sqrt{8.19 \times 10^{16}} = 286.18 \times 10^6 \text{ N}$$

Similarly,

$$\sigma_S^2 = \left(\frac{\partial S}{\partial W_g} \times \sigma_{W_g}\right)^2 + \left(\frac{\partial S}{\partial W_L} \times \sigma_{W_L}\right)^2 + \left(\frac{\partial S}{\partial L} \times \sigma_L\right)^2$$

$$\sigma_S^2 = (6.27 \times 10^6 \times 4.5 \times 10^{-2})^2 + (6.27 \times 10^6 \times 2.40)^2 + (29197.72 \times 354.12)^2 = 3.33 \times 10^{14}$$

$$\therefore \quad \sigma_S = \sqrt{3.33 \times 10^{14}} = 18.25 \times 10^6 \text{ N}$$

Reliability index (β),

$$\beta = \frac{\mu_R - \mu_S}{\sqrt{\sigma_R^2 + \sigma_S^2}} = \frac{(4.04 \times 10^9 - 103.40 \times 10^6)}{\sqrt{(8.19 \times 10^{16} + 3.33 \times 10^{14})}} = 13.73$$

Resistance factor (ϕ),

$$\phi = [1 - 0.75 \times \beta \times (\text{CoV})_R]$$

$$\therefore \quad (\text{CoV})_R = \frac{\sigma_R}{\mu_R} = \frac{286.18 \times 10^6}{4.04 \times 10^9} = 0.07$$

$$\Rightarrow \quad \phi = [1 - 0.75 \times 13.73 \times 0.07] = 0.279$$

Load factor (γ),

$$\gamma = [1 + 0.75 \times \beta \times (\text{CoV})_S]$$

$$\therefore \quad (\text{CoV})_S = \frac{\sigma_S}{\mu_S} = \frac{18.25 \times 10^6}{103.40 \times 10^6} = 0.176$$

$$\Rightarrow \quad \gamma = [1 + 0.75 \times 13.73 \times 0.176] = 2.182$$

Bias resistance factor,

Mean resistance,

$$\mu_R = \mu_{f_y} \times \mu_Z = 272.40 \times 1485.80 \times 10^4 = 4.04 \times 10^9 \text{ N-mm}$$

Characteristic resistance,

$$R_k = f_y \times Z = 250 \times 1363.6 \times 10^4 = 3.40 \times 10^9 \text{ N-mm}$$

$$\therefore \quad b_R = \frac{\mu_R}{R_k} = \frac{4.04 \times 10^9}{3.40 \times 10^9} = 1.188$$

Bias load factor,
Mean load,

$$\mu_S = \frac{(\mu_{W_g} \times \mu_{W_L}) \times \mu_{L^2}}{8} = \frac{(0.45 + 16.04) \times (7082.53)^2}{8} = 103.40 \times 10^6 \text{ N-mm}$$

Characteristic load,

$$S_k = \frac{(W_g \times W_L) \times L^2}{8} = \frac{(0.524 + 20) \times (6500)^2}{8} = 108.39 \times 10^6 \text{ N-mm}$$

$$\therefore \quad b_L = \frac{\mu_S}{S_k} = \frac{103.40 \times 10^6}{108.39 \times 10^6} = 0.953$$

$$\alpha = \frac{\sqrt{\sigma_R^2 + \sigma_S^2}}{\sigma_R + \sigma_S} = \frac{\sqrt{(8.1836 \times 10^{16}) + (3.3458 \times 10^{14})}}{(286.070^2 \times 10^6) + (18.2917 \times 10^6)} = \frac{286.6541 \times 10^6}{304.3619 \times 10^6} = 0.9418$$

Example 10.14 Design by LRFD procedure a tie member of roof truss subjected to 300 kN for the limit state of yield, $g(x) = f_y A - S$ for the steel member.

Sl. No.	Variables	Characteristic value	CoV	Distribution
1	f_y	250 N/mm^2	0.05	Normal
2	A	?	0.05	Normal
3	$S = P$	300×10^3 N	0.2	Normal

Solution:

Equations are taken from AISC-LRFD specifications
Resistance > Load for safety

$$\phi R_k \geq \gamma S_k$$

$$\gamma S_k = 300 \times 10^3 \text{ N (total load)}$$

Yielding of gross cross section

$$\phi R_k = 0.9 \times 250 A$$
$$= 225 A$$

$$\Rightarrow \phi R_k \geq \gamma S_k$$

$225\,A \geq 300 \times 10^3$

$A \geq 1333.33 \text{ mm}^2$

Hence, ISA 100 × 75 × 8 mm section is sufficient and provided area is 1336 mm^2.

Example 10.15 Design by LRFD procedure, a simply supported beam which is loaded with dead, live load. The strength limit state is $g(x) = f_y Z - \dfrac{(W_g + W_L)\,L^2}{8}$

Sl. No.	Variables	Characteristic value	CoV	Distribution
1	W_g	0.524 N/mm	0.10	Normal
2	W_L	20 N/mm	0.15	Normal
3	L	6500 mm	0.05	Normal
4	f_y	250 N/mm^2	0.05	Normal
5	Z	1363.6×10^4 mm^3	0.05	Normal

(i) Determine the section modulus for the beam and (ii) find the allowable load on the beam.

Solution:

LRFD design procedure

Case 1: Design for section modulus by LRFD procedure

$$\phi R_k \geq \gamma S_k$$

$W = 1.2\,W_g + 1.6\,W_L = 1.2 \times 0.524 + 1.6 \times 20 = 32.62$ N/mm

$$\gamma S_k = \frac{WL^2}{8}$$

$$\therefore \quad \gamma S_k = \frac{32.62 \times 6500^2}{8} = 172.27 \times 10^6 \text{ N-mm}$$

$\Rightarrow \phi R_k = 0.9\, f_y\, Z$

$\phi R_k = 0.9 \times 250\, Z$

$= 225\, Z$

$\therefore \quad \phi R_k \geq \gamma S_k$

$225\, Z \geq 172.27 \times 10^6$

$Z \geq 765.66 \times 10^3 \text{ mm}^3$

Case 2: Design for loads

$\gamma S_k \leq \phi R_k$

$$\gamma S_k = \frac{WL^2}{8}, \; W = 1.2\, W_g + 1.6\, W_L$$

$$\Rightarrow \gamma S_k = \frac{W \times 6500^2}{8} = 5.28 \times 10^6\, W$$

$$\Rightarrow \phi R_k = 0.9\, f_y\, Z$$

$$\therefore \; \phi R_k = 0.9 \times 250 \times 1363.6 \times 10^4 = 3.06 \times 10^9$$

$$\gamma S_k \le \varphi R_k$$

$$5.28 \times 10^6\, W \le 3.06 \times 10^9$$

$$W \le 580.9365 \text{ N/mm}$$

$$1.2\, W_g + 1.6\, W_L \le 579.54 \text{ N/mm}$$

Example 10.16 Design by LRFD procedure for the given data. The top chords of bridge truss have an effective length of 4 m and determine the area of cross section. The limit state of buckling $g(x) = P_{cr} - S$; where $P_{cr} = \frac{\pi^2 E A r^2}{L^2}$, $S = P$. S is the load, treat E, A, r, S and L as random variables.

Sl. No.	Variables	Characteristic value	CoV	Distribution
1	E	2×10^5 N/mm^2	0.05	Normal
2	A	?	0.05	Normal
3	r	76.29 mm	0.05	Normal
4	L	4000 mm	0.05	Normal
5	$S = P$	1100×10^3 N	0.20	Normal

Solution: **LRFD design to determine the area of section:**

$$\gamma S = 1.2\, W_g + 1.6\, W_L = 1100 \times 10^3 \text{ N}$$

$$1.2\, W_g + 1.6\, W_L \le 1100 \times 10^3 \text{ N}$$

$$\phi R_k \ge \gamma S_k$$

$$0.85\, \sigma_y\, A_g \ge 1100 \times 10^3 \text{ N}$$

$$A_g \ge 1100 \times 10^3 \text{ N}$$

$$A_g \ge \frac{1100 \times 10^3}{0.85\, \sigma_y}$$

$$\therefore \quad A_g \ge \frac{1100 \times 10^3}{0.85 \times 250} = 5176.47 \text{ mm}^2$$

$$A_g \ge 5176.47 \text{ mm}^2$$

Hence, ISMC – 125 with 2 cover plates of 125 × 8 mm is sufficient

Hence, provide area is 5238 mm^2.

Example 10.17 A simply supported beam loaded at mid-point by a concentrated force P. The length of the beam is L, and the bending capacity at a point along the beam is f_y, Z, where Z is the plastic section modulus and f_y the yield stress. The random variables P, L, f_y and Z are assumed independent.

Sl. No.	Basic Variable	Mean value (μ)	Standard Deviation (σ)
1	f_y	600×10^3 kN/m^2	10^5 kN/m^2
2	Z	100×10^{-6} m^3	2×10^{-5} m^3
3	P	10 kN	2 kN
4	L	8 m	0.1 m

Solution:

The performance function in terms of **strength** g_1 $(P, L, f_y, Z) = f_y Z - PL/4$

The performance function in terms of **stress** g_2 $(P, L, f_y, Z) = f_y - PL/4 \times Z$

The reliability index for strength $\beta_1 = 2.46153$ $P_{f1} = 0.00695$

The reliability index for stress limit state $\beta_2 = 3.48$ $P_{f2} = 0.00025$

Comment: The reliability indexes are different for the two limit states however the limit states are equivalent, it clearly shows lack of invariance by FOSM method and hence it can be overcome by Hasofer-Lind method.

Example 10.18 A simply supported beam is shown in Fig. 10.4(a), is designed as beam as per IS: 800–1984, given f_y = 250 MPa.

A second beam is designed as per AASHTO & AISC is shown in Fig. 10.4(b), given f_y = 250 MPa. Analyze using Hasofer-Lind method.

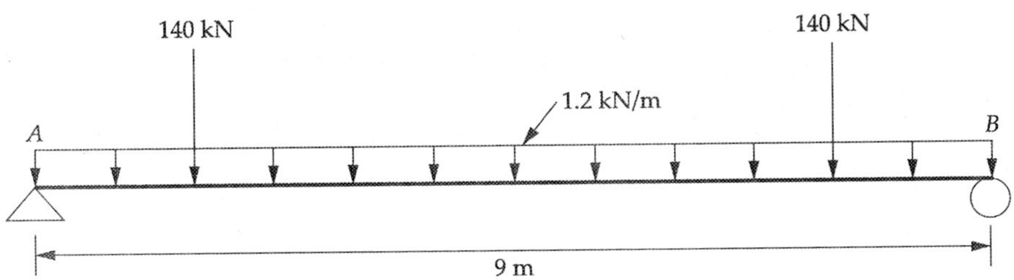

(a) Beam designed as per IS: 800–1984

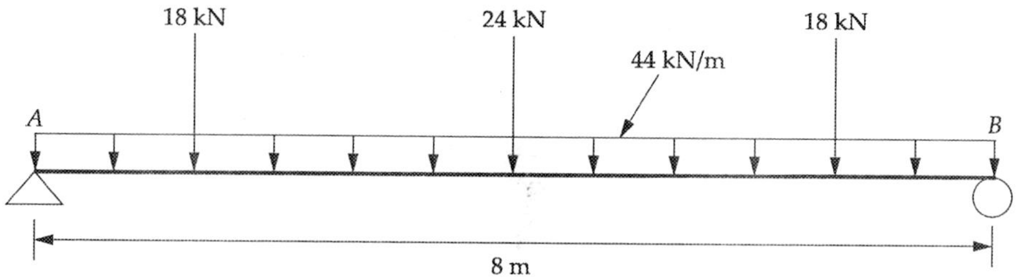

(b) Beam designed as per ASCI and AASHTO

Fig. 10.4

Sl. No.	Variables	Characteristic value	Unit	CoV	Distribution
1	f_y	250	N/mm^2	0.05	Normal
2	Z(IS-800)	3060 × 10^6 ISMB 600 @ 123 kg/m	mm^3	0.05	Normal
3	Z(AISC)	2884.1 × 10^6 W610 @ 113.1	mm^3	0.05	Normal
4	M_{ex} (IS)	433	kN-m	0.20	Normal
5	M_{ex} (AISC)	436	kN-m	0.20	Normal

Sl. No.	β	φ	γ	α_{1fy}	α_{2Z}	α_{3M}
IS Code	4.6777	0.8117	1.7326	–0.4460	–0.4460	0.7759
AASHTO & AISC	5.8152	0.7594	1.9112	–0.4734	–0.4734	0.7430

It is possible to compare different code designs and reliability analysis is performed using Hasofer-Lind method. The beam components calibrated and it is possible to compare two designs. A target reliability of 3.09 for strength can be used as a reference to compare two different designs in order to check for economy and also useful for preparing the tender.

Example 10.19 Show that the LRFD method in split format $\phi R > \gamma S$ from fundamentals, the left side of the inequality is the required strength and the right side is the load combinations.

Solution:

$$\beta = \frac{\mu_R - \mu_S}{\sqrt{(\sigma_R^2 + \sigma_S^2)}}$$

$$\beta\sqrt{(\sigma_R^2 + \sigma_S^2)} = (\mu_R - \mu_S) \quad \text{... (10.16)}$$

A linearised form of equation developed by Rjanitzin and Cornell format (1954) was suggested by Lind. Cross multiplying the terms in Eq. 10.22, the form $\sqrt{(\sigma_R^2+\sigma_S^2)}$ can be reduced to a linear form.

$$\alpha(t)=\frac{\sqrt{1+t^2}}{1+t} \quad \text{... (10.16a)}$$

Which for ($t \geq 0$) becomes $\frac{\sqrt{2}}{2}\leq\alpha(t)\leq 1$ and for $\left(\frac{1}{3}\leq t\leq 3\right)$ becomes $\alpha=0.75$ with an error of 5%.

Substituting $\frac{\sigma_{R_m}}{\sigma_{S_m}}$ for t in Eq. (10.16a)

$$\alpha[\sigma_R+\sigma_S]=\sqrt{(\sigma_R^2+\sigma_S^2)}$$

$$\beta\alpha[\sigma_R+\sigma_S]=[\mu_R-\mu_S]$$

$$\beta\alpha\sigma_R+\beta\alpha\sigma_S=[\mu_R-\mu_S]$$

The standard deviation of resistance is the product of coefficient of variation of resistance and mean resistance.

$$\sigma_R = \sigma_R \times (\text{CoV})_R; \quad \sigma_S = \mu_S \times (\text{CoV})_S$$

Substituting for standard deviation of resistance and load in the above expression and grouping the terms of resistance and load.

$$\beta\,\alpha\,[R\,(\text{CoV})_R + \beta\,\alpha\,[S\,(\text{CoV})_S] = [\mu_R - \mu_S]$$

We obtain the following expression in the split format.

$$R\,[1-\beta\,\alpha\,(\text{CoV})_R] = S\,[1+\beta\,\alpha\,(\text{CoV})_S]$$

$$\phi = [1-\beta\,\alpha\,(\text{CoV})_R];\ \gamma = [1+\beta\,\alpha\,(\text{CoV})_S]$$

$$\phi \times \mu_R = \gamma \times \mu_S$$

Here ϕ is the resistance factor which is always less than the unity and γ is the load factor which is always greater than unity. The split form of the condition of safety format is obtained.

The central safety factor (CSF) = mean resistance/mean load

$$\text{CSF}=\frac{\mu_R}{\mu_S}=\frac{\gamma}{\varphi}=\frac{[1+\beta\alpha(\text{CoV})_S]}{[1-\beta\alpha(\text{CoV})_R]}=\frac{[1+0.75\times\beta(\text{CoV})_S]}{[1-0.75\times\beta(\text{CoV})_R]}$$

Example 10.20 Show that the central safety factor is obtained from fundamentals.

$$C_o=\frac{1+\beta\sqrt{(V_R^2+V_S^2)}}{1-\beta^2 V_R^2}$$

Solution: Reliability index, $\beta = \dfrac{\mu_R - \mu_S}{\sqrt{(\mu_R^2 + \mu_S^2)}}$

where, μ_R = mean of resistance, μ_S = Load σ_R, σ_S are the standard deviation of resistance and load respectively.

$$\beta\sqrt{(\sigma_R^2 + \sigma_S^2)} = \mu_R - \mu_S$$

Squaring on both sides

$$\beta^2[(\sigma_R^2 + \sigma_S^2)] = [\mu_R - \mu_S]^2$$

$$\beta^2[\sigma_R^2] + \beta^2[\sigma_S^2] = [\mu_R^2 + \mu_S^2 - 2\mu_R\mu_S]$$

$$\beta^2\mu_R^2V_R^2 + \beta^2\mu_S^2V_S^2 = [\mu_R^2 + \mu_S^2 - 2\mu_R\mu_S]$$

$$\beta^2\mu_R^2V_R^2 - \mu_R^2 + [\beta^2\mu_S^2V_S^2 - \mu_S^2] + 2\mu_R\mu_S = 0$$

$$\mu_R^2[\beta^2V_R^2 - 1] + \mu_S^2[\beta^2V_S^2 - 1] + 2\mu_R\mu_S = 0$$

$$\mu_R^2[\beta^2V_R^2 - 1] + 2\mu_R\mu_S + \mu_S^2[\beta^2V_S^2 - 1] = 0 \text{ is quadratic equation in } \mu_R$$

$$\mu_R = \frac{-2\mu_S \pm \sqrt{[(2\mu_S)^2 - (4)(\beta^2V_R^2 - 1)\mu_S^2(\beta^2V_S^2 - 1)}}{2\beta^2V_R^2 - 1}$$

$$a = \beta^2V_R^2 - 1,\ b = 2\mu_S;\ c = \mu_S^2[\beta^2V_S^2 - 1]$$

$$\mu_R = \frac{\mu_S[1 \pm \sqrt{[\beta^2(V_R^2 + V_S^2)]}}{2[1 - \beta^2V_R^2]}$$

$$\frac{\mu_R}{\mu_S} = \frac{1 \pm \beta\sqrt{[(V_R^2 + V_S^2)]}}{[1 - \beta^2V_R^2]}$$

$$\frac{\mu_R}{\mu_S} = \frac{1 + \beta\sqrt{[(V_R^2 + V_S^2)]}}{[1 - \beta^2V_R^2]}$$

UNSOLVED PROBLEMS

1. A simply supported steel beam is loaded with a live load and a dead load as shown in Fig. 10.5 determine the reliability index β, load factor γ, resistance factor, φ, and central safety factor, CSF for the following formulations.

   (i) Strength formulation: $g(x) = f_y Z - (W_g + W_L) L^2/8$

   (ii) Stress formulation: $g(x) = f_y - \dfrac{M}{I}y$

(iii) Serviceability formulation : $g(x) = \Delta_{\text{lim}} - \Delta_{\text{act}}$

where $\Delta_{\text{lim}} = \dfrac{L}{360}$ and $\Delta_{\text{act}} = \dfrac{5WL^4}{384EI}$

Shear formulation $g(x) = R - S$ where $R = \tau_s\, t_w\, d$ and $S = 0.5\,(W_g + W_L)$

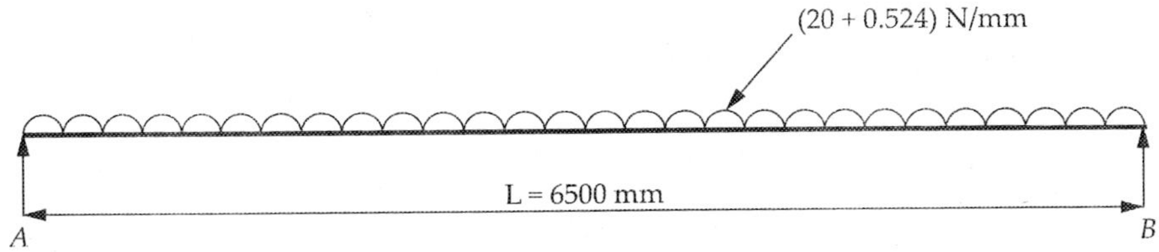

Fig. 10.5 Simply supported beam carrying UDL

Given ISMB 350 @ 0.524 N/mm, $Z = 778.9 \times 10^3 \text{mm}^3$, $I_{xx} = 13630.6 \times 10^4$, Area = 6671 mm^2, $f_y = 250$ N/mm^2, $W_g = 0.524$ N/mm, $W_L = 20$ N/mm, $d = 350$ mm, $t_w =$ 5.10 mm.

2. A segment of brick deck is under axial compression, P_a and prestressing force P_f. The condition for safety corresponding to the stress at top is given by $f_{ck} \geq \dfrac{P_a}{A} + \dfrac{P_f}{A} + \dfrac{P_f \times e \times y}{I} - \dfrac{M \times y}{I}$. The prestressing force P_f can be expressed as product of yield stress, f_y and area of tendons A_{st}. Evaluate the area of cross section A for a reliability index of 3.09, if the characteristic values and coefficient of variation (CoV) are as follows:

Variable	**Characteristic value**	**CoV**
f_{ck}	40 N/mm^2	0.05
f_y	1570 N/mm^2	0.05
P_a	57096000 N	0.20
A_{st}	18780.44 mm^2	0.05
e	42.9 mm	0.05
y	568.4 mm	0.05
M	24651×10^6 N-mm	0.20
A	?	0.05
I	6.2282×10^{12} mm^4	0.05

3. In Problem 1 if characteristic value of A is 12259×10^3 mm^2 then design the deck to evaluate the moment of inertia, I for a reliability index of 3.09.

4. The performance function for buckling of a pylon of a cable stayed bridge is given by $g(x) = \frac{\pi^2 EI}{L^2} - P_a$, where E modulus of elasticity of concrete (M40 Grade), I is the moment of inertia of the pylon, L is the effective length and P_a is the axial load whose characteristic values are

Variable	**Characteristic value**	**CoV**
E	31622.77 N/mm^2	0.05
I	?	0.05
P_a	67221192 N	0.20
L	8322.65 mm	0.05

Design the pylon for its moment of inertia, inertia (I) versus reliability index (β) varying from 3.09 to 10.

Appendix 1

Table A1: Normal Distribution for Finding Probabilities

Z	0.00	0.01	0.02	0.03	0.04	0.05	0.06	0.07	0.08	0.09
–3.0	0.0013	0.0013	0.0013	0.0012	0.0012	0.0011	0.0011	0.0011	0.0010	0.0010
–2.9	0.0019	0.0018	0.0018	0.0017	0.0016	0.0016	0.0015	0.0015	0.0014	0.0014
–2.8	0.0026	0.0025	0.0024	0.0023	0.0023	0.0022	0.0021	0.0021	0.0020	0.0019
–2.7	0.0035	0.0034	0.0033	0.0032	0.0031	0.0030	0.0029	0.0028	0.0027	0.0026
–2.6	0.0047	0.0045	0.0044	0.0043	0.0041	0.0040	0.0039	0.0038	0.0037	0.0036
–2.5	0.0062	0.0060	0.0059	0.0057	0.0055	0.0054	0.0052	0.0051	0.0049	0.0048
–2.4	0.0082	0.0080	0.0078	0.0075	0.0073	0.0071	0.0069	0.0068	0.0066	0.0064
–2.3	0.0107	0.0104	0.0102	0.0099	0.0096	0.0094	0.0091	0.0089	0.0087	0.0084
–2.2	0.0139	0.0136	0.0132	0.0129	0.0125	0.0122	0.0119	0.0116	0.0113	0.0110
–2.1	0.0179	0.0174	0.0170	0.0166	0.0162	0.0158	0.0154	0.0150	0.0146	0.0143
–2.0	0.0228	0.0222	0.0217	0.0212	0.0207	0.0202	0.0197	0.0192	0.0188	0.0183
–1.9	0.0287	0.0281	0.0274	0.0268	0.0262	0.0256	0.0250	0.0244	0.0239	0.0233
–1.8	0.0359	0.0351	0.0344	0.0336	0.0329	0.0322	0.0314	0.0307	0.0301	0.0294
–1.7	0.0446	0.0436	0.0427	0.0418	0.0409	0.0401	0.0392	0.0384	0.0375	0.0367
–1.6	0.0548	0.0537	0.0526	0.0516	0.0505	0.0495	0.0485	0.0475	0.0465	0.0455
–1.5	0.0668	0.0655	0.0643	0.0630	0.0618	0.0606	0.0594	0.0582	0.0571	0.0559
–1.4	0.0808	0.0793	0.0778	0.0764	0.0749	0.0735	0.0721	0.0708	0.0694	0.0681
–1.3	0.0968	0.0951	0.0934	0.0918	0.0901	0.0885	0.0869	0.0853	0.0838	0.0823
–1.2	0.1151	0.1131	0.1112	0.1093	0.1075	0.1056	0.1038	0.1020	0.1003	0.0985
–1.1	0.1357	0.1335	0.1314	0.1292	0.1271	0.1251	0.1230	0.1210	0.1190	0.1170
–1.0	0.1587	0.1562	0.1539	0.1515	0.1492	0.1469	0.1446	0.1423	0.1401	0.1379
–0.9	0.1841	0.1814	0.1788	0.1762	0.1736	0.1711	0.1685	0.1660	0.1635	0.1611
–0.8	0.2119	0.2090	0.2061	0.2033	0.2005	0.1977	0.1949	0.1922	0.1894	0.1867
–0.7	0.2420	0.2389	0.2358	0.2327	0.2296	0.2266	0.2236	0.2206	0.2177	0.2148
–0.6	0.2743	0.2709	0.2676	0.2643	0.2611	0.2578	0.2546	0.2514	0.2483	0.2451

Contd.

Table A1: *Contd.*

–0.5	0.3085	0.3050	0.3015	0.2981	0.2946	0.2912	0.2877	0.2843	0.2810	0.2776
–0.4	0.3446	0.3409	0.3372	0.3336	0.3300	0.3264	0.3228	0.3192	0.3156	0.3121
–0.3	0.3821	0.3783	0.3745	0.3707	0.3669	0.3632	0.3594	0.3557	0.3520	0.3483
–0.2	0.4207	0.4168	0.4129	0.4090	0.4052	0.4013	0.3974	0.3936	0.3897	0.3859
–0.1	0.4602	0.4562	0.4522	0.4483	0.4443	0.4404	0.4364	0.4325	0.4286	0.4247
–0.0	0.5000	0.4960	0.4920	0.4880	0.4840	0.4801	0.4761	0.4721	0.4681	0.4602
0.0	0.5000	0.5040	0.5080	0.5120	0.5160	0.5199	0.5239	0.5279	0.5319	0.5359
0.1	0.5398	0.5438	0.5478	0.5517	0.5557	0.5596	0.5636	0.5675	0.5714	0.5753
0.2	0.5793	0.5832	0.5871	0.5910	0.5948	0.5987	0.6026	0.6064	0.6103	0.6141
0.3	0.6179	0.6217	0.6255	0.6293	0.6331	0.6368	0.6406	0.6443	0.6480	0.6517
0.4	0.6554	0.6591	0.6628	0.6664	0.6700	0.6736	0.6772	0.6808	0.6844	0.6879
0.5	0.6915	0.6950	0.6985	0.7019	0.7054	0.7088	0.7123	0.7157	0.7190	0.7224
0.6	0.7257	0.7291	0.7324	0.7357	0.7389	0.7422	0.7454	0.7486	0.7517	0.7549
0.7	0.7580	0.7611	0.7642	0.7673	0.7704	0.7734	0.7764	0.7794	0.7823	0.7852
0.8	0.7881	0.7910	0.7939	0.7967	0.7995	0.8023	0.8051	0.8078	0.8106	0.8133
0.9	0.8159	0.8186	0.8212	0.8238	0.8264	0.8289	0.8315	0.8340	0.8365	0.8389
1.0	0.8413	0.8438	0.8461	0.8485	0.8508	0.8531	0.8554	0.8577	0.8599	0.8621
1.1	0.8643	0.8665	0.8686	0.8708	0.8729	0.8749	0.8770	0.8790	0.8810	0.8830
1.2	0.8849	0.8869	0.8888	0.8907	0.8925	0.8944	0.8962	0.8980	0.8997	0.9015
1.3	0.9032	0.9049	0.9066	0.9082	0.9099	0.9115	0.9131	0.9147	0.9162	0.9177
1.4	0.9192	0.9207	0.9222	0.9236	0.9251	0.9265	0.9279	0.9292	0.9306	0.9319
1.5	0.9332	0.9345	0.9357	0.9370	0.9382	0.9394	0.9406	0.9418	0.9429	0.9441
1.6	0.9452	0.9463	0.9474	0.9484	0.9495	0.9505	0.9515	0.9525	0.9535	0.9545
1.7	0.9554	0.9564	0.9573	0.9582	0.9591	0.9599	0.9608	0.9616	0.9625	0.9633
1.8	0.9641	0.9649	0.9656	0.9664	0.9671	0.9678	0.9686	0.9693	0.9699	0.9706
1.9	0.9713	0.9719	0.9726	0.9732	0.9738	0.9744	0.9750	0.9756	0.9761	0.9767
2.0	0.9772	0.9778	0.9783	0.9788	0.9793	0.9798	0.9803	0.9808	0.9812	0.9817
2.1	0.9821	0.9826	0.9830	0.9834	0.9838	0.9842	0.9846	0.9850	0.9854	0.9857
2.2	0.9861	0.9864	0.9868	0.9871	0.9875	0.9878	0.9881	0.9884	0.9887	0.9890
2.3	0.9893	0.9896	0.9898	0.9901	0.9904	0.9906	0.9909	0.9911	0.9913	0.9916
2.4	0.9918	0.9920	0.9922	0.9925	0.9927	0.9929	0.9931	0.9932	0.9934	0.9936
2.5	0.9938	0.9940	0.9941	0.9943	0.9945	0.9946	0.9948	0.9949	0.9951	0.9952
2.6	0.9953	0.9955	0.9956	0.9957	0.9959	0.9960	0.9961	0.9962	0.9963	0.9964
2.7	0.9965	0.9966	0.9967	0.9968	0.9969	0.9970	0.9971	0.9972	0.9973	0.9974
2.8	0.9974	0.9975	0.9976	0.9977	0.9977	0.9978	0.9979	0.9979	0.9980	0.9981
2.9	0.9981	0.9982	0.9982	0.9983	0.9984	0.9984	0.9985	0.9985	0.9986	0.9986
3.0	0.9987	0.9987	0.9987	0.9988	0.9988	0.9989	0.9989	0.9989	0.9990	0.9990

Table A2: Normal Distribution for Finding Quartiles

P	0.000	0.001	0.002	0.003	0.004	0.005	0.006	0.007	0.008	0.009
0.00		–3.0902	–2.8782	–2.7478	–2.6521	–2.5758	–2.5121	–2.4573	–2.4089	–2.3656
0.01	–2.3263	–2.2904	–2.2571	–2.2262	–2.1973	–2.1701	–2.1444	–2.1201	–2.0969	–2.0749
0.02	–2.0537	–2.0335	–2.0141	–1.9954	–1.9774	–1.9600	–1.9431	–1.9268	–1.9110	–1.8957
0.03	–1.8808	–1.8663	–1.8522	–1.8384	–1.8250	–1.8119	–1.7991	–1.7866	–1.7744	–1.7624
0.04	–1.7507	–1.7392	–1.7279	–1.7169	–1.7060	–1.6954	–1.6849	–1.6747	–1.6646	–1.6546
0.05	–1.6449	–1.6352	–1.6258	–1.6164	–1.6072	–1.5982	–1.5893	–1.5805	–1.5718	–1.5632
0.06	–1.5548	–1.5464	1.5382	–1.5301	–1.5220	–1.5141	–1.5063	–1.4985	–1.4909	–1.4833
0.07	–1.4758	–1.4684	1.4611	–1.4538	–1.4466	–1.4395	–1.4325	–1.4255	–1.4187	–1.4118
0.08	–1.4051	–1.3984	1.3917	–1.3852	–1.3787	–1.3722	–1.3658	–1.3595	–1.3532	–1.3469
0.09	–1.3408	–1.3346	1.3285	–1.3225	–1.3165	–1.3106	–1.3047	–1.2988	–1.2930	–1.2873
0.10	–1.2816	–1.2759	–1.2702	–1.2646	–1.2591	–1.2536	–1.2481	–1.2426	–1.2372	–1.2319
0.11	–1.2265	–1.2212	–1.2160	–1.2107	–1.2055	–1.2004	–1.1952	–1.1901	–1.1850	–1.1800
0.12	–1.1750	–1.1700	–1.1650	–1.1601	–1.1552	–1.1503	–1.1455	–1.1407	–1.1359	–1.1311
0.13	–1.1264	–1.1217	–1.1170	–1.1123	–1.1077	–1.1031	–1.0985	–1.0939	–1.0893	–1.0848
0.14	–1.0803	–1.0758	–1.0714	–1.0669	–1.0625	–1.0581	–1.0537	–1.0494	–1.0450	–1.0407
0.15	–1.0364	–1.0322	–1.0279	–1.0237	–1.0194	–1.0152	–1.0110	–1.0069	–1.0027	–0.9986
0.16	–0.9945	–0.9904	–0.9863	–0.9822	–0.9782	–0.9741	–0.9701	–0.9661	–0.9621	–0.9581
0.17	–0.9542	–0.9502	–0.9463	–0.9424	–0.9385	–0.9346	–0.9307	–0.9269	–0.9230	–0.9192
0.18	–0.9154	–0.9116	–0.9078	–0.9040	–0.9002	–0.8965	–0.8927	–0.8890	–0.8853	–0.8816
0.19	–0.8779	–0.8742	–0.8705	–0.8669	–0.8633	–0.8596	–0.8560	–0.8524	–0.8488	–0.8452
0.20	–0.8416	–0.8381	–0.8345	–0.8310	–0.8274	–0.8239	–0.8204	–0.8169	–0.8134	–0.8099
0.21	–0.8064	–0.8030	–0.7995	–0.7961	–0.7926	–0.7892	–0.7858	–0.7824	–0.7790	–0.7756
0.22	–0.7722	–0.7688	–0.7655	–0.7621	–0.7588	–0.7554	–0.7521	–0.7488	–0.7454	–0.7421
0.23	–0.7388	–0.7356	–0.7323	–0.7290	–0.7257	–0.7225	–0.7192	–0.7160	–0.7128	–0.7095
0.24	–0.7063	–0.7031	–0.6999	–0.6967	–0.6935	–0.6903	–0.6871	–0.6840	–0.6808	–0.6776
0.25	–0.6745	–0.6713	–0.6682	–0.6651	–0.6620	–0.6588	–0.6557	–0.6526	–0.6495	–0.6464
0.26	–0.6433	–0.6403	–0.6372	–0.6341	–0.6311	–0.6280	–0.6250	–0.6219	–0.6189	–0.6158
0.27	–0.6128	–0.6098	–0.6068	–0.6038	–0.6008	–0.5978	–0.5948	–0.5918	–0.5888	–0.5858
0.28	–0.5828	–0.5799	–0.5769	–0.5740	–0.5710	–0.5681	–0.5651	–0.5622	–0.5592	–0.5563
0.29	–0.5534	–0.5505	–0.5476	–0.5446	–0.5417	–0.5388	–0.5359	–0.533	–0.5302	–0.5273
0.30	–0.5244	–0.5215	–0.5187	–0.5158	–0.5129	–0.5101	–0.5072	–0.5044	–0.5015	–0.4987
0.31	–0.4959	–0.4930	–0.4902	–0.4874	–0.4845	–0.4817	–0.4789	–0.4761	–0.4733	–0.4705
0.32	–0.4677	–0.4649	–0.4621	–0.4593	–0.4565	–0.4538	–0.4510	–0.4482	–0.4454	–0.4427
0.33	–0.4399	–0.4372	–0.4344	–0.4316	–0.4289	–0.4261	–0.4234	–0.4207	–0.4179	–0.4152

Contd.

Table A2: *Contd.*

P	0.000	0.001	0.002	0.003	0.004	0.005	0.006	0.007	0.008	0.009
0.34	–0.4125	–0.4097	–0.4070	–0.4043	–0.4016	–0.3989	–0.3961	–0.3934	–0.3907	–0.3880
0.35	–0.3853	–0.3826	–0.3799	–0.3772	–0.3745	–0.3719	–0.3692	–0.3665	–0.3638	–0.3611
0.36	–0.3585	–0.3558	–0.3531	–0.3505	–0.3478	–0.3451	–0.3425	–0.3398	–0.3372	–0.3345
0.37	–0.3319	–0.3292	–0.3266	–0.3239	–0.3213	–0.3186	–0.3160	–0.3134	–0.3107	–0.3081
0.38	–0.3055	–0.3029	–0.3002	–0.2976	–0.2950	–0.2924	–0.2898	–0.2871	–0.2845	–0.2819
0.39	–0.2793	–0.2767	–0.2741	–0.2715	–0.2689	–0.2663	–0.2637	–0.2611	–0.2585	–0.2559
0.40	–0.2533	–0.2508	–0.2482	–0.2456	–0.2430	–0.2404	–0.2378	–0.2353	–0.2327	–0.2301
0.41	–0.2275	–0.2250	–0.2224	–0.2198	–0.2173	–0.2147	–0.2121	–0.2096	–0.2070	–0.2045
0.42	–0.2019	–0.1993	–0.1968	–0.1942	–0.1917	–0.1891	–0.1866	–0.1840	–0.1815	–0.1789
0.43	–0.1764	–0.1738	–0.1713	–0.1687	–0.1662	–0.1637	–0.1611	–0.1586	–0.1560	–0.1535
0.44	–0.1510	–0.1484	–0.1459	–0.1434	–0.1408	–0.1383	–0.1358	–0.1332	–0.1307	–0.1282
0.45	–0.1257	–0.1231	–0.1206	–0.1181	–0.1156	–0.1130	–0.1105	–0.1080	–0.1055	–0.1030
0.46	–0.1004	–0.0979	–0.0954	–0.0929	–0.0904	–0.0878	–0.0853	–0.0828	–0.0803	–0.0778
0.47	–0.0753	–0.0728	–0.0702	–0.0677	–0.0652	–0.0627	–0.0602	–0.0577	–0.0552	–0.0527
0.48	–0.0502	–0.0476	–0.0451	–0.0426	–0.0401	–0.0376	–0.0351	–0.0326	–0.0301	–0.0276
0.49	–0.0251	–0.0226	–0.0201	–0.0175	–0.0150	–0.0125	–0.0100	–0.0075	–0.0050	–0.0025
0.50	0.0000	0.0025	0.0050	0.0075	0.0100	0.0125	0.0150	0.0175	0.0201	0.0226
0.51	0.0251	0.0276	0.0301	0.0326	0.0351	0.0376	0.0401	0.0426	0.0451	0.0476
0.52	0.0502	0.0527	0.0552	0.0577	0.0602	0.0627	0.0652	0.0677	0.0702	0.0728
0.53	0.0753	0.0778	0.0803	0.0828	0.0853	0.0878	0.0904	0.0929	0.0954	0.0979
0.54	0.1004	0.1030	0.1055	0.1080	0.1105	0.1130	0.1156	0.1181	0.1206	0.1231
0.55	0.1257	0.1282	0.1307	0.1332	0.1358	0.1383	0.1408	0.1434	0.1459	0.1484
0.56	0.1510	0.1535	0.1560	0.1586	0.1611	0.1637	0.1662	0.1687	0.1713	0.1738
0.57	0.1764	0.1789	0.1815	0.1840	0.1866	0.1891	0.1917	0.1942	0.1968	0.1993
0.58	0.2019	0.2045	0.2070	0.2096	0.2121	0.2147	0.2173	0.2198	0.2224	0.2250
0.59	0.2275	0.2301	0.2327	0.2353	0.2378	0.2404	0.2430	0.2456	0.2482	0.2508
0.60	0.2533	0.2559	0.2585	0.2611	0.2637	0.2663	0.2689	0.2715	0.2741	0.2767
0.61	0.2793	0.2819	0.2845	0.2871	0.2898	0.2924	0.2950	0.2976	0.3002	0.3029
0.62	0.3055	0.3081	0.3107	0.3134	0.3160	0.3186	0.3213	0.3239	0.3266	0.3292
0.63	0.3319	0.3345	0.3372	0.3398	0.3425	0.3451	0.3478	0.3505	0.3531	0.3558
0.64	0.3585	0.3611	0.3638	0.3665	0.3692	0.3719	0.3745	0.3772	0.3799	0.3826
0.65	0.3853	0.3880	0.3907	0.3934	0.3961	0.3989	0.4016	0.4043	0.4070	0.4097
0.66	0.4125	0.4152	0.4179	0.4207	0.4234	0.4261	0.4289	0.4316	0.4344	0.4372
0.67	0.4399	0.4427	0.4454	0.4482	0.4510	0.4538	0.4565	0.4593	0.4621	0.4649

Contd.

Table A2: *Contd.*

P	0.000	0.001	0.002	0.003	0.004	0.005	0.006	0.007	0.008	0.009
0.68	0.4677	0.4705	0.4733	0.4761	0.4789	0.4817	0.4845	0.4874	0.4902	0.4930
0.69	0.4959	0.4987	0.5015	0.5044	0.5072	0.5101	0.5129	0.5158	0.5187	0.5215
0.70	0.5244	0.5273	0.5302	0.5330	0.5359	0.5388	0.5417	0.5446	0.5476	0.5505
0.71	0.5534	0.5563	0.5592	0.5622	0.5651	0.5681	0.5710	0.5740	0.5769	0.5799
0.72	0.5828	0.5858	0.5888	0.5918	0.5948	0.5978	0.6008	0.6038	0.6066	0.6098
0.73	0.6128	0.6158	0.6189	0.6219	0.6250	0.6280	0.6311	0.6341	0.6372	0.6403
0.74	0.6433	0.6464	0.6495	0.6526	0.6557	0.6588	0.6620	0.6651	0.6682	0.6713
0.75	0.6745	0.6776	0.6808	0.6840	0.6871	0.6903	0.6935	0.6967	0.6999	0.7031
0.76	0.7063	0.7095	0.7128	0.7160	0.7192	0.7225	0.7257	0.7290	0.7323	0.7356
0.77	0.7388	0.7421	0.7454	0.7488	0.7521	0.7554	0.7588	0.7621	0.7655	0.7688
0.78	0.7722	0.7756	0.7790	0.7824	0.7858	0.7892	0.7926	0.7961	0.7995	0.8030
0.79	0.8064	0.8099	0.8134	0.8169	0.8204	0.8239	0.8274	0.8310	0.8345	0.8381
0.80	0.8416	0.8452	0.8488	0.8524	0.8560	0.8596	0.8633	0.8669	0.8705	0.8742
0.81	0.8779	0.8816	0.8853	0.8890	0.8927	0.8965	0.9002	0.9040	0.9078	0.9116
0.82	0.9154	0.9192	0.9230	0.9269	0.9307	0.9346	0.9385	0.9424	0.9463	0.9502
0.83	0.9542	0.9581	0.9621	0.9661	0.9701	0.9741	0.9782	0.9822	0.9863	0.9904
0.84	0.9945	0.9986	1.0027	1.0069	1.0110	1.0152	1.0194	1.0237	1.0279	1.0322
0.85	1.0364	1.0407	1.0450	1.0494	1.0537	1.0581	1.0625	1.0669	1.0714	1.0758
0.86	1.0803	1.0848	1.0893	1.0939	1.0985	1.1031	1.1077	1.1123	1.1170	1.1217
0.87	1.1264	1.1311	1.1359	1.1407	1.1455	1.1503	1.1552	1.1601	1.1650	1.1700
0.88	1.1750	1.1800	1.1850	1.1901	1.1952	1.2004	1.2055	1.2107	1.2160	1.2212
0.89	1.2265	1.2319	1.2372	1.2426	1.2481	1.2536	1.2591	1.2646	1.2702	1.2759
0.90	1.2816	1.2873	1.2930	1.2988	1.3047	1.3106	1.3165	1.3225	1.3285	1.3346
0.91	1.3408	1.3469	1.3532	1.3595	1.3658	1.3722	1.3787	1.3852	1.3917	1.3984
0.92	1.4051	1.4118	1.4187	1.4255	1.4325	1.4395	1.4466	1.4538	1.4611	1.4684
0.93	1.4758	1.4833	1.4909	1.4985	1.5063	1.5141	1.5220	1.5301	1.5382	1.5464
0.94	1.5548	1.5632	1.5718	1.5805	1.5893	1.5982	1.6072	1.6164	1.6258	1.6352
0.95	1.6449	1.6546	1.6646	1.6747	1.6849	1.6954	1.7060	1.7169	1.7279	1.7392
0.96	1.7507	1.7624	1.7744	1.7866	1.7991	1.8119	1.8250	1.8384	1.8522	1.8663
0.97	1.8808	1.8957	1.9110	1.9268	1.9431	1.9600	1.9774	1.9954	2.0141	2.0335
0.98	2.0537	2.0749	2.0969	2.1201	2.1444	2.1701	2.1973	2.2262	2.2571	2.2904
0.99	2.3263	2.3656	2.4089	2.4573	2.5121	2.5758	2.6521	2.7478	2.8782	3.0902

Table A3: Reliability Index, β, and Corresponding Probability of Failure Values

Reliability index, β	Probability of failure	Reliability index, β	Probability of failure
0	0.500 000 17	2.55	0.00538617
0.05	0.480 060 97	2.60	0.00466122
0.10	0.46817195	2.65	0.00402463
0.15	0.44038215	2.70	0.00346702
0.20	0.42074018	2.75	0.00297982
0.25	0.40129360	2.80	0.00255519
0.30	0.38208852	2.85	0.00218603
0.35	0.36316930	2.90	0.00186588
0.40	0.34457820	2.95	0.00158894
0.45	0.32635515	3.00	0.00134997
0.50	0.30853744	3.05	0.00114428
0.55	0.29115957	3.10	0.00096767
0.60	0.27425299	3.15	0.00081642
0.65	0.25784597	3.20	0.00068720
0.70	0.24196351	3.25	0.00057709
0.75	0.22662722	3.30	0.00048348
0.80	0.21185528	3.35	0.00040411
0.85	0.19766244	3.40	0.00033698
0.90	0.18406005	3.45	0.00028034
0.95	0.17105607	3.50	0.00023267
1.00	0.15865522	3.55	0.00019266
1.05	0.1685905	3.60	0.00015915
1.10	0.13566607	3.65	0.00013115
1.15	0.12507196	3.70	0.00010783
1.20	0.11506971	3.75	0.00008844
1.25	0.10564982	3.80	0.00007237
1.30	0.09680053	3.85	0.00005908
1.35	0.08850803	3.90	0.00004812
1.40	0.08075669	3.95	0.00003909
1.45	0.07352929	4.00	0.00003169
1.50	0.06680722	4.05	0.00002562

Contd.

Table A3: *Contd.*

Reliability index, β	Probability of failure	Reliability index, β	Probability of failure
1.55	0.06057076	4.10	0.00002067
1.60	0.05479928	4.15	0.00001663
1.65	0.04947144	4.20	0.00001335
1.70	0.04456542	4.25	0.00001070
1.75	0.04005911	4.30	0.00000855
1.80	0.03593026	4.35	0.00000681
1.85	0.03215671	4.40	0.00000542
1.90	0.02871649	4.45	0.00000430
1.95	0.02558799	4.50	0.00000340
2.00	0.02275006	4.55	0.00000268
2.05	0.02018214	4.60	0.00000211
2.10	0.01786435	4.65	0.00000166
2.15	0.01577755	4.70	0.00000130
2.20	0.01390340	4.75	0.00000102
2.25	0.01222443	4.80	0.00000079
2.30	0.01072408	4.85	0.00000062
2.35	0.00938669	4.90	0.00000048
2.40	0.00819753	4.95	0.00000037
2.45	0.00714281	5.00	0.00000029
2.50	0.00620968		

Table A4: Consequence of Failure

<table>
<tr><th rowspan="2">Relative Costs of Safety Measures</th><th colspan="4">Consequence of Failure</th></tr>
<tr><th>Small</th><th>Some</th><th>Moderate</th><th>Great</th></tr>
<tr><td>High</td><td>$P_f = 10^0$
$\beta = 0.00$</td><td>$P_f = 10^{-2}$
$\beta = 1.05$</td><td>$P_f = 10^{-2}$
$\beta = 2.3$</td><td>$P_f = 10^{-3}$
$\beta = 3.1$</td></tr>
<tr><td>Moderate</td><td>$P_f = 10^{-1}$
$\beta = 1.3\beta$</td><td>$P_f = 10^{-2}$
$\beta = 2.3$</td><td>$P_f = 10^{-3}$
$\beta = 3.1$</td><td>$P_f = 7 \times 10^{-5}$
$\beta = 3.8$</td></tr>
<tr><td>Low</td><td>$P_f = 10^{-2}$
$\beta = 2.3$</td><td>$P_f = 10^{-3}$
$\beta = 3.1$</td><td>$P_f = 7 \times 10^{-5}$
$\beta = 3.8$</td><td>$P_f = 10^{-5}$
$\beta = 4.3$</td></tr>
</table>

Failure Types	Consequence of Failure		
	Less serious	Serious	Very serious
Ductile failure with reserve strength capacity resulting from strain hardening	$P_f^T = 10^{-3}$ $\beta^T = 3.09$	$P_f^T = 10^{-4}$ $\beta^T = 3.71$	$P_f^T = 10^{-5}$ $\beta^T = 4.26$
Ductile failure with no reserve capacity	$P_f^T = 10^{-4}$ $\beta^T = 3.71$	$P_f^T = 10^{-5}$ $\beta^T = 4.26$	$P_f^T = 10^{-6}$ $\beta^T = 4.75$
Brittle failure and instability	$P_f^T = 10^{-5}$ $\beta^T = 4.26$	$P_f^T = 10^{-6}$ $\beta^T = 4.75$	$P_f^T = 10^{-7}$ $\beta^T = 5.20$

Table A5: Gamma Functions

N	$\Gamma(N)$	N	$\Gamma(N)$	N	$\Gamma(N)$	N	$\Gamma(N)$
1.00	1.00000	1.25	0.90640	1.50	0.88623	1.75	0.91906
1.01	0.99433	1.26	0.90440	1.51	0.88659	1.76	0.92137
1.02	0.98884	1.27	0.90250	1.52	0.88704	1.77	0.92376
1.03	0.98355	1.28	0.90072	1.53	0.88757	1.78	0.92623
1.04	0.97844	1.29	0.89940	1.54	0.88818	1.79	0.92877
1.05	0.97350	1.30	0.89747	1.55	0.88887	1.80	0.93138
1.06	0.96874	1.31	0.89600	1.56	0.88964	1.81	0.93408
1.07	0.96415	1.32	0.89464	1.57	0.89049	1.82	0.93685
1.08	0.95973	1.33	0.89338	1.58	0.89142	1.83	0.93969
1.09	0.95546	1.34	0.89222	1.59	0.89243	1.84	0.94261
1.10	0.95135	1.35	0.89115	1.60	0.89352	1.85	0.94561
1.11	0.94740	1.36	0.89018	1.61	0.89468	1.86	0.94869
1.12	0.94359	1.37	0.88931	1.62	0.89592	1.87	0.95184
1.13	0.93993	1.38	0.88854	1.63	0.89724	1.88	0.95507
1.14	0.93642	1.39	0.88785	1.64	0.89864	1.89	0.95838
1.15	0.93304	1.40	0.88726	1.65	0.90012	1.90	0.96177
1.16	0.92980	1.41	0.88676	1.66	0.90167	1.91	0.96523
1.17	0.92670	1.42	0.88636	1.67	0.90330	1.92	0.96877
1.18	0.92373	1.43	0.88604	1.68	0.90500	1.93	0.97240
1.19	0.92089	1.44	0.88581	1.69	0.90678	1.94	0.97610
1.20	0.91817	1.45	0.88566	1.70	0.90864	1.95	0.97988
1.21	0.91558	1.46	0.88560	1.71	0.91057	1.96	0.98374
1.22	0.91311	1.47	0.88563	1.72	0.91258	1.97	0.98768
1.23	0.91075	1.48	0.88575	1.73	0.91467	1.98	0.99171
1.24	0.90852	1.49	0.88595	1.74	0.91683	1.99	0.99581
						2.00	1.00000

Appendix 2

The problem of selecting the appropriate probability distribution for a random variable is often considered crucial. Attention should be paid to which distribution is to be used in probability modeling of a particular variable. Engineering judgment is necessary in finding the right distribution.

SI. No.	Assigned probability distribution	Given Constraints	Parameters
1	Uniform	2 parameter, symmetric; lower and upper bound. It can be used for modeling when no better information is available.	Minimum and maximum values are known.
2	Exponential	2 parameter, L shaped; positively asymmetric; lower bound values are known.	Only mean values are known. Only average values in stresses and strength.
3	Normal	2 parameter, symmetric; bell shaped no bounds.	Mean and standard deviation are known. Geometric, load and strength.
4	Log-Normal distribution	2 parameter form of log-normal distribution with lower bound equal to zero is commonly used; it is positively asymmetric.	Mean and standard deviation are known. Stresses and strength.
5	Log-Normal distribution	3 parameter, bell shaped; positively and negatively asymmetric.	Effective tool for many problems where asymmetric variables are encountered.

Contd.

6	Beta distribution	4 parameter; bell shaped, J shaped, L shaped, U shaped; lower and upper bound values are known. It is a attractive distribution, but it is difficult to fit data when all four parameters of the distributions. It is effectively used in diverse problems.	Mean, standard deviation, minimum and maximum.
7	Extreme value distributions	3 parameter, in general, positively or negatively asymmetric, bell shaped, lower or upper bound, or no bounds. A family of distributions (maxima or minima) of phenomenon enter model.	Examination of wind load, sea waves load.

References

As far as laws of mathematics refer to reality, they are not certain; and as far as they are certain, they do not refer to reality.
—**Albert Einstein**

1. Maurice Lemaire, *Structural Reliability*, John Wiley, New York, 2009.
2. Augusti G, Baratta A, Casciati F, Probabilistic *Methods in Structural Engineering*, Chapman and Hall, London, 1970.
3. Alfredo H. S. Ang, Wilson H. Tang, *Probability Concepts in Engineering, (Emphasis on Application in Civil and Environmental Engineering)*, 2nd Edition, John Wiley, 2007.
4. Alfredo H. S. Ang, Wilson H. Tang, *Probability Concepts in Engineering Planning and Design, Vol.: I, Basic Principles*, John Wiley, 1975.
5. Alfredo H. S. Ang, Wilson H. Tang, *Probability Concepts in Engineering Planning and Design, Vol.: II, Decision, Risk, & Reliability*, John Wiley, 1984.
6. Milton E. Harr, *Reliability–Based Design in Civil Engineering*, McGraw-Hill Book Company, 1987.
7. Dan Ghiocel, and Dan Lungu, *Wind, Snow and Temperature Effects on structures Based on Probability*, Abacus Press, Turn bridge wells, Kent, 1975.
8. Bilal M. Ayyub, Richard H. McCuen, *Probability, Statistics, and Reliability for Engineers*, CRC Press, New York, 1997.
9. Randolph Kissell, Robert L. Ferry, *Aluminium Structures (A Guide to their Specifications and Design)*, John Wiley & Sons, New York, 1985.
10. Gary C. Hart, *Uncertainity Analysis, Loads, and Safety in Structural Engineering*, Prentice Hall, Inc., Englewood Cliffs, N.J. 1982.
11. Seung_Kyum Choi, Ramana V. Grandhi and Robert A. Canfield, *Reliabiltiy-based Structrual Design*, Springer, 2006.
12. Robert E. Melchers, *Structural Reliability Analysis and Prediction, Second Edition*, John Wiley & Sons, 2002.
13. Robert E. Melchers, *Structural Reliability Analysis and Prediction*, John Wiley & Sons, 1987.
14. K. C. Kapur, L. R. Lamberson, *Reliability in Engineering Design*, John Wiley & Sons, 1977.
15. James N. Siddall, *Probabilistic Engineering Design, Principles and Applications*, Marcel Dekker, Inc., New York, 1983.
16. Palle Thoft-Christensen, Michael J. Baker, *Structural Reliability Theory and Its Applications*, Springer-Verlag, New York, 1982.
17. Palle Thoft-Christensen, Yoshisada Murotsu, *Applicaions of Structural Systems Reliability Theory*, Springer-Verlag, New York, 1986.

18. Jack R. Benjamin, C. Allin Cornell, *Probability, Statistics, and Decision for Civil Engineers*, McGraw-Hill Book Company, New York, 1970.
19. O. Ditlevsen, H. O. Madsen, *Structural Reliability Methods*, John Wiley & Sons, NewYork, 1996.
20. MiliK Tichy, *Applied Methods of Structural Reliability*, Kluwer Academic Publishers, 1993.
21. Yeou-Koung Tung, Ben-Chie Yen, Charles Steven Melching, *Hydrosystems Engineering Reliability Assessment and Risk Analysis*, McGraw-Hill Book Company, New York, 2006.
22. Yeou-Koung Tung, Ben-Chie Yen, *Hydrosystems Engineering Uncertainity Analysis*, McGraw-Hill Book Company, New York, 2005.
23. Achintya Haldar, Sankaran Mahadevan, *Probabilility, Reliability and Statistical Methods in Engineering Design*, John Wiley & Sons, Inc., New York, 2000.
24. Edward B. Haugen, *Probabilistic Approaches to Design*, John Wiley & Sons, 1964.
25. Issac Elishakkoff, *Probabilistic Methods in the Theory of Structures*, John Wiley & Sons, 1983.
26. James N. Siddall, *Analytical Decision-Making in Engineering Design*, Prentice Hall, Inc., New Jersey, 1972.
27. Sir, Alfred Pugsley, *The Safety of Structures*, Edward Arnold Publishers Limited, London, 1966.
28. Gregory B. Baecher, John T. Christian, *Reliability and Statistics in Geotechnical Engineering*, John Wile & Sons, 2003.
29. Andrez S. Nowak, Kevin R. Collins, McGraw Hill Book Company, New York, 2010.
30. Roger D. Leitch, *Reliability Analysis for Engineers*, Oxford University Press, 1995.
31. Ronald H. Doeck, *Measurement Uncertainty, Methods and Applications*, ISA, 2002.
32. Roy Billinton, Ronald N. Allan, *Reliability Evaluation of Engineering Systems, Concepts and Techniques*, 2nd Edition, Springer, 1992.
33. V. V. Bolotin, Sanuel Aroni, *Statistical Methods in Structural Mechanics*, Holden-Day, Inc., San Francisco, 1968.
34. Bilal M Ayyub, *Uncertainity Modeling and Analysis in Civil Engineering*, CRC Press, New York, 1998.
35. R. Ranganathan, *Reliability Analysis and Design of Structures*, Tata McGraw-Hill Publishing Company Limited, New Delhi, 1990.
36. L. S. Srinath, *Concepts in Reliability, with an Introduction to Maintainability and Availability*, East-West Press, 1975.
37. Gaya Prasad Chhalotra, *Reliability Methods Engineering and its Application*, Khanna Publishers, New Delhi, 1986.
38. Guo-Qiang Li, Jin-Jun Li, *Advanced Analysis and Design of Steel Frames*, John Wiley & Sons, Ltd., 2007.
39. Bruce Ellingwood, Theodor V. Galambos, James G. MacGregor, C. Allin Cornell, *Development of a Probability Based Load Criterion for American National Standard A58*, National Bureau of Standards, NBS 577, June 1980.
40. Nathabandu T., Kottegoada, and Renzo Rosso, *Statistics, Probability, and Reliability for Civil and Environmental Engineers*, McGraw Hill International edition, Singapore,1998.
41. Ditlevsen O, *Uncertainty Modelling*, McGraw Hill, New York,1981.
42. Madsen H. O, Krenk S, & Lind N C, *Methods of Structural Safety*, Prentice Hall, Englewood Cliffs, New Jersey, 1986.

43. Blockley D.I., *The Nature of Structural Design and Safety*, Ellis Horwood, Chechester, 1980.
44. Hasofer A. M. & Lind N. C., '*Exact and Invariant Second Moment Code*' *Format, Journal of Engineering Mechanics Division*, ASCE, 100 (EM1), February 1974.
45. *Nordic Committee on Building Regulations, Recommendations for Loading and Safety Regulations for Structural Design*, NKB-Report Nl. 36, November 1978.
46. Kok-Kwang Phoon, *Reliability-Based Design in Geotechnical Engineering, (Computations and Applications*, Taylor & Francis, New York, 2008.
47. Harr, Milton Edward, *Mechanics of Particulate Media*, Mc-Graw Hill Inc, New York, 1977.
48. Mohamed A. El-Reedy, *Reinforced Concrete Structural Reliability*, CRC Press, Boca Raton, New York, 2013.
49. M. E. Criswell, M. D. Vandrbilt, *Reliability Based Desing of Transmission Line Structures*, Electric Power Research Institute, California, EL-4793, Vol:2, 1987.
50. Ravindra M. K. & Galambos T. V., *Load and Resistance Factor Design for Steel, Journal of Structural Division*, ASCE, Vol. 104, No. ST9A, September 1978.

Index